John Dornyan

THE ELECTRONIC STRUCTURE OF ATOMS AND MOLECULES

A Survey of Rigorous Quantum Mechanical Results

HENRY F. SCHAEFER III
University of California, Berkeley

ADDISON-WESLEY PUBLISHING COMPANY
Reading, Massachusetts · Menlo Park, California · London · Don Mills, Ontario

Copyright © 1972 by Addison-Wesley Publishing Company, Inc.
Philippines copyright 1972 by Addison-Wesley Publishing Company, Inc.

All rights reserved. No part of this publication may be reproduced, stored in a retrieval system, or transmitted, in any form or by any means, electronic, mechanical, photocopying, recording, or otherwise, without the prior written permission of the publisher.
Printed in the United States of America. Published simultaneously in Canada. Library of Congress Catalog Card No. 70-190611.

To Karen

PREFACE

This book represents an attempt to familiarize chemists with the very significant advances which have been recently made in that branch of theoretical chemistry which deals with the ab initio quantum mechanical treatment of electronic structure. The term ab initio, as opposed to semi-empirical, implies that no approximations (such as neglect of certain integrals) are used as the wave function is determined by minimization of the total energy. Although the quantum-mechanical principles required for an understanding of electronic structure have been recognized for forty years, it is only during the past five years that rigorous quantum-mechanical investigations have begun to make real contributions to chemistry. The rise to relevance of theory can be attributed to a) the development and availability of high speed digital computers and b) the concurrent development of sophisticated new theoretical and computational methods for the description of the behavior of electrons in atoms and molecules. Innovations in both areas a) and b) will doubtless continue to be made, and we can fully expect that during the next decade the role of rigorous quantum-mechanical calculations in chemistry will continue to grow by leaps and bounds.

An increasing number of introductory texts on quantum mechanics are using simple ab initio calculations to illustrate concepts in valence theory. Although a minimum-basis-set self-consistent-field wave function for water is of considerable pedagogical value, the chemical importance of the calculation is not usually apparent. By such examples theoretical chemistry may come to be thought of as aesthetically pleasing, but certainly not urgent. The present book attempts to highlight the urgency of molecular quantum mechanics. The fact that theoretical chemists discovered methylene (CH_2) to be bent rather than linear (as was supposed) is of chemical importance.

The present book picks up the theory of electronic structure where most texts on quantum chemistry leave off. The book is intended to be simple enough to be useful to the student who has had only a one quarter or semester introductory course in quantum mechanics. As a consequence, I anticipate a few complaints from my theoretical colleagues concerning the absence of detailed discussions concerning theoretical and computational methods. It should be mentioned here that the recent book by McWeeny and Sutcliffe (254), intended primarily for theoreticians, is addressed to the question of methods.

I have organized the book around a series of about fifty theoretical studies, intended to show in a very concrete manner the current "state of the art." The calculations chosen as examples are meant to be illustrative rather than exhaustive, since they represent only a small fraction of the papers published in this field. A nearly complete bibliography of ab initio calculations has been given by Krauss (216) and updated by Allen (6). Another bibliography has recently been compiled by Richards et al. (314). Of necessity, the selection of examples reflects the author's own interests and there is no question but that a number of very important contributions have been omitted.

The discussion of basis sets in chapter I may strike the reader as uninteresting and possibly irrelevant. Such a discussion is necessary, however, as it is otherwise not possible to assess the reliability of a particular calculation.

The reader may notice that I have omitted a discussion of the very accurate and important work of Pekeris (292) on the helium atom and Kolos and Wolniewicz (211) on the hydrogen molecule. The method used in these two-electron calculations, namely explicit use of interparticle coordinates, is potentially very powerful but unlikely in the near future to be applied to systems with more than a few electrons. For this reason, all the calculations discussed herein adopt the orbital approximation, in which the wave function is built up as a linear combination of Slater determinants constructed from one-electron functions.

The writing of this book was made possible by those theoretical chemists who carried out the outstanding work described herein. Thanks are due to the many authors who allowed me to incorporate their tables and figures in the text. I am particularly indebted to each individual who sent me a description of his work prior to its publication.

The typing of the manuscript was cheerfully carried out by Mrs. Sally Olsen and Claudia Redwood. I thank Mr. Peter K. Pearson and Professor Russell M. Pitzer for their careful reading of the manuscript and many helpful suggestions.

Berkeley, California
January 1972

H.F.S.

CONTENTS

I. Methods

    A. Elements of Quantum Mechanics

    B. The Self-Consistent-Field (SCF) Method

Examples; symmetry and equivalence restrictions	6
The energy expression	8
Brillouin's theorem	10
Koopmans' theorem	14
Open shells	19
Extensions of the restricted Hartree-Fock approximation	22

    C. Configuration Interaction

Electron correlation	27
The basic idea; the eigenvalue problem	30
Orbital occupancies, configurations, and symmetry	32
Hamiltonian matrix elements	37
Multiconfiguration SCF method	40
The density matrix and natural orbitals; H_2 as an example	41
The pseudonatural orbital method	48
The iterative natural orbital method	52

    D. Basis Sets

The minimum basis set	58
Double zeta and extended basis sets	63
Gaussian basis sets	67
Contracted functions	70
Polarization functions	75
Comparisons for CO and H_2O	78

II. Atoms

A. Restricted Hartree-Fock Results

Experimental atomic correlation energies; relativistic effects	86
Ionization potentials of neon and argon	89

B. Relaxation of Orbital Restrictions

The Fermi contact interaction in the unrestricted Hartree-Fock approximation	95

C. Variational Consideration of Electron Correlation

Oscillator strengths of boron and carbon	102
Accurate wave functions for Be, C, and Ne	111
Hyperfine structure of first-row atoms	119

D. Pair Correlation Approximations

Spinorbital pairs – Be, Mg, Ar	129
Symmetry-adapted pairs for neon and the question of additivity	134
Treatment of open shells – nitrogen	140

III. Diatomic Molecules

A. Near Hartree-Fock results

Spectroscopic constants and potential curves for N_2 and N_2^+	146
Molecular properties of the alkali halides	153
Paramagnetism predicted for BH	160
Theoretical transition probabilities for N_2^+, C_2^-, CN, CO^+, BO, BF^+, and BeF	168
Exchange splitting of the inner shell ionization potentials of NO	175

B. Multiconfiguration Results

Pair correlations in the first-row hydrides	182
The separated pair approximation: LiH, BH, NH	190
Accurate dipole moment calculations: CO	196
Potential energy curves for the low-lying states of CN	200

Accurate potential energy curves for the ground states
  of F_2 and O_2 205
Interaction potential between two helium atoms 216
Positions of the electronic states of BeO 222
Existence and properties of the NaLi molecule 227
Larger diatomics: KrF and KrF^+ 232

IV. Linear Polyatomic Molecules

  A. Single Configuration Wave Functions

  Dipole moments and dissociation energies 240
  Direct calculation of molecular polarizabilities 243
  Predicted properties of LiOH and Li_2O 250
  Relative stabilities of lithium cyanide and isocyanide 259

  B. Inclusion of Electron Correlation

  The linear HeH_2^+ potential energy surface 266
  The $F + H_2$ chemical reaction 272
  Krypton difluoride 281

V. Nonlinear Polyatomic Molecules

  A. Triatomic Molecules

  Magnetic hyperfine structure of NO_2 290
  Molecular quadrupole moments and other properties
    of sulfur dioxide 294
  Zero-point vibrational effects in the water molecule 298
  Rydberg states of H_2O 304
  Geometry of $^3B_1$ methylene and the singlet-triplet separation .. 309
  The H_3 potential surface 316
  Electron correlation in the water molecule 319

  B. Molecules with Four through Eight Atoms

  Geometry predictions from simple basis sets 325
  1s binding energies of BH_3, CH_4, NH_3, H_2O, and HF ... 331
  The $H_2 + D_2$ exchange reaction 335
  The T and V states of ethylene 338

Approach of two methylenes   344
Dimerization energy of BH_3 348

C. Larger Molecules

Excited states of butadiene   352
The electrocyclic transformation between cyclobutene
   and cis-butadiene   359
Electronic ground states of ortho, meta, and para benzyne   363
Hydrogen bonding in the water dimer and trimer   370
Nickel hexafluoride   378
XeF_2, XeF_4, and XeF_6 383
Naphthalene and azulene   388
The guanine-cytosine base pair   394

D. Other Topics

Barriers to rotation and inversion   398
Localized orbitals   407

References   414

Molecule Index   429

Subject Index   432

I
METHODS

A. ELEMENTS OF QUANTUM MECHANICS

The Schrödinger equation

$$H\Psi = E\Psi \qquad (I,1)$$

provides a theoretical foundation for the solution of virtually all problems in chemistry. Since the analytic solution of the Schrödinger equation has only been possible to date for atoms and molecules with one electron, one must turn to approximate solutions to obtain information of chemical value from theory.

For molecules a first approximation, which is made in all the calculations discussed in this book, is the Born-Oppenheimer approximation (118)*, in which the nuclear and electronic wave functions are considered separable. Using the Born-Oppenheimer method, we first solve the electronic Schrödinger equation

$$H_e \psi_e = E_e \psi_e \qquad (I,2)$$

for all possible positions of the N atoms within the molecule. The resulting electronic energies E_e form a potential energy

---

*Numbers in parentheses are keyed to the References at the end of this text.

surface $V(\vec{R}_1, \vec{R}_2, \ldots \vec{R}_N)$ where $\vec{R}_A$ specifies the coordinates of the Ath atom within the molecule. This leads to a Schrödinger equation describing the motion of the nuclei

$$H_N \psi_N = E \psi_N \qquad (I,3)$$

where

$$H_N = \sum_A \frac{-\nabla_A^2}{2M_A} + V(\vec{R}_1, \vec{R}_2, \ldots \vec{R}_N) \qquad (I,4)$$

in which $-\nabla_A^2/2M_A$ is the kinetic energy operator for the Ath nucleus. Eq. (I,3) may then be solved to yield the approximate total wave function

$$\Psi = \psi_e \psi_N \qquad (I,5)$$

and total energy E of Eq. (I,1).

Our interest in the present volume is in finding approximate solutions to the electronic Schrödinger equation, Eq. (I,2). To this end we make another approximation, that of ignoring relativistic effects and adopting the <u>ordinary electrostatic Hamiltonian</u>

$$H_e = \sum_A \sum_{B>A} \frac{Z_A Z_B}{R_{AB}} + \sum_i \left\{ \frac{-\nabla_i^2}{2} - \sum_A \frac{Z_A}{r_{iA}} \right\} + \sum_i \sum_{j>i} \frac{1}{r_{ij}} \qquad (I,6)$$

The first summation in Eq. (I,6) represents the repulsions between all pairs of nuclei, Z_A being the nuclear charge on atom A, and R_{AB} the distance between nucleus A and nucleus B. The second summation goes over all n electrons in the molecule and includes the one-electron operators for the kinetic energy and the attraction between electrons and nuclei. r_{iA} is the distance between the ith electron and the Ath nucleus. The final term in the Hamiltonian accounts for the repulsions between pairs of electrons, r_{ij} being the distance between electron i and electron j.

In writing the electronic Hamiltonian, Eq. (I,6), we have used <u>atomic units</u>, in which $\hbar$ (Planck's constant divided by 2π), the electronic charge e, and the electron mass m are all unity. In atomic units, distance is given in bohr radii (bohrs), where 1 bohr is 0.52918×10^{-8} cm or 0.52918 Å. The unit of energy is the hartree, equal to 27.21 eV, 627.5 kcal/mole, 2.1948×10^5 cm^{-1}, or 3.1579×10^5 °K, depending on one's preference.

Nearly all current electronic structure calculations are based on the <u>variational principle</u> (290). For any normalized approximate wave function Ψ, the energy is just the expectation value of the Hamiltonian operator

$$E = \int \Psi^* H \Psi \, d\tau. \qquad (I,7)$$

The variational principle is based on the fact that the energy E given in Eq. (I,7) is a rigorous upper bound to the true energy. That is, the energy E calculated for any approximate wave function Ψ will lie <u>above</u> the exact energy of the quantum mechanical system in question. Therefore the "best" wave function of a given form is that for which all parameters have been varied to yield the lowest energy. If the wave function Ψ is capable of completely flexible variation, application of the variational principle will yield the exact solution to the Schrödinger equation. The most frequent application of the variational principle is to the calculation of self-consistent-field wave functions.

B. THE SELF-CONSISTENT-FIELD (SCF) METHOD

The molecular orbital approximation has provided a theoretical framework for the explanation of a wide variety of chemical phenomena (237). The rigorous mathematical expression of the molecular orbital model is the <u>Hartree-Fock (HF) approximation</u> (150, 122). For closed-shell atoms and molecules the Hartree-Fock wave function is of the form

$$\psi_e = A(n) \phi_1(1) \phi_2(2) \ldots \phi_n(n) \qquad (I,8)$$

in which $A(n)$ is the antisymmetrizer for n electrons and the ϕ's are spinorbitals, products of a spatial orbital χ and a one electron spin function α (for $m_s = +\frac{1}{2}$) or β (for $m_s = -\frac{1}{2}$).

$$\phi_i = \begin{array}{c} \chi_i \alpha \\ \text{or} \\ \chi_i \beta \end{array} \qquad (I,9)$$

This same Hartree-Fock wave function is perhaps more concretely written as a <u>Slater determinant</u> (360):

$$\psi_e = \frac{1}{\sqrt{n!}} \begin{vmatrix} \phi_1(1) & \phi_1(2) & \ldots & \phi_1(n) \\ \phi_2(1) & \phi_2(2) & \ldots & \phi_2(n) \\ & \vdots & & \\ \phi_n(1) & \phi_n(2) & \ldots & \phi_n(n) \end{vmatrix} \qquad (I,10)$$

For a given geometry of a particular molecule in a closed-shell state, any number of different Slater determinants may be used as approximate wave functions. However, there is only one Hartree-Fock determinant, namely that for which the orbitals ϕ_i in Eq. (I,8) have been varied to give the lowest possible energy $\int \psi_e^* H_e \psi_e \, d\tau$.

By minimizing the energy resulting from the single determinant wave function in Eq. (I,10) one can derive a rather complicated set of integrodifferential equations, the Hartree-Fock equations. Since the Hartree-Fock equations and methods for their solution have been discussed in considerable detail elsewhere (254,288,296,362), here we need only point out the important physical concept involved. The essential point is that the Hartree-Fock wave function is the best (in the variational sense) wave function which can be constructed by assigning each electron to a separate <u>orbital</u>, or function depending only on the coordinates of that electron.

Only for one-electron systems such as the hydrogen atom can the Hartree-Fock equations be solved in closed form. However, for atoms the HF equations may be solved to a rather high accuracy by numerical integration (152). For molecules, however, one invariably expands the orbitals ϕ_i in terms of a set of analytic basis functions. Since it is never possible to use a mathematically complete set of functions in molecular calculations of a practical nature, one can only obtain approximate solutions of the Hartree-Fock equations. The best (lowest energy) single determinant wave function constructed within a finite basis set is the <u>self-consistent-field</u> (SCF) <u>wave function</u>. Most of the electronic structure calculations reported in the literature during the past ten years are of the SCF variety. Clearly, as the size of a basis set is expanded, the SCF energy and wave function will approach the Hartree-Fock results.

Examples - symmetry and equivalence restrictions. At this point it may be useful to give a few examples. For the neon atom in its ground ^1S electronic state the HF wave function is of the form

$$\psi_e = A(10) \; \chi_{1s}\alpha(1) \; \chi_{1s}\beta(2) \; \chi_{2s}\alpha(3) \; \chi_{2s}\beta(4) \; \chi_{2p_{-1}}\alpha(5) \; \chi_{2p_{-1}}\beta(6)$$

$$\cdot \; \chi_{2p_0}\alpha(7) \; \chi_{2p_0}\beta(8) \; \chi_{2p_{+1}}\alpha(9) \; \chi_{2p_{+1}}\beta(10) \qquad (I,11)$$

As pointed out earlier A(10) is the 10 electron antisymmetrizer and forms a 10×10 determinant from the 10 spinorbitals. The subscripts attached to the different 2p orbitals indicate the values m_ℓ = -1, 0, or +1. The same wave function can be written in abbreviated notation as

$$\psi_e = 1s\alpha \; 1s\beta \; 2s\alpha \; 2s\beta \; 2p_{-1}\alpha \; 2p_{-1}\beta \; 2p_0\alpha \; 2p_0\beta \; 2p_{+1}\alpha \; 2p_{+1}\beta \qquad (I,12)$$

Similarly the HF wave function for the $^1\Sigma^+$ ground state of carbon monoxide is

$$\psi_e = 1\sigma\alpha \; 1\sigma\beta \; 2\sigma\alpha \; 2\sigma\beta \; 3\sigma\alpha \; 3\sigma\beta \; 4\sigma\alpha \; 4\sigma\beta \; 5\sigma\alpha \; 5\sigma\beta \; 1\pi_{-1}\alpha \; 1\pi_{-1}\beta \; 1\pi_{+1}\alpha \; 1\pi_{+1}\beta$$

$$(I,13)$$

where the +1 and -1 subscripts on the π orbitals indicate the value of the orbital angular momentum along the internuclear axis. As a final example consider the $^1A_1$ ground state of the water molecule:

$$\psi_e = 1a_1\alpha\ 1a_1\beta\ 2a_1\alpha\ 2a_1\beta\ 1b_2\alpha\ 1b_2\beta\ 3a_1\alpha\ 3a_1\beta\ 1b_1\alpha\ 1b_1\beta \qquad (I,14)$$

All three of the above examples illustrate the <u>symmetry and equivalence restrictions</u> (270) usually adopted in SCF calculations. Each orbital is assumed to be a symmetry orbital, or more precisely, each orbital must transform according to one of the irreducible representations of the molecule's point group (87). Thus the 1s orbital of neon must be strictly of s($\ell=0$) character, with, for example, no d character mixed in. An example of the equivalence restrictions is that the spatial function $1a_1$ associated with α spin is required to be identical to the $1a_1$ function identified with the $1a_1\beta$ spin-orbital. A second equivalence restriction is that the $1\pi_+$ and $1\pi_-$ orbitals of CO differ only in their dependence on the angle ϕ about the internuclear axis. For the $1\pi_+$ orbital this ϕ dependence is $e^{i\phi}$, while for $1\pi_-$ it is $e^{-i\phi}$. The solution of the Hartree-Fock equations including symmetry and equivalence restrictions yields the <u>restricted Hartree-Fock (RHF)</u> wave function.

The molecules discussed above - Ne, CO, and H_2O - belong to the point groups K_h, $C_{\infty v}$, and C_{2v}. For a clear discussion of molecular symmetry in general and the symmetries of individual molecular orbitals, the reader is referred to the monographs by Herzberg (157,161).

The energy expression.

It is usually the case that all the orbitals in the Slater determinant (I,10) are orthogonal

$$\int \phi_i(1)\phi_j(1)\,dv(1) = \delta_{ij} \tag{I,15}$$

where dv(1) indicates integration over the space and spin coordinates of electron 1. Given orbital orthogonality, a very simple expression can be obtained (362) for the energy (excluding the nuclear repulsions) of a single determinant wave function

$$E = \int \psi_e^* H_e \psi_e \, d\tau$$

$$= \sum_i \int \phi_i^*(1) \left\{ \frac{-\nabla_1^2}{2} - \sum_A \frac{Z_A}{r_{1A}} \right\} \phi_i(1)\,dv(1)$$

$$+ \sum_{ij} \left[\iint \phi_i^*(1)\phi_j^*(2)\frac{1}{r_{12}}\phi_i(1)\phi_j(2)\,dv(1)\,dv(2) \right. \tag{I,16}$$

$$\left. - \iint \phi_i^*(1)\phi_j^*(2)\frac{1}{r_{12}}\phi_j(1)\phi_i(2)\,dv(1)\,dv(2) \right]$$

The spin integrations in (I,16) drop out immediately since none of the terms in our electrostatic Hamiltonian are spin dependent and

$$\int \alpha(1)^*\alpha(1)\,ds(1) = \int \beta(1)^*\beta(1)\,ds(1) = 1 \tag{I,17}$$

$$\int \alpha(1)^*\beta(1)\,ds(1) = \int \beta(1)^*\alpha(1)\,ds(1) = 0 \tag{I,18}$$

The first summation in Eq. (I,16) goes over the one-electron integrals, which we abbreviate as $I(i|j)$

$$I(i|j) = \int \phi_i^*(1) \left\{ \frac{-\nabla_1^2}{2} - \sum_A \frac{Z_A}{r_{1A}} \right\} \phi_j(1) \, dv(1) \qquad (I,19)$$

The electron repulsion integrals or two-electron integrals may be abbreviated as (288)

$$(ij|k\ell) = \int \phi_i^*(1) \phi_j^*(2) \frac{1}{r_{12}} \phi_k(1) \phi_\ell(2) \, dv(1) \, dv(2) \qquad (I,20)$$

In this simplified notation the energy of a single determinant wave function is

$$E = \sum_i I(i|i) + \sum_i \sum_j [(ij|ij) - (ij|ji)] \qquad (I,21)$$

Two-electron integrals of the type $(ij|ij)$ are generally called <u>coulomb integrals</u>, while those of type $(ij|ji)$ are exchange integrals.

As an example, consider the ground $^1\Sigma^+$ state of boron hydride, for which the Restricted Hartree-Fock wave function is

$$\psi_e = 1\sigma\alpha \; 1\sigma\beta \; 2\sigma\alpha \; 2\sigma\beta \; 3\sigma\alpha \; 3\sigma\beta \qquad (I,22)$$

From Eq. (I,21) the RHF energy is, after carrying out spin integrations,

$$\begin{aligned} E = {}& 2I(1\sigma|1\sigma) + 2I(2\sigma|2\sigma) + 2I(3\sigma|3\sigma) + (1\sigma1\sigma|1\sigma1\sigma) \\ & + 4(1\sigma2\sigma|1\sigma2\sigma) - 2(1\sigma2\sigma|2\sigma1\sigma) + (2\sigma2\sigma|2\sigma2\sigma) \\ & + 4(1\sigma3\sigma|1\sigma3\sigma) - 2(1\sigma3\sigma|3\sigma1\sigma) \\ & + 4(2\sigma3\sigma|2\sigma3\sigma) - 2(2\sigma3\sigma|3\sigma2\sigma) + (3\sigma3\sigma|3\sigma3\sigma) \end{aligned} \qquad (I,23)$$

By varying the forms of the 1σ, 2σ, and 3σ orbitals to minimize the energy E given by (I,23), Cade and Huo (66) have calculated an SCF wave function of nearly Hartree-Fock accuracy for BH. From the Cade-Huo wave function the author has computed the values of the integrals in Eq. (I,23) and these are given in Table I-1. It is seen, as is usually the case, that the coulomb integrals $(ij|ij)$ are significantly larger than the exchange integrals $(ij|ji)$. The Cade-Huo calculation was carried out at an internuclear separation of 2.336 bohrs. Adding the nuclear repulsion term

$$\frac{Z_A Z_B}{R_{AB}} = \frac{5 \cdot 1}{2 \cdot 336} \qquad (I,24)$$

to the energy obtained using the numerical values of Table I-1 and the expression of Eq. (I,23) we obtain -25.1314 hartrees, the total electronic energy of BH. This SCF energy probably lies less than 0.001 hartree above the true Hartree-Fock energy (66).

Brillouin's theorem. This theorem applies in its most general form to self-consistent-field wave functions for closed-shell atoms and molecules. Brillouin's theorem (51) states that matrix elements of the Hamiltonian operator between the Hartree-Fock determinant and all singly excited determinants are zero. A singly excited determinant is one which differs by a single spinorbital from the Hartree-Fock determinant.

Table I-1. Electronic integrals arising in the SCF energy expression for the ground $^1\Sigma^+$ state of boron hydride, BH. The numerical values are given in hartrees, the atomic unit for energy.

One-electron integrals

$I(1\sigma\|1\sigma)$	-12.8667
$I(2\sigma\|2\sigma)$	-3.0684
$I(3\sigma\|3\sigma)$	-2.6539

Two-electron integrals

Coulomb integrals

$(1\sigma1\sigma\|1\sigma1\sigma)$	2.8911
$(1\sigma2\sigma\|1\sigma2\sigma)$	0.5848
$(2\sigma2\sigma\|2\sigma2\sigma)$	0.5694
$(1\sigma3\sigma\|1\sigma3\sigma)$	0.5822
$(2\sigma3\sigma\|2\sigma3\sigma)$	0.3845
$(3\sigma3\sigma\|3\sigma3\sigma)$	0.4552

Exchange integrals

$(1\sigma2\sigma\|2\sigma1\sigma)$	0.0248
$(1\sigma3\sigma\|3\sigma1\sigma)$	0.0201
$(2\sigma3\sigma\|3\sigma2\sigma)$	0.0630

As an example, consider the NH_3 molecule (point group C_{3v}), for which the RHF wave function is of the form (161)

$$\psi_{RHF} = 1a_1\alpha\ 1a_1\beta\ 2a_1\alpha\ 2a_1\beta\ 1e_x\alpha\ 1e_x\beta\ 1e_y\alpha\ 1e_y\beta\ 3a_1\alpha\ 3a_1\beta \qquad (I,25)$$

If we replace the $2a_1\beta$ spin orbital by a higher orbital $na_1\beta$ (n>3) of the same symmetry, from Brillouin's theorem the singly-excited determinant ψ_S

$$\psi_S = 1a_1\alpha\ 1a_1\beta\ 2a_1\alpha\ 1e_x\alpha\ 1e_x\beta\ 1e_y\alpha\ 1e_y\beta\ 3a_1\alpha\ 3a_1\beta\ na_1\beta \qquad (I,26)$$

has a zero Hamiltonian matrix element with ψ_{RHF}

$$\int \psi_{RHF}^* H_e \psi_S d\tau = 0 \qquad (I,27)$$

The reader may note that in the above example the statement that the RHF spinorbital removed was replaced by a higher spinorbital of the *same* symmetry. What if the β spinorbital had been replaced by a spinorbital of some other spatial symmetry, for example, $1a_2\beta$ or $2e_x\beta$? In this case the H matrix element between the singly excited determinant ψ_S and ψ_{RHF} would again be zero, but by a much stronger theorem than that of Brillouin. In this case the overall spatial symmetry of ψ_S would be different from that of the RHF wave function. If $2a_1\beta$ were replaced by $1a_2\beta$, the spatial symmetry of ψ_S would be A_2, while that of ψ_{RHF} is A_1. And it is true in general that H matrix elements between determinants of different overall symmetries are identically zero.

For exact Hartree-Fock wave functions Brillouin's theorem

guarantees that Eq. (I,27) holds for any singly excited determinant ψ_S of the same symmetry. Brillouin's theorem also holds for SCF wave functions, but in a more limited sense. Here we must remember that the SCF wave function is the single determinant of lowest energy within a finite basis set. To obtain Hartree-Fock results, of course, one must use a complete set of functions. For SCF wave functions ψ_{SCF} Brillouin's theorem states that the H matrix element

$$\int \psi_{SCF}^* H_e \psi_S d\tau = 0 \qquad (I,28)$$

<u>only</u> if the single spinorbital in ψ_S not in ψ_{SCF} can be constructed from the finite basis set in which the SCF calculation was carried out.

An interesting and useful corollary to Brillouin's theorem is the <u>theorem of Møller and Plesset</u> (262) which states that one-electron properties calculated from Hartree-Fock wave functions are correct to "first order." A one-electron property is obtained as the expectation value of an operator which (aside from terms involving nuclei only) is a sum of n terms, each depending only on the coordinates of a single electron. The electric dipole moment of a molecule is a well-known example of a one-electron property. The operator for the x-component of the dipole moment vector is

$$\mu_x = \sum_{A=1}^{N} Z_A x_A - \sum_{i=1}^{n} x_i \qquad (I,29)$$

where the first sum is over nuclei and the second over electrons. x_A is the x-coordinate of the Ath nucleus and x_i is the x-coordinate of the ith electron. The theorem of Møller and Plesset gives us hope that SCF wave functions of near Hartree-Fock accuracy may be of chemical value. However, as we will see, just how accurate "first-order" accuracy is must be determined for each one-electron property by detailed calculations on a variety of atoms and molecules.

Many experimental studies are carried out for the stated purpose of obtaining information about the nature of the electron distribution in a molecule or series of molecules. For this reason it is especially noteworthy that Brillouin's theorem implies that Hartree-Fock electron densities should be correct to first order. During the past few years many theoretical studies based on SCF wave functions have used electron density contour maps (and, very recently, three-dimensional computer-generated plots) to illustrate interesting features of molecular electronic structure. For example, Figure I-1 shows some of the results of a theoretical study (108) of the formaldehyde molecule, H_2CO. The basis set used by Dunning and Winter (108) is sufficiently large that the electron distribution shown in Figure 1 is not likely to be qualitatively different from the Hartree-Fock electron distribution.

<u>Koopmans' theorem</u>. This is another very useful theorem which applies to closed-shell SCF wave functions. In the derivation of the Hartree-Fock equations, Lagrange multipliers ε_{ij} are introduced

Contour and three dimensional perspective plots of the total molecular density

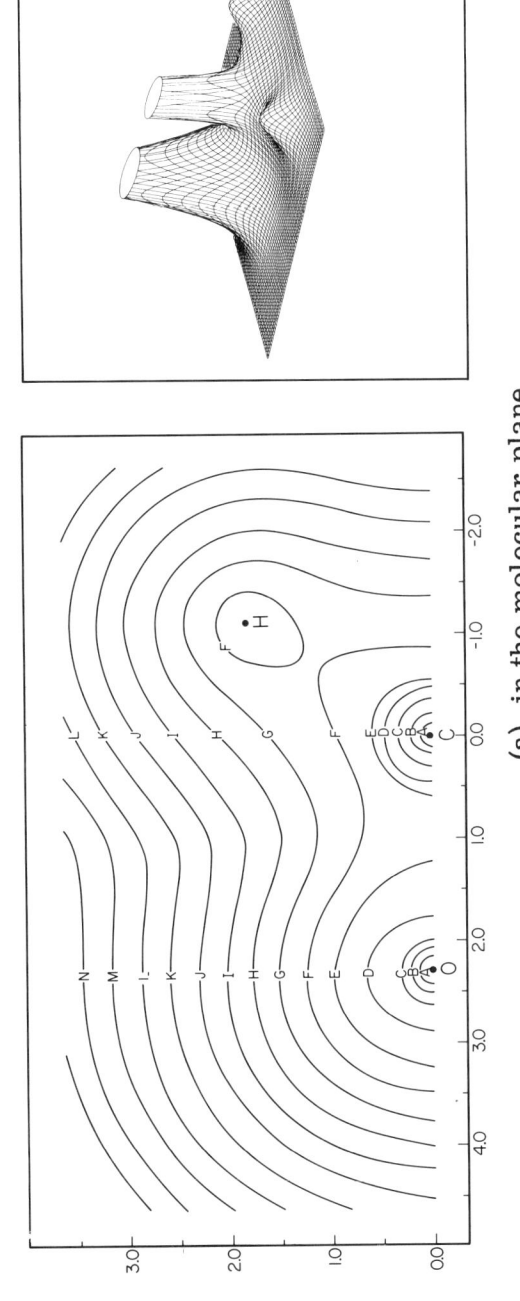

(a) in the molecular plane

Figure I-1. Total electron density plots for the formaldehyde molecule. The traditional contour map has axes marked to show distances in bohrs from the carbon atom. The contour key is given on the next page.

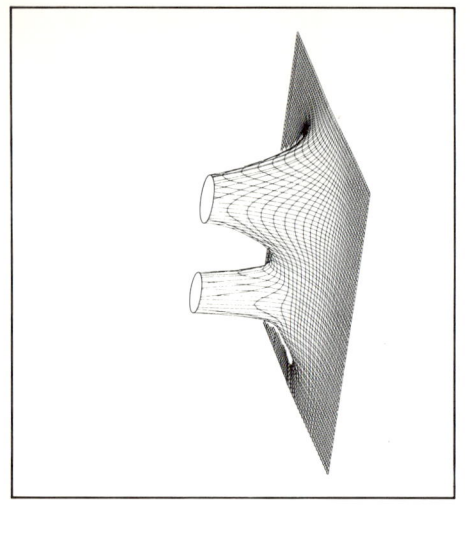

(b) in the pi plane

Letter	Magnitude[a]		Letter	Magnitude[a]
A	64.7837		H	0.0633
B	16.1959		I	0.0316
C	4.0490		J	0.0158
D	1.0122		K	0.0079
E	0.5061		L	0.0040
F	0.2531		M	0.0020
G	0.1265		N	0.0010

[a] Units are electrons/(atomic unit of length)^3

to guarantee that all pairs of Hartree-Fock orbitals i and j will be orthogonal (288). Upon solution of the HF equations only the diagonal elements ε_{ii}, which we hereafter designate ε_i and call orbital energies, have nonzero values. <u>Koopmans' theorem</u> (212) states that these orbital energies ε_i may be associated with the ionization potentials of the atom or molecule for which the SCF wave function has been obtained. For a more precise definition of Koopmans' theorem we note that

$$\varepsilon_i = I(i|i) + \sum_j [(ij|ij) - (ij|ji)] \qquad (I,30)$$

where the sum of coulomb minus exchange integrals goes over all n spinorbitals occupied in the SCF wave function. Now consider a two step process:

(1) Calculate an SCF wave function for a neutral molecule and evaluate its energy using Eq. (I,16).

(2) Using the SCF orbitals from the above neutral molecule calculation, remove one spinorbital and evaluate the single determinant energy of the resulting positive ion.

Koopmans showed that the energy difference or ionization potential calculated in this way is just the orbital energy ε_i given by Eq. (I,30).

As an example, consider the LiH molecule, for which the ground state Hartree-Fock wave function is of the form

$$\psi_{RHF} = 1\sigma\alpha \ 1\sigma\beta \ 2\sigma\alpha \ 2\sigma\beta \qquad (I,31)$$

According to Eq. (I,30) the 1σ orbital energy is

$$\varepsilon_{1\sigma} = I(1\sigma|1\sigma) + (1\sigma 1\sigma|1\sigma 1\sigma)$$
$$+ 2(1\sigma 2\sigma|1\sigma 2\sigma) - (1\sigma 2\sigma|2\sigma 1\sigma) \qquad (I,32)$$

From Eq. (I,16), the HF energy for LiH is

$$E(\text{LiH}) = 2I(1\sigma|1\sigma) + 2I(2\sigma|2\sigma)$$
$$+ (1\sigma 1\sigma|1\sigma 1\sigma)$$
$$+ 4(1\sigma 2\sigma|1\sigma 2\sigma) - 2(1\sigma 2\sigma|2\sigma 1\sigma) \qquad (I,33)$$
$$+ (2\sigma 2\sigma|2\sigma 2\sigma) + 3/R_{\text{LiH}}$$

For that positive ion LiH^+ in which a 1σ electron has been removed the RHF wave function takes the form

$$\psi_{\text{RHF}} = 1\sigma\alpha \; 2\sigma\alpha \; 2\sigma\beta \qquad (I,34)$$

and the resulting energy is

$$E(\text{LiH}^+) = I(1\sigma|1\sigma) + 2I(2\sigma|2\sigma)$$
$$+ 2(1\sigma 2\sigma|1\sigma 2\sigma) - (1\sigma 2\sigma|2\sigma 1\sigma) \qquad (I,35)$$
$$+ (2\sigma 2\sigma|2\sigma 2\sigma) + 3/R_{\text{LiH}}$$

We see by inspection that the energy difference $E(\text{LiH}) - E(\text{LiH}^+)$, the ionization potential, is in fact just the 1σ orbital energy given by Eq. (I,32).

Perhaps the most important feature of Koopmans' theorem is that the ionization potential it refers to is that obtained from a calculation on the positive ion using orbitals variationally determined for the neutral molecule. In general one would expect a better result to be obtained by carrying out independent SCF calculations on the neutral molecule and the particular state of interest of the positive ion. We will return to this point in later sections of the book.

$\underset{\sim\sim\sim\sim\sim\sim\sim\sim\sim\sim}{\text{Open shells}}$. In general it is not possible to express the Hartree-Fock wave function for an open-shell atom or molecule as a single determinant. However, the HF wave functions for the ground states of most radicals (open-shell species) can be written as single determinants. In particular any electron configuration with only a single electron outside closed shells will be described as a single determinant. Consider the ground states of the C (carbon), CH, CH_2 (methylene), and CH_3 (methyl) radicals, which belong to point groups K_h, $C_{\infty v}$, C_{2v}, and D_{3h}. The forms of the HF wave functions are

$$1s\alpha\ 1s\beta\ 2s\alpha\ 2s\beta\ 2p_0\alpha\ 2p_{+1}\alpha \qquad ^3P \quad C \qquad (I,36)$$

$$1\sigma\alpha\ 1\sigma\beta\ 2\sigma\alpha\ 2\sigma\beta\ 3\sigma\alpha\ 3\sigma\beta\ 1\pi_+\alpha \qquad ^2\Pi \quad CH \qquad (I,37)$$

$$1a_1\alpha\ 1a_1\beta\ 2a_1\alpha\ 2a_1\beta\ 1b_2\alpha\ 1b_2\beta\ 3a_1\alpha\ 1b_1\alpha \qquad ^3B_1 \quad CH_2 \qquad (I,38)$$

$$1a_1'\alpha\ 1a_1'\beta\ 2a_1'\alpha\ 2a_1'\beta\ 1e_x'\alpha\ 1e_x'\beta\ 1e_y'\alpha\ 1e_y'\beta\ 1a_2''\alpha \qquad ^2A_2'' \quad CH_3 \qquad (I,39)$$

We see that all four of the above open-shell Hartree-Fock wave functions can be written as single determinants.

The excited states of atoms and molecules are frequently described by RHF wave functions which must be written as a linear combination of Slater determinants. This is necessary to enable the RHF wave function to have the same spin and spatial symmetry as the excited state in question. As an example, consider the lowest ^1S state of the carbon atom, for which

$$\psi_{RHF} = \frac{1}{\sqrt{3}} \; 1s\alpha \; 1s\beta \; 2s\alpha \; 2s\beta \; 2p_{-1}\alpha \; 2p_{+1}\beta$$

$$- \frac{1}{\sqrt{3}} \; 1s\alpha \; 1s\beta \; 2s\alpha \; 2s\beta \; 2p_{-1}\beta \; 2p_{+1}\alpha \qquad (I,40)$$

$$- \frac{1}{\sqrt{3}} \; 1s\alpha \; 1s\beta \; 2s\alpha \; 2s\beta \; 2p_0\alpha \; 2p_0\beta$$

For the CH radical an excited state of interest is the C $^2\Sigma^+$ state, arising from the $1\sigma^2 2\sigma^2 3\sigma 1\pi^2$ electron configuration. The RHF wave function for this state is of the form

$$\frac{1}{\sqrt{2}} \; 1\sigma\alpha \; 1\sigma\beta \; 2\sigma\alpha \; 2\sigma\beta \; 3\sigma\alpha \; 1\pi_+\alpha \; 1\pi_-\beta$$
$$\qquad\qquad\qquad\qquad\qquad\qquad\qquad\qquad (I,41)$$
$$- \frac{1}{\sqrt{2}} \; 1\sigma\alpha \; 1\sigma\beta \; 2\sigma\alpha \; 2\sigma\beta \; 3\sigma\alpha \; 1\pi_+\beta \; 1\pi_-\alpha$$

A third example is provided by the $\tilde{b} \; ^1B_1$ state of methylene:

$$RHF = \frac{1}{\sqrt{2}} \; 1a_1\alpha \; 1a_1\beta \; 2a_1\alpha \; 2a_1\beta \; 1b_2\alpha \; 1b_2\beta \; 3a_1\alpha \; 1b_1\beta$$
$$\qquad\qquad\qquad\qquad\qquad\qquad\qquad\qquad (I,42)$$
$$- \frac{1}{\sqrt{2}} \; 1a_1\alpha \; 1a_1\beta \; 2a_1\alpha \; 2a_1\beta \; 1b_2\alpha \; 1b_2\beta \; 3a_1\beta \; 1b_1\alpha$$

Excited states of chemical interest are usually represented by Hartree-Fock wave functions with just a few electrons outside closed shells. For this reason, as seen above, the number of determinants required to give the proper spin and spatial symmetry to the RHF function is usually quite small. And the coefficients of these Slater determinants are generally numbers like $\frac{1}{\sqrt{2}}$ or $\frac{1}{\sqrt{3}}$.

However, when configuration interaction is taken into account, as we shall see in Section IC, very complicated linear combinations of determinants arise.

In principle it should be no more difficult to obtain SCF wave functions for open-shell systems than for closed-shell systems. In both cases, we take a symmetry-determined linear combination of determinants and vary the orbitals to minimize the calculated total energy. In our discussion of natural orbitals we will discuss briefly a procedure for obtaining SCF wave functions which is no different for open- than for closed-shell cases. The Nesbet procedure (270) for solving the SCF equations is also conceptually the same for open and closed shells, but for some open-shell cases (e.g., the $1s^2 2s 2p$ ^1P state of beryllium) approximations are made so that the wave functions obtained yield a total energy somewhat above the SCF energy. The most commonly used procedure for solving the SCF equations in terms of a finite basis set of analytic functions is that of Roothaan (317-319). The Roothaan procedure is considerably simpler for closed- (317) than for open-shell (318) systems and as formulated (318,319) is restricted to no more than one open shell of a particular symmetry. A more general method recently introduced by Hunt, Dunning, and Goddard (175) is being increasingly used to obtain open-shell SCF wave functions.

Both Brillouin's theorem and Koopmans' theorem apply with no restrictions only to closed-shell systems. Sometimes, however, they apply in a limited sense to open shells. For example, <u>some classes</u> of singly excited configurations do have zero matrix elements with open-shell SCF wave functions. It is also <u>sometimes</u> true that orbital energies obtained from open-shell SCF calculations do correspond to ionization potentials in the sense of Koopmans' theorem. The NH_2 radical in its $^2B_1$ ground state may be taken as an example

$$\psi_{RHF} = 1a_1\alpha \; 1a_1\beta \; 2a_1\alpha \; 2a_1\beta \; 1b_2\alpha \; 1b_2\beta \; 3a_1\alpha \; 3a_1\beta \; 1b_1\alpha \qquad (I,43)$$

All determinants formed by replacing the $1b_1\alpha$ spinorbital with a higher spinorbital $2b_1\alpha$, $3b_1\alpha$, $4b_1\alpha$,... have an identically zero Hamiltonian matrix element with the RHF wave function. In addition the $1b_1$ orbital energy ε_{1b_1} will in fact correspond to the difference between the energy obtained from wave function (I,43) and the positive ion wave function which results from removing the $1b_1\alpha$ spinorbital. However, the other types of determinants formed by replacing one spinorbital of (I,43) <u>will</u> have nonzero H matrix elements with the RHF wave function. Furthermore, the $1a_1$, $2a_1$, $1b_2$, and $3a_1$ orbital energies do <u>not</u> correspond to ionization potentials of NH_2 obtained as differences between the energy of wave function (I,43) and the energies of the positive ion NH_2 wave functions constructed using the same orbitals.

Extensions of the restricted Hartree-Fock approximation.
By removing the symmetry and equivalence restrictions placed on RHF wave functions, single-determinant wave functions of lower

energy can frequently be obtained. In particular for atoms there have been a considerable number of calculations which go beyond the restricted Hartree-Fock approximation but retain the one-spinorbital-per-electron concept. We shall use the ^4S ground state of the nitrogen atom to compare the different "extended" Hartree-Fock approximations. For reference the RHF wave function for ^4S N is of the form

$$\psi_{RHF} = 1s\alpha \; 1s\beta \; 2s\alpha \; 2s\beta \; 2p_{-1}\alpha \; 2p_0\alpha \; 2p_{+1}\alpha \tag{I,44}$$

and the RHF energy is found to be -54.4009 hartrees (71).

The first equivalence restriction usually removed is that by which the spatial parts of each closed-shell pair of spinorbitals (e.g., $1s\alpha$ and $1s\beta$) are identical. Removal of this equivalence yields what is usually called the unrestricted Hartree-Fock (UHF) wave function (305). However, since the only equivalence removed is that between orbitals with the same spatial quantum numbers but different values of m_s, it seems more reasonable to call this wave function <u>spin unrestricted Hartree-Fock</u> (SUHF). For the nitrogen atom ground state the SUHF wave function is of the form

$$\psi_{SUHF} = 1s\alpha \; 1s'\beta \; 2s\alpha \; 2s'\beta \; 2p_{-1}\alpha \; 2p_0\alpha \; 2p_{+1}\alpha \tag{I,45}$$

where the 1s and 1s' orbitals are no longer identical, but the spinorbitals $1s\alpha$ and $1s'\beta$ remain orthogonal due to the orthogonality of the α and β spin functions, Eq. (I,18). Varying the forms of the 1s, 1s', 2s, 2s', and 2p orbitals to minimize the energy expression (I,16) yields (13) an SUHF energy of -54.4046 hartrees for ^4S N.

The energy lowering over the RHF is seen to be rather small, of the order of one-tenth of an electron volt.

The frequently raised and primary objection to the SUHF method (127) is that the resulting single-determinant wave function is not an exact eigenfunction of the spin operator $\underset{\sim}{S}^2$. The correct electronic wave function must of course satisfy

$$\underset{\sim}{S}^2 \psi_e = S(S+1) \psi_e \qquad (I,46)$$

In all fairness it should be pointed out that the SUHF wave functions are expected to nearly satisfy Eq. (I,46). For example, the SUHF wave function of Bagus and Liu (11) for the ^5D state of the iron atom was found to have $S(S+1) = 6.018$, as opposed to the correct value, $2 \cdot 3 = 6$.

The most straightforward way of making the SUHF wave function an exact eigenfunction of $\underset{\sim}{S}^2$ is to apply a spin projection operator O_S (235). The resulting wave function may be referred to as the projected unrestricted Hartree-Fock (PUHF) wave function. For ^4S N this wave function is of the form

$$\psi_{PUHF} = O_S \psi_{SUHF} \qquad (I,47)$$

and the resulting total energy is -54.4058 hartrees (136). However, the PUHF wave function is not strictly variational--that is the energy was not minimized for the wave function of form (I,47), but rather for the SUHF wave function (I,45). For the <u>spin-extended Hartree-Fock</u> (SEHF) wave function, the variational principle is applied _after_ the spin projection. For nitrogen we have

$$\psi_{SEHF} = O_S \; 1s\alpha \; 1s'\beta \; 2s\alpha \; 2s'\beta \; 2p_{-1}\alpha \; 2p_0\alpha \; 2p_{+1}\alpha \qquad (I,48)$$

and the total energy has been calculated (136) to be -54.4064 hartrees.

An even further extension of the RHF method can be obtained by realizing for example, that the SEHF wave function (I,48) is just one of _six_ linear combinations of determinants arising from the electron configuration 1s 1s' 2s 2s' $2p^3$ and having L=0, S=3/2. For a general discussion of this "spin degeneracy" problem the reader is referred to the book by Pauncz (291). We shall refer to that six configuration ^4S N wave function for which the 1s, 1s', 2s, 2s', and 2p orbitals have been variationally determined as the _spin-optimized Hartree-Fock_ (SOHF) wave function (192,219). For nitrogen, the SOHF energy is -54.4217 hartrees (193), which unlike the other types of extended HF functions, _does_ represent a very significant lowering of the RHF energy. The SOHF method is about as far as one can go while retaining the independent electron interpretation (one orbital per electron). However, the SOHF method abandons orbital orthogonality, Eq. (I-15), and as a result the energy expression, unlike Eq. (I,16), is extremely complicated. Because of the nonorthogonality problem (363) it is unlikely that SOHF wave functions will be available for atoms and molecules with more than 10 electrons in the near future.

Goddard and coworkers (219, 45) have shown that for molecules the SOHF wave functions are more closely related to valence bond ideas than to the molecular orbital picture theoretical chemists generally work within. That is, the SOHF orbitals in molecules tend to be of a localized and frequently hybridized nature.

As an example, consider the BH molecule for which the SOHF wave function is the optimum linear combination of the five singlet (S=1) spin functions which can be constructed (291) under the restriction that each of the 1σ, $1\sigma'$, 2σ, $2\sigma'$, 3σ, and $3\sigma'$ spatial orbitals is singly occupied. As was the case for the nitrogen atom, the SOHF wave function yields a markedly lower energy than the RHF wave function. The Cade-Huo (66) near RHF energy was -25.1314 hartrees, while an SCF calculation using a smaller basis set but within the SOHF formulation gives an energy of -25.1664 hartrees (45). The nature of the spin coupling in the approximate SOHF wave function for BH is particularly interesting since it is found (45) that one of the five spin functions dominates the wave function. The only six-electron spin function of any importance for BH is that in which the 1σ and $1\sigma'$ orbitals are singlet coupled. Similarly the 2σ and $2\sigma'$ orbitals are singlet coupled, as are the 3σ $3\sigma'$ pair in this valence-bond-like spin function.

Thus far the BH molecular orbitals have been constrained to have σ symmetry, that is to be cylindrically symmetric about the internuclear axis. Blint and Goddard (46) have shown that by removing this symmetry restriction, an even lower energy, -25.1801 hartrees, is obtained. Since the six spatial orbitals are now allowed to have π character, the final molecular orbitals may be referred to as 1a, 1b, 2a, 2b, 3a, and 3b. Figure I-2 shows the valence orbitals as a function of internuclear separation for the calculation of Blint and Goddard (46) described above. The 2a and 2b orbitals seen in Figure I-2 are nonbonding sp hybrids localized on the boron atom. Orbitals 2a and 2b differ only in that 2a points in the positive x direction (out of the book) while 2b is oppositely directed. The 3a and 3b orbitals are radically different and in the simplest picture are related to the $2p_z$ boron and 1s hydrogen orbitals which are coupled in the traditional valence bond picture (89).

C. CONFIGURATION INTERACTION

Electron correlation. In the Hartree-Fock approximation, the motion of each electron is solved for in the presence of the average potential created by the remaining (n-1) electrons. As such the HF approximation neglects the instantaneous (rather than averaged) repulsions between pairs of electrons. The contribution to the total energy due to instantaneous repulsions is called the correlation energy (234), since the motion of the electrons is correlated in that two electrons are unlikely to get very close to each other, given Coulomb's law. To be more specific the

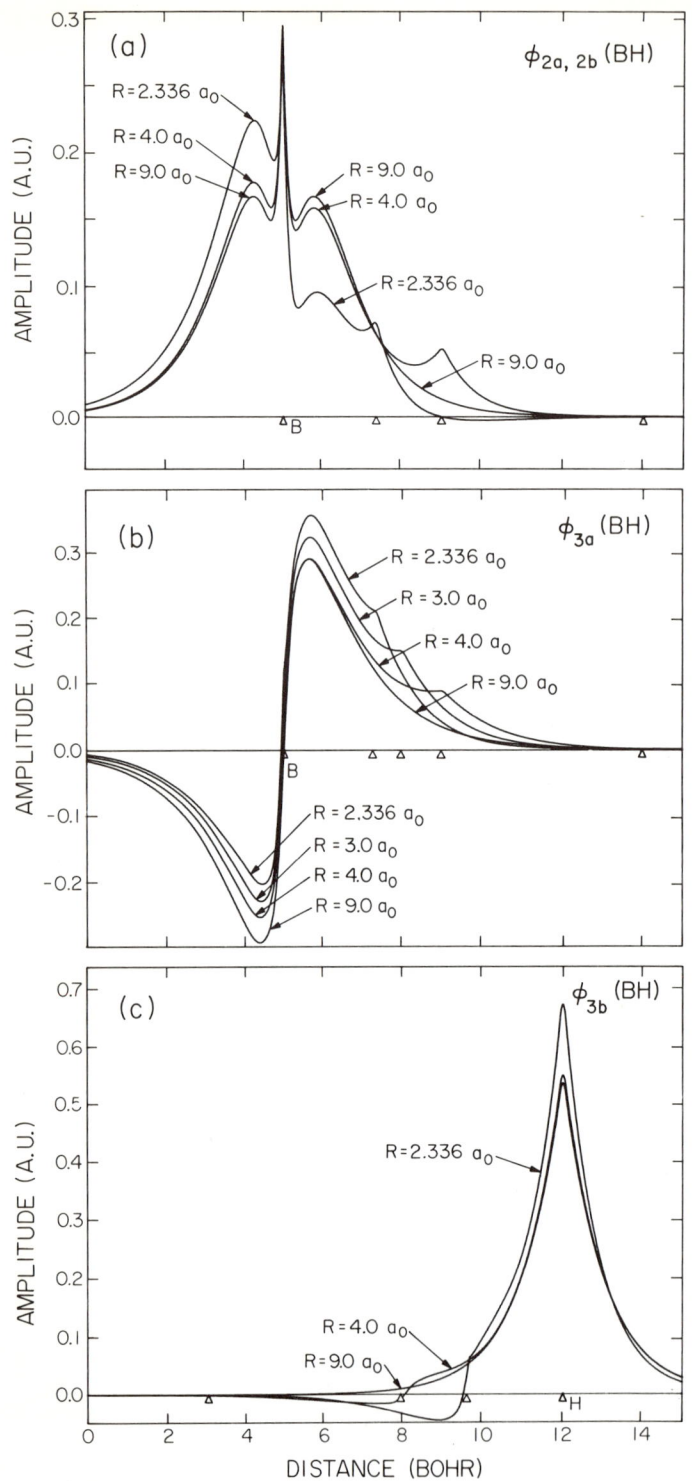

Figure I-2. Amplitude of the extended SCF orbitals of Blint and Goddard (46) for the ground state of BH as a function of internuclear separation. In (a) and (b) the boron atom position is fixed and the small triangles denote the position of hydrogen. In (c) the hydrogen atom position is fixed and the triangles denote the position of the boron.

correlation energy E_c is usually accepted (72) to be the difference between the restricted Hartree-Fock energy and the exact non-relativistic energy of a particular system.

The correlation energy is usually a relatively small percentage of the total energy of an atom or molecule. For example, for the neon atom the correlation energy, 0.38 hartrees, is only 0.3% of the total energy -129.06 hartrees (377). This small percentage might lead an unsuspecting observer to conclude that the RHF approximation, which neglects electron correlation, is adequate. And for many purposes the RHF approximation is an excellent one; for example, molecular geometries, some one-electron properties, and ionization potentials predicted from RHF wave functions are frequently in good agreement with experiment. However, though 0.3% is a small percentage, the actual value of the correlation energy of neon is more than 10 eV, which is larger than the amount of energy required to break most chemical bonds.

Since energy differences are of primary importance to chemists, one might hope the correlation energy to be constant as a function of molecular geometry. Were this the case, accurate dissociation energies and potential energy surfaces could be obtained from Hartree-Fock wave functions. For chemical reactions involving only closed-shell species, Snyder and Basch (364) have in fact found SCF calculations to give fair agreement with experiment for heats of reaction. However, for most diatomic molecules the molecular RHF wave functions do not dissociate to RHF wave functions describing the correct states of the separated atoms (363). The F_2 molecule is predicted to have a negative dissociation energy in the restricted Hartree-Fock approximation (384) and the RHF dissociation energy for O_2 is slightly more than one-fourth of the experimental value (342). The barrier height or activation energy for the simple chemical reaction $Cl + H_2 \rightarrow HCl + H$ is predicted from accurate SCF calculations (325) to be three or more times greater than experiment. In general, perhaps the most serious drawback of the Hartree-Fock approximation is its inability to describe molecule formation and dissociation.

Recent theoretical work has shown other serious deficiencies in the self-consistent-field approach. Despite the theorem of Møller and Plesset (262) the RHF dipole moment of carbon monoxide has been shown to have the incorrect sign (178). The ^4S ground state of the nitrogen atom has a hyperfine structure which is well known experimentally but vanishes in the RHF approximation. The list of examples could be extended but the essential point is clear: in many circumstances the theoretician must go beyond the RHF approximation in order to make a contribution to the solution of problems of chemical interest.

The basic idea; the eigenvalue problem. The most frequently used method for approaching the electron correlation problem is configuration interaction (185), abbreviated CI. For any atom or molecule, there are an infinite number of orbitals in addition to the Hartree-Fock orbitals. These higher orbitals can of course be used to construct other (after the RHF) configurations. A CI wave function is just a linear combination of such configurations with coefficients variationally determined. The CI method is in principle exact because, as the basis set of one-electron functions (orbitals) approaches completeness and we include in our wave function all configurations which can be constructed from these orbitals, we approach the exact solution to the Schrödinger equation. Although in practice it is never possible to use a complete set of orbitals, the CI method is a very reasonable procedure for going beyond Hartree-Fock.

More precisely, the CI wave function is of the form

$$\psi_e = \sum_i c_i \Phi_i \tag{I,49}$$

where the Φ's are an orthonormal set of n electron configurations, to be described in more detail in the following section. The coefficients c_i are determined to minimize the energy $\int \psi_e^* H_e \psi_e d\tau$. Application of the variation principle leads to the well-known eigenvalue problem (165)

$$(\underset{\sim}{H} - E \underset{\sim}{1}) \underset{\sim}{C} = 0 \tag{I,50}$$

$\underset{\sim}{H}$ is composed of matrix elements between configurations

$$H_{ij} = \int \Phi_i^* H_e \Phi_j d\tau \tag{I,51}$$

The eigenvalue problem (I,50) is usually solved by one of several iterative techniques (396). If M configurations are included in the CI calculations, solution of equation (I,50) yields M distinct energies, or eigenvalues. From the variational principle the lowest of the calculated energies is an upper bound to the true energy of the ground state. More generally, MacDonald's theorem (238) states that the kth lowest eigenvalue of Eq. (I,50) is a rigorous upper bound to the kth lowest exact energy. With each energy E is associated an eigenvector or set of coefficients c_i which define the corresponding CI wave function.

Hamiltonian matrix elements H_{ij} between configurations i and j of different symmetries are zero. Therefore the secular equation is greatly simplified by only considering configurations which have the total symmetry of the particular electronic state being investigated. Frequently one is only interested in the wave function and energy of the lowest state of a particular symmetry. In this case one need not solve for all M eigenvalues and eigenvectors of Eq. (I,50). In fact theoretical chemists have developed special methods (274,349) for obtaining the lowest eigenvalue and corresponding eigenvector of large, real symmetric matrices. These methods, due to Nesbet (274) and Shavitt (349), make possible the consideration of much larger CI problems than are usually considered practical. For the ethylene molecule, for example, a 12,077 configuration wave function has been obtained using these methods (349).

Orbital occupancies, configurations, and symmetry. It is

necessary here to introduce the notation to be used throughout in the discussion of configuration interaction results. An <u>orbital occupancy</u> is just a collection of orbitals, written without regard to those quantum numbers which do not affect the orbital energy. For example, the m_ℓ and m_s quantum numbers do not affect the orbital energy of a 2p orbital in the RHF approximation ($\varepsilon_{2p_{-1}\alpha} = \varepsilon_{2p_{-1}\beta} = \varepsilon_{2p_0\alpha} = \varepsilon_{2p_0\beta} = \varepsilon_{2p_{+1}\alpha} = \varepsilon_{2p_{+1}\beta}$). Therefore the orbital occupancy for any state of an atom does not specify m_ℓ or m_s values. The ground state of neon is thus represented by the orbital occupancy $1s^2 2s^2 2p^6$. For the oxygen molecule the lowest orbital occupancy is $1\sigma_g^2 1\sigma_u^2 2\sigma_g^2 2\sigma_u^2 3\sigma_g^2 1\pi_u^4 1\pi_g^2$, while that for the methyl radical is $1a_1'^2 2a_1'^2 1e'^4 1a_2''$. Note that for the methyl radical the orbital occupancy does not distinguish between $1e_x'$ and $1e_y'$ orbitals but merely states that four electrons occupy $1e'$. What we call the orbital occupancy is frequently referred to as the <u>electron configuration</u> elsewhere.

We define a <u>configuration</u> (not to be confused with electron configuration) as a symmetry-adapted linear combination of Slater determinants. The phrase <u>symmetry-adapted</u> implies that the chosen linear combination of determinants possesses all the symmetry of the molecular state being described by the approximate wave function. A Slater determinant D_i is defined as in Eq. (I,10) and thus the configuration Φ is written

$$\Phi = \sum_i c_i D_i \qquad (I,52)$$

Many examples of configurations have already been given, the least trivial of which are (I,40), (I,41), and (I,42). It is often found that a single orbital occupancy gives rise to several different configurations. To determine what configurations arise from a given orbital occupancy it is only necessary to consider the electrons outside closed shells. Thus, for the lowest orbital occupancy of O_2, $1\sigma_g^2 \, 1\sigma_u^2 \, 2\sigma_g^2 \, 2\sigma_u^2 \, 3\sigma_g^2 \, 1\pi_u^4 \, 1\pi_g^2$, only the last two $1\pi_g$ electrons are pertinent. It turns out (363) that three configurations are possible for $1\pi_g^2$ which can be written as

$$\Phi_1 = 1\pi_{g-}\alpha \, 1\pi_{g+}\alpha \qquad\qquad {}^3\Sigma_g^- \qquad (I,53)$$

$$\Phi_2 = 1\pi_{g+}\alpha \, 1\pi_{g+}\beta \qquad\qquad {}^1\Delta_g \qquad (I,54)$$

$$\Phi_3 = \frac{1}{\sqrt{2}} 1\pi_{g-}\alpha \, 1\pi_{g+}\beta$$

$$\qquad\qquad\qquad\qquad\qquad {}^1\Sigma_g^+ \qquad (I,55)$$

$$\quad - \frac{1}{\sqrt{2}} 1\pi_{g-}\beta \, 1\pi_{g+}\alpha$$

and each corresponds to an electronic state of different symmetry.

CI calculations can be greatly simplified if they are carried out in terms of configurations, rather than Slater determinants. For example, a recent calculation (381) on the neon atom included 8392 distinct Slater determinants. It is clear that the diagonalization of a matrix of order 8392 is a difficult task. However, only 434 configurations of ${}^1S$ atomic symmetry can be constructed as linear combinations of the 8392 determinants and the 434 configuration calculation was relatively straightforward. Diagonalization of the 8392 × 8392 matrix would have given exactly the same result.

For simple cases (e.g., ground state RHF wave functions) symmetry-adapted configurations can be constructed by inspection or by trial and error (362,363). In general, however, this is not possible. Two methods of general applicability involve (a) the use of projection operators (235) and (b) the direct diagonalization (186) of the appropriate symmetry operators (e.g., $\underset{\sim}{S}^2$). For systems of highest symmetry, particularly atoms and diatomic molecules, construction of symmetry-adapted functions is most difficult. For atoms and diatomics both the projection operator (320,214) and direct diagonalization (334,339) methods have been implemented on large-scale computers to handle essentially any case of interest. Table I-2 illustrates a nontrivial example, the three $^3\Sigma_g^-$ configurations arising from the $1\sigma_g^2\ 1\sigma_u^2\ 2\sigma_g^2\ 2\sigma_u^2\ 3\sigma_g\ 3\sigma_u\ 1\pi_u^3\ 1\pi_g^3$ orbital occupancy. For molecules of symmetry lower than $C_{\infty v}$ but containing a threefold or higher axis of rotation (87) the problem of constructing symmetry-adapted configurations is still difficult due to doubly degenerate irreducible representations (e.g., e'_x and e'_y in the CH_3 radical), but at least two practical approaches have been implemented (134,58). The simplest cases are those arising in molecules belonging to point group D_{2h} or lower. In these cases each determinant has a trivially obtained spatial symmetry and only the well-studied spin degeneracy problem (291) need be considered.

Table I-2. The three linearly independent $^3\Sigma_g^-$ configurations Φ_1, Φ_2, and Φ_3 which arise from the orbital occupancy $1\sigma_g^2\,1\sigma_u^2\,2\sigma_g^2\,2\sigma_u^2\,3\sigma_g^2\,3\sigma_u\,1\pi_u^3\,1\pi_g^3$ of the oxygen molecule O_2. Since the $1\sigma_g$, $1\sigma_u$, $2\sigma_g$, and $2\sigma_u$ orbitals are doubly occupied in all determinants, they are not explicitly included.

	Φ_1	Φ_2	Φ_3
$D_1 = 3\sigma_g\alpha\;3\sigma_u\alpha\;1\pi_{u-}\alpha\;1\pi_{u+}\alpha\;1\pi_{u-}\beta\;1\pi_{g-}\beta\;1\pi_{g+}\beta$	-0.19811	-0.57937	0.00930
$D_2 = 3\sigma_g\alpha\;3\sigma_u\alpha\;1\pi_{u-}\alpha\;1\pi_{u+}\beta\;1\pi_{u-}\beta\;1\pi_{g-}\alpha\;1\pi_{g+}\beta$	0.57174	0.02375	0.21805
$D_3 = 3\sigma_g\alpha\;3\sigma_u\alpha\;1\pi_{u-}\alpha\;1\pi_{u+}\alpha\;1\pi_{u-}\beta\;1\pi_{g-}\beta\;1\pi_{g+}\beta$	0.19811	0.57937	-0.00930
$D_4 = 3\sigma_g\alpha\;3\sigma_u\alpha\;1\pi_{u-}\beta\;1\pi_{u+}\alpha\;1\pi_{u-}\alpha\;1\pi_{g-}\alpha\;1\pi_{g+}\beta$	-0.57174	-0.02375	-0.21805
$D_5 = 3\sigma_g\alpha\;3\sigma_u\beta\;1\pi_{u-}\alpha\;1\pi_{u+}\beta\;1\pi_{u-}\beta\;1\pi_{g-}\alpha\;1\pi_{g+}\alpha$	-0.36579	0.34642	0.34813
$D_6 = 3\sigma_g\alpha\;3\sigma_u\beta\;1\pi_{u-}\alpha\;1\pi_{u+}\alpha\;1\pi_{u-}\beta\;1\pi_{g-}\beta\;1\pi_{g+}\alpha$	0.36579	-0.34642	-0.34813
$D_7 = 3\sigma_g\beta\;3\sigma_u\alpha\;1\pi_{u-}\beta\;1\pi_{u+}\alpha\;1\pi_{u-}\alpha\;1\pi_{g-}\alpha\;1\pi_{g+}\beta$	-0.00785	0.20920	-0.57548
$D_8 = 3\sigma_g\beta\;3\sigma_u\alpha\;1\pi_{u-}\alpha\;1\pi_{u+}\alpha\;1\pi_{u-}\beta\;1\pi_{g-}\alpha\;1\pi_{g+}\alpha$	0.00785	-0.20920	0.57548

It is often found that a CI wave function is dominated (normalized coefficient ≥ 0.9) by a single configuration, the RHF wave function. In such cases it is found that 95% or more of the correlation energy can be accounted for by configurations which differ by one or two orbitals from the RHF or reference configuration. Therefore we introduce a shorter excitation notation to specify the orbital occupancies from which these configurations arise. Take the BH_2 molecule as an example,

$$\psi_{RHF} = 1a_1\alpha \; 1a_1\beta \; 2a_1\alpha \; 2a_1\beta \; 1b_2\alpha \; 1b_2\beta \; 3a_1\alpha \quad {}^2A_1 \qquad (I,56)$$

where the RHF orbital occupancy is $1a_1^2 \; 2a_1^2 \; 1b_2^2 \; 3a_1$. A singly excited configuration (or <u>single excitation</u>) is one arising from an orbital occupancy differing from the RHF by one orbital. For the ${}^2A_1$ ground state of BH_2 the possible types of single excitations and the shorthand notation are given in Table I-3. Note that for several orbital occupancies there are three unpaired electrons, which give rise to two linearly independent doublet (S=1/2) spin configurations. Similarly a <u>double excitation</u> is a configuration arising from an orbital occupancy differing by two orbitals from the RHF orbital occupancy. The shorthand notation $2a_1 \; 1b_2 \rightarrow 4a_1 \; 2b_2$ is equivalent to the orbital occupancy $1a_1^2 \; 2a_1 \; 1b_2 \; 3a_1 \; 4a_1 \; 2b_2$ for the ${}^2A_1$ state of BH_2.

<u>Hamiltonian matrix elements</u>. To carry out CI calculations, matrix elements H_{ij} of the type given in Eq. (I,51) must be computed.

Table I-3. Singly-excited configurations for the ground state of BH_2.

Orbital Occupancy	Shorthand Notation	Number of $^2A_1$ Configurations per Orbital Occupancy
$1a_1\ 2a_1^2\ 1b_2^2\ 3a_1^2$	$1a_1 \to 3a_1$	1
$1a_1\ 2a_1^2\ 1b_2^2\ 3a_1\ na_1\ (n>3)$	$1a_1 \to na_1$	2
$1a_1^2\ 2a_1\ 1b_2^2\ 3a_1^2$	$2a_1 \to 3a_1$	1
$1a_1^2\ 2a_1\ 1b_2^2\ 3a_1\ na_1\ (n>3)$	$2a_1 \to na_1$	2
$1a_1^2\ 2a_1^2\ 1b_2\ 3a_1\ nb_2\ (n>1)$	$1b_2 \to nb_2$	2
$1a_1^2\ 2a_1^2\ 1b_2^2\ na_1\ (n>3)$	$3a_1 \to na_1$	1

The most straightforward way of doing this is to express the configurations Φ_i and Φ_j as linear combinations of determinants D and evaluate matrix elements of the type

$$\int D_k^* H_e D_\ell \, d\tau \tag{I,57}$$

As long as the orbitals used to build up determinants D_k and D_ℓ form an orthonormal set, (I,57) may be easily evaluated. In particular, if $k=\ell$, the formula (I,16) for the energy of a single-determinant wave function may be used.

If D_k and D_ℓ differ by a single spinorbital, ϕ_p in D_k and ϕ_q in $D\ell$,

$$D_k = A(n) \; \phi_1(1) \; \phi_2(2) \ldots \phi_p(i) \ldots \phi_n(n) \qquad (I,58)$$

$$D_\ell = A(n) \; \phi_1(1) \; \phi_2(2) \ldots \phi_q(i) \ldots \phi_n(n) \qquad (I,59)$$

it is easy to show (362) that

$$\int D_k^* \; H_e \; D_\ell \; d\tau = I(p|q) + \sum_{r \neq p}^{n} [(pr|qr) - (pr|rq)] \qquad (I,60)$$

Similarly, if D_k and D_ℓ differ by two spinorbitals,

$$D_k = A(n) \; \phi_1(1) \; \phi_2(2) \ldots \phi_p(i) \; \phi_q(j) \ldots \phi_n(n) \qquad (I,61)$$

$$D_\ell = A(n) \; \phi_1(1) \; \phi_2(2) \ldots \phi_r(i) \; \phi_s(j) \ldots \phi_n(n) \qquad (I,62)$$

the matrix element has an especially simple form:

$$\int D_k^* \; H_e \; D_\ell \; d\tau = (pq|rs) - (pq|sr) \qquad (I,63)$$

Within an orthonormal set of orbitals, if two determinants differ by three or more spinorbitals the H matrix element between these two determinants is identically zero. It is essentially for this reason that when the RHF configuration dominates the wave function, only the singly and doubly excited configurations will give significant contributions to the wave function. A configuration Φ_T differing by three orbitals from the RHF will have a zero matrix element

$$\int \Phi_{RHF} \; H_e \; \Phi_T \; d\tau = 0 \qquad (I,64)$$

with the RHF configuration. Using perturbation theory and the RHF wave function as the zero-order wave function, (I,64) implies that Φ_T makes no contribution to the energy through second order.

Multiconfiguration SCF method. Suppose we want to carry out a configuration interaction calculation on the helium atom. First we must decide what configurations to include in the wave function. Suppose a three configuration wave function including $1s^2$, $2s^2$, and $2p^2$ is decided upon

$$\psi_e = c_1 \; 1s\alpha \; 1s\beta + c_2 \; 2s\alpha \; 2s\beta \qquad (I,65)$$
$$+ c_3 \left[\frac{1}{\sqrt{3}} 2p_{-1}\alpha \; 2p_{+1}\beta - \frac{1}{\sqrt{3}} 2p_{-1}\beta \; 2p_{+1}\alpha - \frac{1}{\sqrt{3}} 2p_0\alpha \; 2p_0\beta \right]$$

Solution of the eigenvalue problem (I,50) will yield the optimum values of c_1, c_2, and c_3. However, the wave function has not been determined in a completely variational manner unless the forms of 1s, 2s, and 2p orbitals have also been varied to minimize the total energy. For a specified set of configurations, the <u>multi-configuration Hartree-Fock</u> wave function is the best (lowest energy) wave function that can be obtained by simultaneously varying both the orbitals ϕ and the CI coefficients c. The first multiconfiguration self-consistent-field (MCSCF) calculations were performed by Hartree, Hartree and Swirles (151) on the oxygen atom.

Sabelli and Hinze (329) have calculated an MCSCF wave function of the form (I,63) for He and obtain 87% of the correlation energy. An interesting feature of these MCSCF results for helium is that the optimum 2s and 2p orbitals are very different from the orbitals which describe the 1s2s ^{2}S and 1s2p ^{2}P excited states He. The latter orbitals are quite diffuse, with an average distance from the nucleus of the order of 5.0 bohr radii. On the contrary, the 2s and 2p orbitals optimum for describing electron correlation in the He ground state have values of <r> close to that for the 1s orbital, ∼1.0 bohr. This result has a general validity, namely that the optimum orbitals for describing electron correlation are physically located in the vicinity of the RHF orbitals being correlated (388).

In principle all CI calculations should be of the MCSCF variety. Unfortunately, solution of the MCSCF equations is very difficult and to date practical calculations have been limited to ten or fewer configurations. However, for many purposes a small number of configurations may be adequate. In particular Wahl and his coworkers (385) have implemented the MCSCF method and solved a number of problems of chemical interest. In later chapters we will discuss some of these results.

The density matrix and natural orbitals; H_2 as an example. The concept of natural orbitals, introduced by Löwdin (233) provides a practical approach to the calculation of CI wave functions (with optimized orbitals) including large numbers of configurations. We will give a qualitative discussion of the central ideas; for an excellent detailed description of the density matrix and natural orbitals, the reader is referred to the book by McWeeny and Sutcliffe (254).

For any electronic wave function ψ_e the one-electron density function γ is given by

$$\gamma = n \int \psi_e^* [v(1'),v(2),\ldots v(n)] \, \psi_e[v(1),v(2),\ldots v(n)] \, dv(2)\ldots dv(n) \quad (I,66)$$

in which we have integrated over the coordinates of all but the first electron. The physical interpretation of (I,66) is that γdv gives the probability of finding an electron in volume element dv. For a CI wave function constructed from orbitals ϕ, γ is of the form

$$\gamma = \sum_i \sum_j a_{ij} \phi_i^* \phi_j \quad (I,67)$$

where the a_{ij} are a set of numbers which form the <u>density matrix</u>. The <u>natural orbitals</u> reduce the density matrix γ to diagonal form,

$$\gamma = \sum_k b_k \phi_k^* \phi_k \quad (I,68)$$

and can be obtained by diagonalizing the matrix of coefficients $\underset{\sim}{a}$. The coefficients b_k are called <u>occupation numbers</u> and indicate the degree of importance of each orbital.

The reason for the importance of natural orbitals is that in a certain mathematical sense (233, 81) they give the most rapidly convergent CI expansion. If a CI calculation is carried out in terms of an arbitrary basis set, one expects to find many configurations which contribute significantly to the wave function.

However, if the density matrix a_{ij} is diagonalized to give the natural orbitals and the same calculation is repeated in terms of these natural orbitals (NO's), only those configurations built up from NO's with large occupation numbers will be important.

The above ideas may be somewhat clearer in terms of an example, the hydrogen molecule. This example also illustrates some of the unusual properties of the natural orbitals for two-electron atoms and molecules. Hagstrom and Shull (143) have used a basis set of 15 functions to study the natural orbitals of $^1\Sigma_g^+$ H_2. Within this basis set, they carried out a full configuration interaction, including the orbital occupancies given in Table I-4. Orbital occupancies such as $1\sigma_g 1\sigma_u$ are not included since they are of u (ungerade) inversion symmetry while the $^1\Sigma_g^+$ ground state of H_2 has g inversion symmetry. It is always true that a two-electron orbital occupancy can give rise to only a single configuration of any particular symmetry. Each of the 33 orbital occupancies in Table I-4 yields one $^1\Sigma_g^+$ configuration.

After carrying out this 33 configuration calculation, Hagstrom and Shull (143) found that all 33 configurations had nonvanishing coefficients c_i in the wave function (I,49). Then they set up and diagonalized the density matrix (I,67). Then using the new set of orbitals (I,68), the natural orbitals, the calculation was repeated. The interesting result is that in this second calculation, 18 configurations had identically zero coefficients. This second

Table I-4. Orbital occupancies included in the Hagstrom-Shull wave function for H_2.

$1\sigma_g^2$	$1\sigma_u^2$	$1\pi_u^2$	$1\pi_g^2$	$1\delta_g^2$
$1\sigma_g\,2\sigma_g$	$1\sigma_u\,2\sigma_u$	$1\pi_u\,2\pi_u$	$1\pi_g\,2\pi_g$	$1\delta_g\,2\delta_g$
$1\sigma_g\,3\sigma_g$	$1\sigma_u\,3\sigma_u$	$1\pi_u\,3\pi_u$	$2\pi_g^2$	$2\delta_g^2$
$1\sigma_g\,4\sigma_g$	$2\sigma_u^2$	$2\pi_u^2$		
$1\sigma_g\,5\sigma_g$	$2\sigma_u\,3\sigma_u$	$2\pi_u\,3\pi_u$		
$2\sigma_g^2$	$3\sigma_u^2$	$3\pi_u^2$		
$2\sigma_g\,3\sigma_g$				
$2\sigma_g\,4\sigma_g$				
$2\sigma_g\,5\sigma_g$				
$3\sigma_g^2$				
$3\sigma_g\,4\sigma_g$				
$3\sigma_g\,5\sigma_g$				
$4\sigma_g^2$				
$4\sigma_g\,5\sigma_g$				
$5\sigma_g^2$				

wave function is given in Table I-5. Table I-5 shows that only the configurations x^2, in which an orbital is doubly occupied, contribute to the wave function. Configurations xy are eliminated when the wave function is recomputed in terms of the natural orbitals. The total energy -1.1731 hartrees (corresponding to 99% of the dissociation energy of H_2) is the same for both the 33 and 15 configuration wave functions since in fact the two total wave functions are identical.

For this simple two-electron case each occupation number b_k (I,68) is just 2 c_k^2, where c_k is the coefficient of the configuration in which the kth natural orbital appears. The occupation numbers provide a useful index of the importance of each orbital. It is seen that the $1\sigma_g$ orbital is by far the most important and it turns out that this $1\sigma_g$ orbital is rather similar to the RHF $1\sigma_g$ obtained by minimizing the energy of the single configuration $1\sigma_g^2$. The most important configuration in a CI wave function based on natural orbitals is called the <u>first natural configuration</u> and, near the equilibrium geometry of a molecule, is frequently very similar to the SCF wave function.

The motivation for the use of natural orbitals was the feeling that a rapidly convergent CI expansion would be obtained (233, 81). For H_2 Hagstrom and Shull (143) have investigated the effect of including only those configurations constructed from orbitals with large occupation numbers in the final wave function. For reference

Table I-5. CI wave function for H_2 in terms of natural orbitals. There is a one to one correspondence between orbital occupancies and configurations for this simple system.

Configuration	Coefficient
$1\sigma_g^2$	0.9909
$1\sigma_u^2$	0.1008
$1\pi_u^2$	0.0659
$2\sigma_g^2$	0.0551
$1\pi_g^2$	0.0126
$3\sigma_g^2$	0.0103
$2\pi_u^2$	0.0095
$2\sigma_u^2$	0.0094
$1\delta_g^2$	0.0093
$4\sigma_g^2$	0.0074
$2\pi_g^2$	0.0033
$5\sigma_g^2$	0.0029
$3\pi_u^2$	0.0027
$3\sigma_u^2$	0.0025
$2\delta_g^2$	0.0021

the RHF energy for H_2 is -1.1336 hartrees (209). The energy of the first natural configuration in Table I-5 is -1.1334 hartrees. The wave function including the first three configurations in Table I-5 yields energy -1.1607 hartrees or 66% of the correlation energy. And the five-configuration wave function based on natural orbitals gives E = -1.1704 hartrees or 90% of the correlation energy. For comparison, the full 33 configuration calculation corresponds to slightly less than 97% of the correlation energy.

The natural orbital results for the ground state of H_2 tend to be somewhat misleading as regards the general usefulness of natural orbitals. In particular, for systems with more than two electrons, it does not appear possible to rigorously eliminate any configurations from the CI expansion by redoing the calculation in terms of the natural orbitals. However, the transformation to natural orbitals does ensure that the most important configurations will be among those constructed from the natural orbitals with highest occupation numbers. Another very serious problem which may not be immediately apparent from the H_2 discussion concerns the method of obtaining the natural orbitals. The method adopted by Hagstrom and Shull (143) is the most straightforward: carry out a full CI calculation and diagonalize the density matrix. However, for larger systems this procedure is not practical due to the enormous numbers of configurations which must be considered. Furthermore, if one did have the capability of doing such a full CI calculation, the resulting wave function would already contain all the desired information about the molecule and, as in the H_2 calculations discussed, the natural orbitals would only serve an interpretive purpose, albeit a very valuable one. Although direct calculation (omitting the CI procedure) of natural orbitals is an alluring thought and preliminary approximate schemes may be promising (312), rigorous direct methods for obtaining natural orbitals appear to be possible only for two-electron systems (1).

The pseudonatural orbital method. Introduced to take advantage of density matrix concepts while avoiding the problems raised in the previous paragraph, the <u>pseudonatural orbital</u> approach (112) exploits the unusual properties of natural orbitals for two-electron systems. The basic idea, due to Edmiston and Krauss (112) is to calculate the natural orbitals for a single pair of SCF orbitals while holding the remaining (n-2) electrons fixed in their SCF spinorbitals.

Edmiston and Krauss use as an example the He_2^+ molecular ion, for which the ground state RHF wave function is of the form

$$\psi_{RHF} = 1\sigma_g \alpha \; 1\sigma_g \beta \; 1\sigma_u \alpha \qquad (I,69)$$

They calculated the pseudonatural orbitals for the $1\sigma_g^2$ pair within the full orbital occupancy $1\sigma_g^2 \, 1\sigma_u$. To do this, a calculation is first carried out including orbital occupancies of the type xy $1\sigma_u$, given in Table I-6 for the basis set of Edmiston and Krauss. Note that the $1\sigma_u$ SCF orbital (the pseudonatural orbital (PSNO) calculation is always preceded by an SCF calculation) remains singly occupied in all configurations.

Many of the orbital occupancies in Table I-6 give rise to <u>two</u> $^2\Sigma_u^+$ configurations. Only one of these configurations is included in the PSNO calculation, namely that in which the first two orbitals (e.g., $n\pi_u\, m\pi_u$ in $n\pi_u\, m\pi_u\, 1\sigma_u$) are coupled to give $^1\Sigma_g^+$ symmetry and the $1\sigma_u\alpha$ spinorbital provides the total $^2\Sigma_u^+$ symmetry. With this restriction 93 configurations are included in the Edmiston-Krauss He_2^+ PSNO calculation.

After carrying out this 93 configuration calculation Edmiston and Krauss diagonalized the resulting density to obtain the PSNO's. If they had then repeated the 93 configuration calculation, only the 23 configurations of types $n\sigma_g^2\, 1\sigma_u$, $n\sigma_u^2\, 1\sigma_u$, $n\pi_u^2\, 1\sigma_u$, and $n\pi_g^2\, 1\sigma_u$ would have nonzero coefficients in the CI expansion. However, this is <u>not</u> the purpose of the PSNO procedure. It is rather to obtain a nearly optimum set of orbitals with which to calculate as much of the correlation energy as possible, not just the $1\sigma_g^2$ pair correlation energy.

In the simplest approximation the $1\sigma_g$ SCF orbital for He_2^+ is just $1s_A + 1s_B$ while the $1\sigma_u$ orbital is $1s_A - 1s_B$. The $1\sigma_g$ and $1\sigma_u$ orbitals thus lie in the same general region of space and Edmiston and Krauss reasonably argued that the $1\sigma_g^2$ PSNO's should also be appropriate for describing electron correlation arising from the unpaired $1\sigma_u$ orbital. Since the occupation numbers order the

Table I-6. Orbital occupancies used to calculate the CI wave function required to obtain pseudonatural orbitals for He_2^+. For the basis set of Edmiston and Krauss (112), n and m go from 2 through 9 for σ_g and σ_u, while n and m go from 1 to 3 for π_u and π_g. In all cases n>m.

$1\sigma_g^2 1\sigma_u$			
$1\sigma_g n\sigma_g 1\sigma_u$			
$n\sigma_g^2 1\sigma_u$	$n\sigma_u^2 1\sigma_u$	$n\pi_u^2 1\sigma_u$	$n\pi_g^2 1\sigma_u$
$n\sigma_g m\sigma_g 1\sigma_u$	$n\sigma_u m\sigma_u 1\sigma_u$	$n\pi_u m\pi_u 1\sigma_u$	$n\pi_g m\pi_g 1\sigma_u$

Table I-7. Coefficients of the seven most important configurations in the final He_2^+ wave function of Edmiston and Krauss (112) based on pseudonatural orbitals.

	Orbital Occupancy	Coefficient
1.	$1\sigma_g^2 1\sigma_u$	0.9919
2.	$1\sigma_g 1\sigma_u 2\sigma_g$	0.0827
3.	$1\sigma_g 1\sigma_u 3\sigma_g$	0.0417
4.	$1\sigma_u^2 2\sigma_u$	0.0362
5.	$1\sigma_g 1\pi_u 1\pi_g$	0.0345
6.	$1\pi_u^2 1\sigma_u$	0.0333
7.	$1\sigma_g 2\sigma_u 3\sigma_g$	0.0292

PSNO's in their importance in describing the $1\sigma_g^2$ correlation, at this stage Edmiston and Krauss truncate their basis to include only the more important orbitals $1\sigma_g - 4\sigma_g$, $1\sigma_u - 3\sigma_u$, $1\pi_u$, $2\pi_u$, $1\pi_g$, and $2\pi_g$. With this small basis they then select 45 important configurations, which are included in the final wave function for He_2^+. The six most important configurations in this final wave function for He_2^+ near its equilibrium internuclear separation are seen in Table I-7. Although the second configuration $1\sigma_g 2\sigma_g 1\sigma_u$ may appear to be one of those eliminated by the PSNO procedure, in fact it corresponds to the second spin coupling for this orbital occupancy (the $1\sigma_g$ and $2\sigma_g$ orbitals are coupled to give $^3\Sigma_g^+$ and this two electron spin function is recoupled to the $1\sigma_u$ orbital to give the necessary $^2\Sigma_u^+$ symmetry).

It should be clear that the purpose of the PSNO method is simply to generate a set of orbitals which will result in a rapidly convergent CI expansion. In systems larger than He_2^+, the SCF orbitals may be localized in different regions of space and in such cases it is necessary to calculate PSNO's for more than one pair and merge the different sets. In a recent calculation on the neon atom (381), for example, PSNO's were calculated for the $1s^2$ and $2s^2$ pairs, the latter PSNO's being assumed to also describe correlation involving the 2p orbitals. Finally, it should be pointed out that the PSNO approach is similar in spirit to the work of Kutzelnigg and coworkers (2), who directly (no CI calculation such as that outlined in Table I-6) obtain approximate natural orbitals for each independent pair of spinorbitals.

The Iterative Natural Orbital Method. Perhaps the most powerful method for carrying out large scale CI calculations with optimized orbitals is the iterative natural orbital (INO) method introduced by Bender and Davidson (27). In its simplest form a set of configurations is chosen, a CI calculation carried out, and the density matrix diagonalized to give the natural orbitals for that wave function. Using the same set of configurations the calculation is repeated, but in terms of the natural orbitals found from the first calculation or iteration. Normally the CI energy on the second iteration is significantly lower than the first. The iterative process is repeated until the energy reaches a minimum. The primary objection to the INO method is that one must carry out several (four is a typical number) CI calculations before the final result is obtained. However, the final energy is likely to be lower than would be possible with any single CI calculation which is less than complete.

There are certain types of CI expansions for which the INO procedure is particularly well-defined. As earlier implied, SCF wave functions can be calculated via the INO method. Starting with any reasonable guess for the SCF orbitals, for a closed-shell system all singly excited configurations are added to form a CI. Of course, if the SCF solution itself were guessed, from Brillouin's theorem the single excitations would not contribute to the wave function. Nevertheless, by carrying out natural orbital iterations on the CI consisting of the SCF guess plus all single excitations, after several iterations it is found that the coefficients of the singly excited configurations vanish and the first natural configuration becomes the SCF wave function. For open shells the procedure is identical except that only certain classes of single excitations can be annihilated, in particular those which retain the open shell angular momentum coupling of the SCF orbitals outside closed shells.

The original Bender-Davidson calculation on LiH (27) provides an excellent example of the INO method. Bender and Davidson found first of all that the effectiveness of the procedure was greatly enhanced by performing an SCF calculation prior to the natural orbital iterations. The final 1σ and 2σ natural orbitals are essentially unchanged from the starting SCF orbitals. The INO procedure began by the choice of 50 configurations thought to be most important. As the iterations proceeded, those configurations found unimportant were deleted and new configurations added. The final wave function contains 45 configurations and has a total energy of -8.0606 hartrees, corresponding to about 87% of the correlation energy. Perhaps the strongest tribute that can be made to the Bender-Davidson LiH calculation (27) is that six years later (with computers two orders of magnitude faster) it is still the most accurate calculation to be found in the literature on the most frequently studied molecule other than H_2. For comparison, a straightforward calculation including SCF plus all (938) single and double excitations yielded an energy of -8.0604 hartrees for LiH (29).

Table I-8 shows the most important configurations and occupation numbers for LiH at its experimental bond distance. This information sheds considerable light on the electronic structure of LiH. First of all it is seen that the SCF configuration is a very good approximation to the exact wave function. This is seen by the large value of c^2 for the first natural configuration and also by the fact that the 1σ and 2σ occupation numbers are nearly 2. The other occupation numbers, representing correlation effects, are much smaller. The CI expansion is in fact rather rapidly convergent--the eight configurations seen in Table I-8 account for nearly 90% of the calculated correlation energy. The configurations are ranked by their contribution to the total energy of the wave function, not by their coefficients c_i. This is a good idea since configurations involving replacements of the $1\sigma^2$ pair are seen to make a large energy contribution but nevertheless have a small coefficients c. Finally, we note that four of the first eight configurations involve replacement of the $1\sigma^2$ pair and the remaining four involve replacement of the $2\sigma^2$ pair. None involve replacement of the $1\sigma 2\sigma$ intershell pair.

Table I-8. Important configurations and occupation numbers from the Bender-Davidson LiH calculation (27). The energy E_k associated with the kth configuration is that obtained from the CI including all configurations through the kth.

	Configuration	c^2	E
1.	$1\sigma^2\ 2\sigma^2$	0.9718	-7.9871
2.	$1\sigma^2\ 3\sigma^2$	0.0114	-8.0023
3.	$2\sigma^2\ 2\pi^2$	0.0010	-8.0161
4.	$1\sigma^2\ 1\pi^2$	0.0104	-8.0289
5.	$2\sigma^2\ 5\sigma^2$	0.0008	-8.0391
6.	$2\sigma^2\ 6\sigma^2$	0.0005	-8.0451
7.	$1\sigma^2\ 4\sigma^2$	0.0029	-8.0494
8.	$2\sigma^2\ 4\sigma\ 5\sigma$	0.0002	-8.0514

Occupation Numbers	
1σ	1.9941
2σ	1.9498
3σ	0.0227
4σ	0.0063
5σ	0.0022
1π	0.0207
2π	0.0022
1δ	0.0002

This is because in LiH the 1σ and 2σ SCF orbitals are rather localized--1σ corresponding to the 1s lithium orbital and 2σ to the bonding orbital. Since the 1σ and 2σ orbitals occur in different regions of space the $1\sigma 2\sigma$ correlation is quite small.

The INO procedure has proven useful in solving a variety of problems of chemical interest, some of which will be discussed in later chapters.

D. BASIS SETS

Advances in high speed computers and in theoretical and computational methods have made it currently feasible to calculate from first principles electronic wave functions and properties for molecules containing as many as 29 atoms (80). However, the mere fact that the calculation can be done by no means guarantees that the predicted molecular properties will be reliable, i.e., in agreement with reality. A primary consideration in evaluating the reliability of an electronic structure calculation is the basis set. In this section we point out some of the known characteristics of the most frequently used types of basis sets. Other examples of basis set dependence (e.g., in the calculation of the inversion barrier for ammonia) will be given in later chapters.

Molecular calculations are generally carried out in terms of basis functions centered on each atom in the molecule. Exponential or gaussian functions are most frequently employed.

The use of exponential functions was first suggested by Slater (361) and functions of the type

$$A\, r^{n-1} e^{-\zeta r} \qquad (I,70)$$

are usually called Slater functions or Slater-type orbitals (STO's). In (I,70) A is a normalization factor, n is the principal quantum number, and ζ is the orbital exponent or screening parameter. Since r is the distance from the atom on which the particular Slater function is centered, (I,70) only gives the radial dependence of the function. Angular dependence is usually introduced by multiplying (I,70) by a spherical harmonic $Y_{\ell m}(\theta,\phi)$.

The use of gaussian functions in electronic structure calculations was first suggested by Boys (49). In analogy with (I,70) the radial dependence of a gaussian function may be written

$$B\, r^n e^{-\alpha r^2} \qquad (I,71)$$

In (I,71) n is the analog of the principal quantum number in the Slater function case, and can take the values ℓ, $\ell+2$, $\ell+4$, etc. Angular dependence may be introduced by a factor $Y_{\ell m}(\theta,\phi)$. However, for gaussians the angular dependence is more frequently introduced by a function of the form

$$C\, x^p y^q z^s e^{-\alpha r^2} \qquad (I,72)$$

where p, q, and s are integers. Functions of the type (I,72) are usually called <u>cartesian gaussians</u>. If, for example, p=q=0 and s=1, the function is of p_z symmetry. Another way of using gaussian functions was developed independently by Preuss (308) and Whitten (395). The <u>gaussian lobe</u> method (308,395) uses only 1s gaussians $Be^{-\alpha r^2}$ but these 1s functions are not required to be atom-centered. Linear combinations of these lobe functions are then chosen to simulate s, p, d, etc. orbitals.

In principle one would probably prefer to use Slater functions in all molecular calculations. For the neon atom, for example, a basis of four s and two p Slater functions yields an SCF energy which cannot be improved upon with fewer than ten s and six p gaussian functions (182). However, for nonlinear polyatomic molecules the two-electron integrals (I,20) in terms of a Slater basis set are extremely difficult to compute, while the same integrals in terms of gaussians may be evaluated comparatively simply and rapidly using closed analytic expressions.

The minimum basis set. A minimum basis set includes one function for each SCF -occupied atomic orbital with distinct n and ℓ quantum numbers. For the sulfur atom the minimum basis thus consists of 1s, 2s, 3s, 2p, and 3p functions. The functions in a minimum basis set are usually Slater functions.

The minimum basis set occupies a unique position in the history of molecular quantum mechanics. Prior to 1960 virtually all electronic structure calculations were envisaged in terms of the minimum basis. All the currently popular semi-empirical methods

(167,101,306) are in principle based on a minimum basis set of Slater functions. In fact there are some concepts in physical organic chemistry which can only be discussed in terms of a minimum basis set (370). In light of the above, it is clear that, no matter how trivial it becomes to carry out calculations with extended basis sets, the minimum basis set will continue to be, at the very least, of interpretive value.

The first question which naturally arises relative to a minimum basis set is that of choosing the orbital exponents ζ in (I,70). Long ago Slater suggested

$$\zeta = \frac{(Z - s)}{n^*} \qquad (I,73)$$

in which Z is the actual charge on the nucleus, s is a screening constant, and n^* is an effective quantum number. Slater's rules (361), derived from empirical considerations, provide values of s and n^* for atomic orbitals with principal quantum numbers n = 1, 2, 3, 4, 5, 6.

More recently Clementi and coworkers (73) have attacked the orbital exponent problem more directly. For the ground electronic state of each atom, the ζ values are varied in repeated atomic SCF calculations until a set of ζ's is obtained yielding the lowest possible minimum basis SCF energy. Table I-9 reproduces these results for the atoms helium through fluorine. In most cases the energy-optimized exponents are not a great deal different from

Table I-9. Optimized orbital exponents ζ from atomic SCF calculations using minimum basis sets (73).

	He	Li	Be	B	C	N	O	F	Ne
1s	1.6875	2.6906	3.6848	4.6795	5.6727	6.6651	7.6579	8.6501	9.6421
2s		0.6396	0.9560	1.2881	1.6083	1.9237	2.2458	2.5638	2.8792
2p				1.2107	1.5679	1.9170	2.2266	2.5500	2.8792

those obtained by Slater's rules. For example, for oxygen the values from Slater's rules are 1s (7.7), 2s (2.275), and 2p (2.275) compared to 7.6579, 2.2458, and 2.2266.

The atom-optimized minimum basis sets may be used directly in molecular calculations. Essentially this type of calculation has been carried out for cyclopropane (C_3H_6) and yields an SCF energy of -116.7516 hartrees (48). A lower energy may be obtained by optimizing some or all of the exponents by repeated molecular SCF calculations. Clearly this can become a very expensive procedure. However, for cyclopropane Stevens, et.al. (368) have found the following optimized exponents: 1s (H) = 1.2096 and 2p (C) = 1.7118. In this calculation (368) the carbon 1s (5.68) and 2s (1.74) exponents were taken from fully optimized calculations on smaller hydrocarbons. The resulting total energy -116.8322, is 0.0806 hartrees below the calculation (48) with atom optimized exponents. The carbon 2s and 2p exponents optimized for C_3H_6 are seen to be significantly _larger_ than the values in Table I-9, 1.6083 and 1.5679. This is generally the case and it is not unreasonable that the valence atomic orbitals should contract somewhat in the molecular environment. The orbital occupancy for the ground state of cyclopropane is

$$1a_1'^2 1e'^4 2a_1'^2 2e'^4 1a_2'^2 3a_1'^2 1e''^4 3e'^4 \qquad (I,74)$$

and the SCF wave function of Stevens, et.al. (368) is seen in Table I-10. Each molecular orbital is of course seen to be a linear combination of atomic orbitals (LCAO). Figure I-3 shows the resulting total electron density in the CCC plane.

We have already mentioned the extreme difficulties which arise in the calculation of two-electron integrals required for polyatomic calculations in terms of Slater functions. To bypass these difficulties while retaining the concept of a minimum basis set of Slater functions, many workers expand each Slater function in terms of a linear combination of gaussians (123). Pople and coworkers (154) have carried out extensive calculations to test the efficacy of this scheme. In particular Hehre, et.al. (154) have investigated the convergence to the correct Slater function results as the numbers of gaussians used in each least-squares expansion is increased. Some of their results are seen in Table I-11. It is seen that even when each Slater function is expanded in terms of six gaussians, the total molecular SCF energies are almost 1 eV above the exact Slater function energies. However, in

Table I-10. SCF orbitals for cyclopropane constructed from an optimized minimum basis set of Slater functions (368). Orbital exponents ζ are given in the text. The $z=0$ plane contains the three carbon atoms. H_1, H_2, and H_3 are bonded to C_1, C_2, and C_3 and lie above (positive) the CCC plane while H_4, H_5, and H_6 lie below the plane.

	$1a_1'$	$1e_x'$	$1e_y'$	$2a_1'$	$2e_x'$	$2e_y'$	$1a_2'$	$3a_1'$	$1e_x''$	$1e_y''$	$3e_x'$	$3e_y'$
$1s(H_1)$	-0.0028	-0.0038	0.0	0.0704	0.2385	0.0	0.1945	-0.1828	0.3801	0.0	0.1592	0.0
$1s(H_2)$	-0.0028	0.0019	-0.0033	0.0704	-0.1192	0.2065	0.1945	-0.1828	-0.1901	0.3292	-0.0796	0.1379
$1s(H_3)$	-0.0028	0.0019	0.0033	0.0704	-0.1192	-0.2065	0.1945	-0.1828	-0.1901	-0.3292	-0.0796	-0.1379
$1s(H_4)$	-0.0028	-0.0038	0.0	0.0704	0.2385	0.0	-0.1945	-0.1828	-0.3801	0.0	0.1592	0.0
$1s(H_5)$	-0.0028	0.0019	-0.0033	0.0704	-0.1192	0.2065	-0.1945	-0.1828	0.1901	-0.3292	-0.0796	0.1379
$1s(H_6)$	-0.0028	0.0019	0.0033	0.0704	-0.1192	-0.2065	-0.1945	-0.1828	0.1901	0.3292	-0.0796	-0.1379
$1s(C_1)$	0.5747	0.8122	0.0	-0.1361	-0.1641	0.0	0.0	0.0072	0.0	0.0	0.0154	0.0
$2s(C_1)$	0.0104	0.0224	0.0	0.3532	0.5099	0.0	0.0	-0.0313	0.0	0.0	-0.1001	0.0
$2p_z(C_1)$	0.0	0.0	0.0	0.0	0.0	0.0	-0.3197	0.0	-0.4661	0.0	0.0	0.0
$2p_x(C_1)$	0.0015	-0.0027	0.0	-0.1161	0.1413	0.0	0.0	-0.3761	0.0	0.0	0.3240	0.0
$2p_y(C_1)$	0.0	0.0	-0.0026	0.0	0.0	0.1524	0.0	0.0	0.0	0.0	0.0	-0.6244
$1s(C_2)$	0.5747	-0.4061	0.7033	-0.1361	0.0820	-0.1421	0.0	0.0072	0.0	0.0	-0.0077	0.0133
$2s(C_2)$	0.0104	-0.0112	0.0194	0.3532	-0.2549	0.4415	0.0	-0.0313	0.0	0.0	0.0501	-0.0867
$2p_z(C_2)$	0.0	0.0	0.0	0.0	0.0	0.0	-0.3197	0.0	0.2331	-0.4037	0.0	0.0
$2p_x(C_2)$	-0.0007	-0.0026	0.0001	0.0581	0.1496	0.0048	0.0	0.1881	0.0	0.0	-0.3873	-0.4107
$2p_y(C_2)$	-0.0013	0.0001	-0.0027	-0.1006	0.0048	0.1441	0.0	-0.3257	0.0	0.0	-0.4107	0.0869
$1s(C_3)$	0.5747	-0.4061	-0.7033	-0.1361	0.0820	0.1421	0.0	0.0072	0.0	0.0	-0.0077	-0.0133
$2s(C_3)$	0.0104	-0.0112	-0.0194	0.3532	-0.2549	-0.4415	0.0	-0.0313	0.0	0.0	0.0501	0.0867
$2p_z(C_3)$	0.0	0.0	0.0	0.0	0.0	0.0	-0.3197	0.0	0.2331	0.4037	0.0	0.0
$2p_x(C_3)$	-0.0007	-0.0026	-0.0001	0.0581	0.1496	-0.0048	0.0	0.1881	0.0	0.0	-0.3873	0.4107
$2p_y(C_3)$	-0.0013	-0.0001	-0.0027	0.1006	-0.0048	0.1441	0.0	0.3257	0.0	0.0	0.4107	0.0869

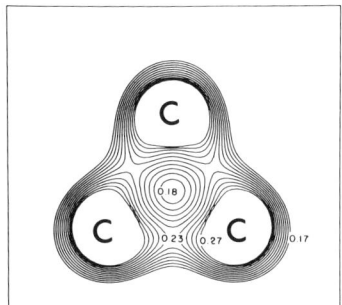

Figure I-3. Results of Stevens, et al. (368) for the total electron density in the three-carbon plane of cyclopropane. Contours are spaced at intervals of 0.01 electrons/bohrs^3. Contours near each carbon atom are omitted.

---

chemistry, energy <u>differences</u> are more important than absolute energies. And the atomization energies (SCF energy of the molecule minus the sum of the SCF energies of the separated atoms) are very close to the Slater function results when a four gaussian expansion is used. Table I-11 also shows that dipole moments reach essentially the correct values when four gaussians are used. It can probably be concluded that, for chemical purposes, four gaussian expansions are suitable replacements for Slater functions.

<u>Double zeta and extended basis sets.</u> We shall see that minimum basis sets yield SCF energies moderately far to very far above the

Table I-11. The approximation of a minimum basis set of Slater functions as linear combinations of gaussians obtained by least-squares fits. The integers above each column indicate the number of gaussians used to fit each Slater function. All quantities are in atomic units.

SCF Total Energies

	2	3	4	5	6	Slater functions
NH_3	-53.8350	-55.4536	-55.8490	-55.9553	-55.9874	-56.0050
HCN	-88.9456	-91.6622	-92.3275	-92.5058	-92.5608	-92.5903
H_2CO	-109.0037	-112.3295	-113.1362	-113.3497	-113.4146	-113.4496

SCF Atomization Energies

	2	3	4	5	6	Slater functions
NH_3	0.3482	0.3085	0.2999	0.3003	0.3000	0.2999
HCN	0.2879	0.2396	0.2270	0.2270	0.2271	0.2266
H_2CO	0.4129	0.3491	0.3383	0.3387	0.3384	0.3381

Dipole Moments

	2	3	4	5	6	Slater functions
NH_3	1.617	1.742	1.769	1.766	1.765	1.76
HCN	1.858	2.099	2.122	2.122	2.121	2.11
H_2CO	0.545	0.941	1.003	1.008	1.008	1.006

Hartree-Fock energies. For this reason electronic structure calculations are frequently carried out with larger basis sets. A particularly popular basis set is the <u>double zeta</u> set, which includes exactly twice as many functions as the minimum basis. For the carbon atom, then, the double zeta basis includes two 1s Slater functions (with different orbital exponents ζ), two 2s functions and two sets (p_x, p_y, and p_z) of 2p functions. For the first, second, and third row atoms, double zeta basis sets have been determined by successive ground state SCF calculations until the best possible set of ζ values is obtained (74, 184). Table I-12 gives the optimized double zeta sets for helium through neon.

If one wishes to obtain SCF energies very close to Hartree-Fock, larger than double zeta basis sets must be used. We refer to any basis set of Slater functions larger than double zeta as an <u>extended basis set</u>. Two frequently used extended basis sets for the oxygen atom are seen in Table I-13. For the ^3P ground state of O, the first extended set (75) yields an SCF energy of -74.80936 hartrees while the SCF energy obtained from the second basis (12) is -74.80938. The author estimates the true Hartree-Fock energy for ^3P O to be not lower than -74.8095 hartrees.

Table I-14 gives a comparison of the atomic SCF energies obtained from basis sets of Slater functions. The fact that many theoretical chemists have strong reservations about the use

Table I-12. Orbital exponents ζ from the optimized double zeta basis sets of Huzinaga and Arnau (184) for helium and the first-row atoms.

	1s	1s'	2s	2s'	2p	2p'
He	2.9156	1.4546				
Li	4.6875	2.4823	1.9757	0.6716		
Be	5.5430	3.3476	1.0090	0.5886		
B	6.5450	4.2448	1.4142	0.8788	2.2116	1.0044
C	7.4831	5.1117	1.8366	1.1635	2.7238	1.2549
N	8.5276	5.9990	2.2523	1.4148	3.2390	1.4961
O	9.5507	6.8758	2.6709	1.6603	3.6856	1.6555
F	10.5136	7.7159	3.1202	1.9333	4.1746	1.8470
Ne	11.5976	8.6081	3.5247	2.1619	4.6784	2.0530

of minimum basis sets can be partly understood in terms of Table I-14. For the oxygen atom, for example, the minimum basis set SCF energy lies 0.269 hartrees (169 kcal/mole) above the near Hartree-Fock energy obtained from the extended basis. The popularity of the double zeta basis set is also understandable in light of the SCF energies. The double zeta basis is relatively

Table I-13. Extended basis sets for the oxygen atom.

First Extended Basis (75)		Second Extended Basis (12)	
Type	Exponent	Type	Exponent
1s	13.324	1s	12.418
1s	7.616	1s	6.995
2s	5.945	3s	8.681
2s	4.283	2s	2.922
2s	2.563	2s	1.818
2s	1.758	2p	8.450
2p	7.907	2p	3.744
2p	3.438	2p	2.121
2p	1.796	2p	1.318
2p	1.154		

modest in size but yields an SCF energy for oxygen only 0.0051 hartrees or 3 kcal/mole above the RHF energy.

Gaussian basis sets. Carefully optimized gaussian basis sets of several sizes have been reported in the literature (182,183,378,382, 394). For the ground state of each atom, the gaussian exponents α of Eq. (I, 71) are varied until a minimum is reached in the computed

Table I-14. Ground state self-consistent-field energies for the first row atoms using three Slater function basis sets described in the text.

	Minimum (73)	Double Zeta (184)	Extended (12)
Li	-7.4185	-7.4327	-7.4327
Be	-14.5567	-14.5724	-14.5730
B	-24.4984	-24.5279	-24.5291
C	-37.6224	-37.6868	-37.6886
N	-54.2689	-54.3980	-54.4009
O	-74.5404	-74.8043	-74.8094
F	-98.9421	-99.4013	-99.4093
Ne	-127.8122	-128.5351	-128.5471

SCF energy. For the first-row atoms, a comparison of the resulting SCF energies is given in Table I-15. By comparing Tables I-14 and I-15, one can see that in order to equal the minimum (2s 1p) basis of Slater functions, a gaussian basis of size (7s 4p) is required for B, C, and N. For O and F, the (5s 3p) gaussian set yields an energy below the (2s 1p) Slater set. To match the double zeta (4s 2p) Slater results, a (10s 6p) gaussian basis must be used.

Table I-15. SCF energies obtained for the ground states of first-row atoms using basis sets (182,183,394) with optimized gaussian exponents α. The notation (7s 4p) means that seven 1s gaussians and four 2p gaussians were used in the calculations.

	B	C	N	O	F
(2s 1p)	-20.7667	-32.0162	-46.2839	-63.6089	-84.4439
(4s 2p)	-24.3359	-37.3557	-53.8620	-73.9734	-98.1789
(5s 3p)	-24.4646	-37.5791	-54.2299	-74.5514	-99.0396
(7s 4p)	-24.5185	-37.6694	-54.3725	-74.7620	-99.3387
(9s 5p)	-24.5271	-37.6853	-54.3953	-74.8003	-99.3956
(10s 6p)	-24.5283	-37.6873	-54.3989	-74.8063	-99.4049
(11s 7p)	-24.5287	-37.6881	-54.4001	-74.8081	-99.4076
Estimated Hartree-Fock	-24.5291	-37.6886	-54.4009	-74.8094	-99.4094

For the hydrogen atom the comparison between Slater and gaussian functions is not representative, since a single 1s Slater function with exponent 1.0 is the exact solution to the non-relativistic Schrödinger equation. However, gaussian sets including 1 through 6 functions yield energies -0.42441, -0.48581, -0.49698, -0.49928, -0.49981, and -0.49994 hartrees. The exact result of course is -0.5 hartree.

As has been mentioned, the primary reason theoretical chemists utilize gaussian functions is the great speed with which the multicenter (involving orbitals on different atoms) two-electron integrals (I,20) may be computed. This point is clearly illustrated by a recent comparative calculation (170) on H_2O using the most efficient computer programs currently available for evaluation of the two types of integrals. A Slater double zeta basis was used for the oxygen (4s 2p) and hydrogen (2s) atoms. The gaussian basis used was (9s 5p) on oxygen and (4s) on hydrogen. Despite the facts that (a) the number of gaussian functions is more than twice the number of Slater functions and (b) the total number of two-electron integrals is proportional to n^4, where n is the number of basis functions, the integrals over gaussians were computed using a total of 49 seconds while those over Slater functions required 569 seconds (Control Data Corporation 6400 computer). In addition the SCF energy obtained with the gaussian set was -76.0093 hartrees, below the Slater function result -76.0053 hartrees.

Contracted functions. We have seen that the two-electron integrals required to carry out a polyatomic calculation can (using currently available programs) be most readily evaluated within a gaussian basis. Unfortunately, the electronic structure calculation is not over when the integrals have been computed. In most calculations the other time consuming step is the solution of the SCF equations via the Roothaan procedure (317,318). This procedure requires an amount of time proportional to the fourth power of the number of basis functions. In addition, the Roothaan procedure is iterative and large basis sets may necessitate more iterations in order to attain convergence. For these reasons, it turns out that calculations carried out directly in terms of _primitive_ gaussians (I,72) become uneconomical due to the enormous amounts of time required to solve the SCF equations.

The above considerations suggest the use of _contracted gaussians_, linear combinations of gaussians with fixed coefficients (395,77). In solving the SCF equations, then, only the coefficients in each SCF orbital of the contracted functions must be determined. This approach allows one to exploit the analytic properties of gaussians in integral evaluation and yet keep the time required for the SCF iterations at a reasonable level. Whitten (395) was among the first to make effective use of contracted gaussians and in particular he optimized a basis set of ten s and five p (10s5p) _specifically_ to be contracted to three s and one p functions. We denote this a (10s5p/3s1p) basis and Table I-16 shows the s functions in this set for carbon.

The contraction of gaussian basis sets should be carried out very carefully (105). Perhaps the most obvious type of contraction would be a complete contraction to atomic SCF orbitals. For the (11s7p) gaussian set, such a contraction would be designated (11s7p/2s1p). One would then carry out molecular calculations in terms of a minimum basis set of near Hartree-Fock atomic orbitals

Table I-16. Two contractions of the Whitten (10s) gaussian basis set for the carbon atom (395). The notation (10s 5p/4s 2p) implies that four fixed linear combinations of ten s type gaussians are taken and two linear combinations of the five p type gaussians are used. Wiggly lines separate the different contracted functions.

1s (α=2548.7256)	0.0324	0.0324
1s (α=781.6495)	0.1344	0.1344
1s (α=159.6274)	1.0000	1.0000
	~~	~~~~~
1s (α=41.8427)	0.6308	0.6308
1s (α=17.1893)	1.0000	1.0000
1s (α=7.0591)	1.8651	1.8651
1s (α=2.5269)	1.1090	1.1090
	~~	~~~~~
1s (α=4.9344)	-0.1585	-0.1585
1s (α=0.4735)	1.1412	1.1412
		~~
1s (α=0.1480)	1.0000	1.0000

(HFAO's) 1s, 2s, and 2p for first-row atoms. However, the use of such a basis set would be a serious mistake since it is drastically overcontracted. Although a minimum basis set of HFAO's describes the atoms perfectly (in the single configuration approximation), such a basis does not have the flexibility required to describe the rearrangement in the electron distribution accompanying molecule formation.

As an example, consider some recent full CI calculations (338) on F_2 and Cl_2 using a minimum basis of near Hartree-Fock atomic orbitals. The results are seen in Table I-17, and for F_2 comparison is made with analogous CI calculations (146) which use a minimum basis set of simple Slater functions. Several points can be made:

(1) For F_2 the total energy obtained with the Slater functions is much worse (higher) than the near HFAO energy.

(2) For both F_2 and Cl_2 the near HFAO bond distance and dissociation energy are in poor agreement with experiment.

(3) The Slater function bond distance and dissociation energy are in much better agreement with experiment for F_2 than are the near HFAO results.

Table I-17 leads us to a perhaps surprising conclusion--for properties likely to be of interest to chemists, e.g., bond distances and dissociation energies, a minimum basis set of Slater functions is preferable to the energetically superior minimum basis of HFAO's. And in fact the primary characteristics of over-contracted functions are predictions of unrealistically large bond distances and unrealistically small dissociation energies.

Perhaps the greatest amount of effort as regards the contraction of functions has gone towards the goal of obtaining a

Table I-17. Full configuration interaction calculations (338) on F_2 and Cl_2 using minimum basis sets of near Hartree-Fock atomic orbitals. For F_2 results are also shown with a minimum basis of Slater functions.

	F_2		
	E (hartrees)	r_e (bohrs)	D_e (eV)
Slater functions	-197.9600	2.61	2.04
Near Hartree-Fock atomic orbitals	-198.8303	3.13	0.32
Experiment	--	2.68	~1.65
	Cl_2		
	E (hartrees)	r_e (bohrs)	D_e (eV)
Near Hartree-Fock atomic orbitals	-918.9896	4.42	0.71
Experiment	--	3.76	2.52

contracted gaussian basis comparable to the double zeta Slater function sets (74, 184). The Whitten (10s 5p/3s 1p) basis is, as such, overcontracted. However, by freeing the coefficients of the most diffuse s and p gaussians the basis becomes (10s 5p/4s 2p) and is of the gaussian double zeta variety. The Whitten (10s/4s) contraction is demonstrated for carbon in Table I-16. However, the most effective double zeta gaussian basis sets are the (9s 5p/4s 2p) sets of Dunning (105) for boron through fluorine. Perhaps the most important principle in devising a contraction procedure is to retain maximum flexibility in the valence region. That is, the coefficients of the most diffuse gaussian functions should be determined independently in each molecular calculation.

The Dunning basis set for carbon is seen in Table I-18. The contracted SCF energy for the ground state of carbon is -37.6845 hartrees, 0.0007 hartrees above the completely uncontracted (9s5p) energy of Huzinaga (182) given in Table I-15. This contracted energy is 0.0023 hartrees above the double zeta Slater function SCF energy seen in Table I-14. The utility of the Dunning contraction becomes apparent when molecular calculations are considered. For the water molecule, SCF calculations using (a) the Dunning contraction, (b) a double zeta Slater basis and (c) the uncontracted (9s 5p) gaussian set yield energies -76.0093, 76.0053, and -76.0133 hartrees. Similar results have been found for the nitrogen molecule (105) and it seems safe to conclude that for molecular calculations, an optimally contracted double zeta basis can be quite competitive with an optimized Slater function double zeta set.

Polarization functions. For the ground states of atoms through argon, the exact RHF energy is obtained using only a basis of $s(\ell=0)$ and $p(\ell=1)$ functions. However, when one brings two nitrogen atoms together forming the nitrogen molecule, functions with higher ℓ values (d,f,...) must be added to the

Table I-18. Optimally contracted gaussian double zeta s basis sets for the ground states of carbon and hydrogen (105). Orbital exponents α are given for each gaussian function in parentheses. The wiggly lines separate the different contracted functions.

Carbon (9s/ 4s)		Hydrogen (4s/2s)	
Function	Coefficient	Function	Coefficient
1s (4232.61)	0.00203	13.3615	0.03283
1s (634.882)	0.01554	2.0133	0.23121
1s (146.097)	0.07541	0.4538	0.81724
1s (42.4974)	0.25712	0.1233	1.0
1s (14.1892)	0.59656		
1s (1.9666)	0.24252		
1s (5.1477)	1.0		
1s (0.4962)	1.0		
1s (0.1533)	1.0		

basis in order to approach the RHF energy limit. The importance of such functions with higher ℓ values, or <u>polarization functions</u>, was first emphasized by Nesbet (271). For the hydrogen atom, the ground state description requires only s functions, so p,d,... functions centered on H for a molecular calculation are considered polarization functions.

We shall see that the reliable calculation of many properties of chemical interest, e.g., dissociation energies and dipole moments, necessitates the inclusion of polarization functions in the basis set. To give some idea of their energetic importance we note that an SCF calculation on the $^1\Sigma_g^+$ ground state of nitrogen with a large s, p basis gives an energy -108.9025 (65). However, the true molecular Hartree-Fock energy of N_2 lies slightly lower than -108.9928 hartrees. Most of the energy difference, 2.6 eV, can be accounted for by the addition of 3d functions on each nitrogen atom. In general one finds that after the basis set for a first row atom has reached the double zeta level (4s 2p), the next most important contributor to the total energy will be a set of 5 (corresponding to the 5 m_ℓ values associated with $\ell=2$) 3d functions. Almost equally important for molecules containing hydrogen is the inclusion of $2p_x$, $2p_y$, and $2p_z$ functions on each H atom. A basis including (4s 2p 1d) on each first row atom and (2s1p) on each hydrogen is referred to as <u>double zeta plus polarization</u> (273).

It is not usually economical to optimize the exponents of the chosen polarization functions in the course of each molecular calculation. Fortunately calculated total energies are not too sensitive to small variations in exponents and one can usually make reasonable choices based on optimized small molecule calculations (251,316,324,107). For hydrogen, a reasonable value of the Slater exponent ζ (2p) is 2.0, while a reasonable value of the gaussian exponent α (2p) is 1.0. For first-row atoms reasonable values are ζ (3d) = 2.0 and α (3d) = 0.8, although for beryllium a Slater function as small as 1.5 might be somewhat more appropriate while for neon ζ (3d) as large as 3.0 might prove more effective. The general trend is for the optimum 3d exponents of both Slater and gaussian functions to increase in going from Be to Ne.

Comparisons for CO and H_2O. The reliability of any basis set which is less than complete must be determined by comparison with other basis sets or with experiment. The last four chapters of this book emphasize comparison with experiment. In this section we evaluate different basis sets by comparison of calculated SCF properties with results thought to be very close to the Hartree-Fock limit. For this purpose we consider two simple molecules, CO and H_2O, for which many different SCF calculations have been carried out. In chapter IV we present a third detailed comparison of basis sets, for the lithium hydroxide molecule.

Table I-19 shows a comparison of SCF properties for CO. There are several other calculations on CO which are less complete in the number of properties reported but fill some gaps in Table I-19. A minimum basis set calculation with orbital exponents optimized for the C and O atoms (see Table I-9) yields energy -112.3260 hartrees and dipole moment -0.592 debyes (310). Thus the energy lowering obtained by specifically optimizing the minimum basis for CO is 0.0650 hartrees. A much greater energy

Table I-19. Self-consistent-field properties computed with different basis sets for carbon monoxide. Dissociation energies are obtained with respect to C and O atom calculations at the same level of basis set.

Basis Set	(2s 1p)	(4s 2p 1d)	(10s 5p 2d/5s 3p 2d)	(5s 4p 1d 1f)
Description	Minimum basis of Slater functions optimized for CO	Double zeta plus polarization Slater functions	Extended gaussian	Near Hartree-Fock Slater functions
Reference	178	273	282	250
Total energy E(hartrees)	-112.3910	-112.7588	-112.7622	-112.7891
Dissociation energy (eV)	6.21	7.30	8.18	7.93
Orbital energies (hartrees)				
1σ	-20.6679	-20.6937	-20.6614	-20.6643
2σ	-11.2856	-11.3913	-11.3605	-11.3597
3σ	-1.4812	-1.5493	-1.5213	-1.5210
4σ	-0.7279	-0.8252	-0.8031	-0.8038
5σ	-0.4842	-0.5704	-0.5543	-0.5544
1π	-0.5582	-0.6531	-0.6378	-0.6395
Dipole moment (debyes)	-0.464	0.361	0.245	0.280
Quadrupole moment (10^{-26} esu)	-1.72	-1.806	-2.089	-2.204
Electric field gradient at carbon nucleus (a.u.)	-1.02	-1.17	-1.135	-1.175
Electric field gradient at oxygen nucleus (a.u.)	-1.16	-0.627	-0.697	-0.724

lowering is obtained by going to a double zeta set of Slater functions, yielding energy -112.6758 hartrees and dipole moment 0.60 debyes (250). Another result not seen in Table I-19 is the double zeta plus polarization calculation of McLean and Yoshimine (250) yielding energy -112.7700 hartrees and dipole moment 0.14 debyes. The above energy is lower than that of Nesbet (273) with the same size basis due to optimization of the 3d exponents by McLean and Yoshimine.

Several trends may be seen in Table I-19. The most general is that the three larger basis sets usually yield similar results while the minimum basis results are for some properties quite different. The minimum basis set dissociation energy 6.21 eV is 1.72 eV below the near RHF value. The minimum basis dipole moment is nearly twice as large as the RHF value and of the opposite sign. Three other calculations on diatomic molecules show the same problem: (a) for BF the minimum basis dipole moment is -2.16 debyes (310) while the near RHF value of McLean and Yoshimine (250) is -0.88; (b) for LiF the minimum basis dipole moment is 2.94 debyes, the near RHF value 6.30; (c) for HF the two values are 0.878 and 1.934 debyes (249). Another reflection of the inadequacy of the minimum basis for CO is the electric field gradient at the nucleus, -1.16 atomic units, compared to the near RHF value -0.724 atomic units. McLean and Yoshimine (250) have estimated the true Hartree-Fock energy of CO to be -112.790 hartrees.

The comparison for water, seen in Table I-20, shows trends similar to those for CO. The minimum basis dipole moment is in close agreement with those calculations using the largest basis sets, but in light of the diatomic examples mentioned above, this agreement must be considered fortuitous. This is borne out by the poor agreement with larger calculations of the minimum basis quadrupole moment tensor. In general then, we must conclude that a minimum basis set, however, carefully optimized, is <u>not capable</u> of yielding reliable values of sensitive one-electron properties such as dipole moments, quadrupole moments, and electric field gradients. For certain properties, however, minimum basis sets may yield qualitatively correct results of chemical importance. Two such examples to be discussed in later chapters are the prediction of molecular geometries (chapter V) and of potential energy curves via full configuration interaction (chapter III).

The double zeta basis set represents a significant improvement over the minimum basis. This is apparent from the results of Table I-20 for both total energy, dissociation energy and most properties. In particular the elements Θ_{xx} and Θ_{yy} of the quadrupole moment tensor are qualitatively correct within the double zeta basis. However, the double zeta dipole moment is more than 25% (or ∿0.6 debyes) larger than the near Hartree-Fock values. This is in fact a common observation, that double zeta sets may yield unreliable values for sensitive molecular properties. By the time

Table I-20. Properties of H_2O within the self-consistent-field approximation using a variety of basis sets. All calculations were carried out at the experimentally known geometry. Unless indicated properties are given in atomic units.

Basis Set	E	Dissociation Energy(eV)	μ(debyes)	Quadrupole Moments		Electric Field Gradients				
				Θ_{xx}	Θ_{yy}	$q_{xx}(O)$	$q_{yy}(O)$	$q_{xx}(H)$	$<1/r_O>$	$<1/r_H>$
O(2s 1p) H(1s) [a]	-75.7033	4.43	1.921	1.110	-1.104	-2.325	2.616	0.271	23.398	5.834
O(9s 5p/4s 2p) [b] H(4s/2s)	-76.0093	5.73	2.683	1.859	-1.732	-1.808	2.052	0.311	23.443	5.738
O(9s 5p 2d/4s 2p 1d) [c] H(4s 1p/2s 1p)	-76.0491	6.81	2.079	1.864	-1.830	-1.704	1.931	0.297	23.440	5.773
O(10s 6p 2d) [d] H(4s 2p)	-76.0594	6.89	1.995	1.882	-1.801	-1.679	1.852	0.294	23.445	5.795
O(11s 7p 2d/6s 5p 2d) [e] H(5s 1p/3s 1p)	-76.0621	6.91	2.084	1.924	-1.875	-1.686	1.885	0.282	23.438	5.769
O(5s 4p 1d) [f] H(3s 1p)	-76.0631	6.90								
Estimated Hartree-Fock	-76.066±0.002									

a. Slater function minimum basis optimized for H_2O(9).
b. Gaussian double zeta basis(105).
c. Gaussian double zeta plus polarization (106).
d. Uncontracted gaussian set(281).
e. Large contracted gaussian set (106).
f. Large set of Slater functions (299).

one reaches the double zeta plus polarization level, however, the calculated properties, as illustrated in Table I-20, are usually quite close (within a few per cent) of the RHF values. Since the estimated RHF energy of H_2O is -76.066 hartress (106,299), the most extended calculations in Table I-20 should be indicative of the Hartree-Fock values of the reported properties.

II
ATOMS

There are at least two reasons why a chemist might be interested in the electronic structure of atoms:

(a) Viewed from a broad perspective, an atom is just a simple molecule. Calculations on atoms may reveal important insights which are applicable to larger molecules.

(b) There are a number of atomic problems which are of inherent interest to chemists as well as physicists. Two such examples are atomic hyperfine structure and atomic electron affinities.

With reference to point (a) above, it should be noted that the most accurate electronic structure calculations carried out to date are atomic calculations. This is primarily due to i) the ease of calculating two-electron integrals (I,20) for atoms and ii) other simplifications brought about by the high degree of symmetry of an atom. In this chapter we discuss the implications of recent ab initio calculations of electronic wave functions for atoms.

A. RESTRICTED HARTREE-FOCK RESULTS

Using highly sophisticated computer programs (319,128) it is now possible to obtain very-close-to Hartree-Fock wave functions for essentially any atom of interest. In fact an accurate calculation, including relativistic effects, was recently reported (293) on the hypothetical superheavy element Z=164. These calculations may be carried out either numerically (152,128) or in terms of a basis set of analytic functions (319). A large number of atomic SCF calculations are reviewed in the book by Fraga and Malli (125).

Experimental atomic correlation energies; relativistic effects. A logical starting point for the investigation of electron correlation in molecules involves the consideration of the atomic problem. Recall that the correlation energy is usually defined as the difference between the restricted Hartree-Fock energy and the nonrelativistic exact energy. As pointed out above, the RHF atom of any atomic state of interest may be obtained straightforwardly. The most direct approach to finding the nonrelativistic exact energy would be to solve the nonrelativistic Schrödinger equation to the required accuracy. However, this is not currently feasible except for atoms through lithium and possibly beryllium. A more practical approach, taken by Clementi (72, 377) is to combine theoretical and experimental information in order to estimate the nonrelativistic exact energy and by deduction the correlation energy.

The exact energy of an atom may be obtained if all the ionization potentials are known experimentally. For example, the exact total energy of Be is just the sum of the ionization potentials of Be, Be^+, Be^{++}, and Be^{+++}. For at least the first-row atoms through neon, rather accurate values of the total energy can be thus obtained. However, to evaluate the nonrelativistic exact energy one must be able to determine the relativistic corrections to the energy. Although the many-electron relativistic Hamiltonian is not known, Clementi and Hartmann (149) have used the approximate Breit-Pauli Hamiltonian to calculate relativistic corrections for atoms through argon. By subtracting the calculated relativistic corrections from the experimental total energies, then, Clementi (72, 377) obtains a reasonable estimate of the nonrelativistic exact energies. The estimated correlation energy is then finally found by subtracting the known Hartree-Fock energies (corrected for the finite mass of the nucleus) from the estimated nonrelativistic exact energies.

The results of the above correlation energy analysis (377) are seen in Table II-1. Also included in Table II-1 are estimates of the Lamb shift (a quantum electrodynamic effect) for these atoms. However, the Lamb correction has been ignored in obtaining the correlation energies seen in Table II-1. Table II-1 shows that the correlation energy increases rather steadily with increasing number of electrons. A perhaps surprising result is the magnitude of the relativistic effects. In fact, for atoms larger

Table II-1. Correlation energy analysis of Veillard and Clementi (377) for the ground states of atoms through argon. All energies are in hartrees.

Atomic	State	RHF Energy	Relativistic Energy	Correlation Energy	Lamb Correction
He	^1S	−2.861680	−0.00007	−0.0420	0.000022
Li	^2S	−7.432726	−0.00055	−0.0454	0.000106
Be	^1S	−14.57302	−0.00220	−0.0940	0.000323
B	^2P	−24.52905	−0.00603	−0.1240	0.000740
C	^3P	−37.68861	−0.01381	−0.1551	0.001439
N	^4S	−54.40091	−0.02732	−0.1861	0.002500
O	^3P	−74.80937	−0.04940	−0.2539	0.004000
F	^2P	−99.40928	−0.08289	−0.3160	0.006015
Ne	^1S	−128.54701	−0.13121	−0.381	0.008614
Na	^2S	−161.85889	−0.20021	−0.386	0.011856
Mg	^1S	−199.61458	−0.29505	−0.428	0.015791
Al	^2P	−241.87665	−0.42062	−0.459	0.020460
Si	^3P	−288.85426	−0.58351	−0.494	0.025887
P	^4S	−340.71866	−0.79111	−0.521	0.032085
S	^3P	−397.50475	−1.05076	−0.595	0.039051
Cl	^2P	−459.48187	−1.37168	−0.667	0.046765
Ar	^1S	−526.81734	−1.76094	−0.732	0.055190

than aluminum, the relativistic corrections to the energy are
larger than the correlation energy itself. This might lead one to
believe that for molecules containing silicon and larger atoms,
theoretical chemists should evaluate relativistic effects and
secondly consider the correlation problem.

Fortunately, however, nearly all of the energy correction from
relativistic effects is due to the electrons described by inner
shell orbitals. Within the RHF approximation, Hartmann and Clementi
find the following contributions (in hartrees) to the relativistic
energy of argon: 1s (-1.226), 2s (-0.235), 2p (-0.257), 3s (-0.025)
and 3p (-0.022). The valence orbitals 3s and 3p contribute only
a small fraction of the relativistic corrections to the energy.
As chemists we are primarily interested in the behavior of the
"valence electrons". For example, we might want to predict
whether or not the hypothetical ArF_2 molecule exists. This requires us to determine to a high precision the energy difference
between ArF_2 and Ar + F + F. It seems very reasonable to assume
that the relativistic energies of these two entities will be
virtually identical since the inner shells are likely to be very
similar in the molecule and the separated atoms. Therefore it
would appear that even for molecules containing larger atoms,
the correlation problem will be the primary stumbling block in
the path of the theoretical chemist attempting to go beyond the
nonrelativistic Hartree-Fock approximation.

   Ionization potentials of neon and argon. In section IB it

was pointed out that orbital energies ε obtained in SCF calculations may be identified with ionization potentials. As an example, consider the argon atom, for which ε_{1s} (the 1s orbital energy) is 3227.4 eV (10). The experimental ionization potential is 21.1 eV smaller. The primary reason for this deviation is clear if one recalls the sense in which Koopmans' theorem is intended. That is, the Koopmans' ionization potential for argon is given by

$$E_{RHF}(Ar) - E(Ar^+) \qquad (II,1)$$

in which $E(Ar^+)$ is the energy obtained from a single determinant wave function corresponding to orbital occupancy $1s2s^22p^63s^23p^6$. But the orbitals used in the Ar^+ wave function are exactly the Hartree-Fock orbitals obtained for the ground state of the neutral argon atom. Clearly this is not an even-handed arrangement--the ionization potential is given as the difference between a variational energy for the neutral atom and a non-variational energy for the positive ion.

Clearly a more sensible (but much more difficult) approach to the calculation of ionization potentials would be to carry out comparable calculations on both the neutral atom and the positive ion, which in the example given above is referred to as the 1s hole state of Ar. In particular, it would be desirable to carry out a Hartree-Fock calculation on the ^2S state of Ar^+ corresponding to the orbital occupancy $1s2s^22p^63s^23p^6$. The general problem discussed here was first investigated using nonempirical SCF wave functions by Bagus (10).

Bagus (10) discussed in some detail a rather thorny problem which arises in the theoretical consideration of species such as the 1s hole state of argon. A variational wave function corresponding to the kth lowest eigenvalue of (I,50) is by MacDonald's theorem (238) a rigorous upper bound to the energy of the kth lowest state of the symmetry being considered. There are many ^2S states of Ar^+ with energies below the 1s hole state energy. For example, the states corresponding to orbital occupancies $1s^2 2s^2 2p^6 3s 3p^6$, $1s^2 2s^2 2p^6 3p^6 4s$, $1s^2 2s^2 2p^6 3p^6 5s$, and so on all give rise to ^2S states of Ar^+ which lie below the energy of the 1s hole state. Therefore it would appear that an electronic calculation on the 1s hole state of Ar would involve at best a very large CI calculation for ^2S symmetry and the possibly tricky problem (373) of determining which of the higher eigenvalues represents the 1s hole state. Actually, the problem is formally even more difficult than stated above since the 1s hole state of Ar lies in the Ar^{++} continuum. That is, the energy of the 1s hole state lies above the energy of the ground state of Ar^{++} plus an electron at infinite separation.

As it turns out, however, Bagus uses <u>only</u> the single determinant wave function corresponding to $1s 2s^2 2p^6 3s^2 3p^6$ to describe the 1s hole state.

The only constraint imposed to obtain the wave function is the specification of the electronic configuration. That is, the

SCF equations are solved under the assumption that the 1s orbital is occupied by only one electron. Another way of looking at the problem is in terms of the fact that the 1s hole state wave function will yield an upper bound to the true energy if this wave function is <u>orthogonal</u> to all lower states of the same symmetry. And the physical model which Bagus expects to guarantee near orthogonality of the single determinant 1s hole state function to the lower states is just the orbital or shell structure of the atom (10).

Probably the strongest evidence to support the above qualitative arguments of Bagus are the quantitive results seen in Table II-2 for the hole states of neon and argon. In each case the direct hole state SCF calculation necessarily gives a lower energy than could be obtained from the neutral atom SCF orbitals. For this reason the directly calculated ionization potentials are always less than those obtained from Koopmans' theorem. The most striking improvements over the Koopmans' theorem results are obtained for inner shell electrons where relativistic effects are not too large, e.g., 1s neon and 2p argon. The difference between the calculated and experimental 1s Ne ionization potentials is 21.4 eV from the orbital energy but only 1.7 eV using the direct SCF calculation on Ne^+. For the 2p argon hole state, the Koopmans' theorem deviation is 11.9 eV while the direct calculation is only 0.4 eV from experiment.

Table II-2. Ionization potentials (in eV) of Ne and Ar obtained (a) from neutral atom orbital energies (Koopmans' theorem) and (b) by subtracting the near Hartree-Fock energy of the neutral atom from the near Hartree-Fock energy of the appropriate state of the positive ion (10).

			Ionization Potentials		
Designation	Orbital Occupancy	Ion	Koopmans' Theorem	Direct SCF Calculations	Experiment
1s hole	$1s 2s^2 2p^6$	Ne^+	891.7	868.6	870.3
2s hole	$1s^2 2s 2p^6$	Ne^+	52.5	49.3	48.5
2p hole	$1s^2 2s^2 2p^5$	Ne^+	23.1	19.8	21.6
1s hole	$1s 2s^2 2p^6 3s^2 3p^6$	Ar^+	3227.4	3195.2	3206.3
2s hole	$1s^2 2s 2p^6 3s^2 3p^6$	Ar^+	335.3	324.8	--
2p hole	$1s^2 2s^2 2p^5 3s^2 3p^6$	Ar^+	260.4	248.9	248.5
3s hole	$1s^2 2s^2 2p^6 3s 3p^6$	Ar^+	34.8	33.2	29.2
3p hole	$1s^2 2s^2 2p^6 3s^2 3p^5$	Ar^+	16.1	14.8	15.8

For our original example, the 1s hole state of argon, Table II-2 shows the direct calculation to be in error with respect to experiment by 11.1 eV, which is only about a factor of 2 better than the Koopmans' theorem result. Bagus has shown (10) that this apparently large deviation is due primarily to relativistic effects. That is, the state of Ar$^+$ missing a 1s electron has substantially lower relativistic corrections to its energy than does neutral argon, with the 1s orbital doubly occupied in the RHF approximation. Taking this relativistic correction into account, the directly calculated 1s ionization potential differs by experiment by only 1.3 eV. This remaining 1.3 eV deviation is presumably due to the difference between the two states in the correlation energies, which are of course ignored in RHF calculations.

It is perhaps surprising to note that the first ionization potentials (2p for Ne and 3p for Ar) of both atoms are in better agreement with experiment by Koopmans' theorem than by the direct calculations. Since relativistic effects should be small for these outer shell electrons, the correlation energy must be responsible for the deviations between the direct calculations and experiment. One would normally expect the correlation energy of the ground state of Ar$^+$ to be less than that of Ar, since Ar has an additional electron. And because the calculated outermost ionization potentials are less than experiment, inclusion of correlation will in fact lower the energy of Ar relative to Ar^+ and bring the theoretical results into agreement with experiment.

The above line of reasoning breaks down in a consideration of the Ne 2s and Ar 3s ionization potentials. Both of these ionization potentials are from the direct calculations found to be <u>larger</u> than experiment, the deviation for Ar 3s being 4 eV. This implies that the correlation energy of these states of Ar^+ is <u>greater</u> than that of neutral argon. As pointed out by Bagus (10), this result is of particular interest because there has been some feeling among theoreticians (5) that pair correlations (see section IID) may be approximately separable and transferable. Were this strictly the case the correlation energy of any state of Ar^+ would of necessity be less than that of Ar.

Direct hole state SCF calculations of the Bagus type have also been carried out for molecules and will be discussed in chapters III and V. Theoretical calculations on positive ion states have become increasingly of interest in light of new experimental techniques, in particular ESCA (350) and photoelectron spectroscopy (375), for the measurement of ionization potentials.

B. RELAXATION OF ORBITAL RESTRICTIONS

The Fermi contact interaction in the unrestricted Hartree-Fock approximation. The Fermi contact interaction or (unpaired) spin density at the nucleus is an important contributor to the hyperfine structure of atoms. For molecules the spin density frequently dominates the hyperfine structure. Two commonly used measures of the spin density are χ and $|\psi(0)|^2$

$$\chi = \frac{4\pi}{S} \int \psi_e^* \left\{ \sum_i \delta(\vec{r}_i) \, s_{zi} \right\} \psi_e \, d\tau \tag{II,2}$$

$$|\psi(0)|^2 = 2 \int \psi_e^* \left\{ \sum_i \delta(\vec{r}_i) \, s_{zi} \right\} \psi_e \, d\tau \tag{II,3}$$

in which S is the total spin of the atom, $\delta(\vec{r}_i)$ is the Dirac delta function with respect to the nucleus, and s_{zi} is a one-electron spin operator having the effect

$$s_z \alpha = +\tfrac{1}{2}\alpha \qquad s_z \beta = -\tfrac{1}{2}\beta \tag{II,4}$$

when operating on an individual spinorbital.

A serious deficiency of the restricted Hartree-Fock approximation is that for many open-shell atoms it predicts an identically zero Fermi contact interaction, while the experimental spin density is known to be nonzero. Perhaps the most commonly referred to example of this problem is the ^4S ground state of nitrogen, the RHF wave function for which is (I,44). Only s orbitals can contribute to the spin density of an atom, since p, d, f, ... orbitals have zero amplitude at the atomic nucleus on which they are centered. Inspection of (I,45) and (II,2) shows that for nitrogen the 1sα and 1sβ orbitals will give equal and opposite contributions to the spin density, as will the 2sα and 2sβ pair. Thus the RHF spin density is identically zero for ^4S N and will also be zero for any atom described by an RHF wave function with no singly-occupied s orbitals.

The primary motivation for the use of spin unrestricted Hartree-Fock (SUHF) wave functions is the ability of the SUHF wave function for systems such as ^4S N to predict a nonzero spin density (127). Inspection of the SUHF wave function (I,46) shows that the difference between 1s spatial orbital associated with α spin and the 1s' spatial orbital with β spin will result in a non-zero spin density. The actual SUHF spin density χ for ^4S N can be written

$$\chi = \frac{8\pi}{3} \{1s(0) - 1s'(0) + 2s(0) - 2s'(0)\} \qquad (II,5)$$

where 1s(0) is the value of the 1s spatial orbital at the nucleus. From accurate calculations (13) the above four contributions (in atomic units) to χ are 409.8094, -412.9225, 21.7487, and -17.8504, yielding a total χ = 0.7852.

A stumbling block in the serious evaluation of the SUHF approximation has been the fact that the spin density is a property at one point in space, namely the nucleus. For this reason, finding an analytic basis set capable of yielding the true SUHF spin densities is a difficult task. With a set of Slater functions, for example, only 1s functions can contribute to the spin density and a relatively small change in one of the 1s orbital exponents may drastically alter the calculated spin density while leaving the total energy virtually unchanged. Although Slater basis sets have been constructed which give reliable spin densities (136,335), this basis set problem is perhaps most straightforwardly solved using numerical orbitals (152, 128), with which the accuracy required may be attained by simply using a finer grid size.

By way of numerical SUHF calculations it is then possible to make an unambiguous comparison with experiment. Such a comparison, from the calculations of Bagus, Liu, and Schaefer (13), is seen in Table II-3. The reader's first impression of Table II-3 may be that the agreement between SUHF and experimental spin densities is very poor. However, it is a fact (335) that the theoretical prediction of atomic spin densities is <u>exceedingly</u> difficult. In this light some useful correlations can be made between SUHF and experiment. For this purpose the atoms in Table II-3 may be placed in five distinct groups (13):

(a) Alkali atoms. The SUHF spin densities for Li, Na, and K, as earlier pointed out by Goodings (138), are in rather good agreement with experiment (97, 86, and 77%, respectively). The largest contribution to the spin density comes from the unpaired 2s, 3s, or 4s electron.

(b) First row atoms. The SUHF spin densities of these atoms increase monotonically from boron to fluorine. For N, O, and F the SUHF values are 193%, 172%, and 186% of the experimental values. It is probably safe to assume that the B and C spin densities are also of the order of twice the (unknown) experimental values.

(c) Second row atoms. From Al to Cl the SUHF spin densities increase monotonically in increments of ∼0.4 atomic units. The experimental spin density is known only for phosphorous and the SUHF value is of the incorrect sign. This serious discrepancy was a cause of considerable concern among early advocates of the SUHF method (43). Despite the poor agreement for ^4S P, it does seem probable that the true spin densities, like the SUHF values, increase in an orderly fashion from aluminum to chlorine.

(d) Transition metal atoms. For Ti, V, Mn, and Co, the SUHF spin densities are 82, 75, 72, and 67% of experiment. This represents the best agreement obtained between SUHF and experiment for any group of atoms. The SUHF spin densities are all negative, decrease monotonically from Sc to Cu, and confirm the earlier qualitative conclusions of Freeman and Watson (127).

(e) Third row atoms gallium through bromine. The SUHF values of χ follow the same pattern seen for the atoms of the first and second row filling up the 2p and 3p

Table II-3. Fermi contact interaction, indicated by the spin density χ, within the spin-unrestricted Hartree-Fock (SUHF) approximation (13). Values of χ are given in atomic units.

Atom	Open-Shell Occupation	χ(SUHF)	χ(Experiment)
Li(^{2}S)	2s	2.823	2.9062
B(^{2}P)	2p	0.215	--
C(^{3}P)	2p^2	0.485	--
N(^{4}S)	2p^3	0.785	0.4071
O(^{3}P)	2p^4	1.228	0.715
F(^{2}P)	2p^5	1.675	0.901
Na(^{2}S)	3s	8.136	9.42
Al(^{2}P)	3p	-1.370	--
Si(^{3}P)	3p^2	-0.972	--
P(^{4}S)	3p^3	-0.584	0.3824
S(^{3}P)	3p^4	-0.055	--
Cl(^{2}P)	3p^5	0.425	--
K(^{2}S)	4s	10.727	13.91
Sc(^{2}D)	3d	-0.301	--
Ti(^{3}F)	3d^2	-0.423	-0.508
V(^{4}F)	3d^3	-0.487	-0.646
Cr(^{5}D)	3d^4	-0.537	--
Mn(^{6}S)	3d^5	-0.598	-0.826
Fe(^{5}D)	3d^6	-0.648	--
Co(^{4}F)	3d^7	-0.682	-1.02
Ni(^{3}F)	3d^8	-0.726	--
Cu(^{2}D)	3d^9	-0.794	--
Ga(^{2}P)	4p	-4.353	--
Ge(^{3}P)	4p^2	-3.324	--
As(^{4}S)	4p^3	-2.501	-1.085
Se(^{3}P)	4p^4	-1.495	--
Br(^{2}P)	4p^5	-0.663	--

orbitals. That is, the spin densities monotonically increase. Only for arsenic is the experimentally value known, and the SUHF value is a factor of 2.3 larger.

To summarize these results, it must be said that the SUHF approximation is a poor theoretical tool for predicting one particular spin density. However, there are several trends among the SUHF spin densities, and where experimental values are available, the same trends are qualitatively observed experimentally. The above atomic experience should probably be given serious consideration by those carrying out SUHF calculations on polyatomic free radicals. In addition to errors inherent in the SUHF method, polyatomic hfs calculations (257) may be subject to further uncertainties of a severe nature due to basis set incompleteness.

C. VARIATIONAL CONSIDERATION OF ELECTRON CORRELATION

As pointed out above, the most direct procedure for evaluating the correlation energy in many-electron systems is via large scale configuration interaction. To date this has only been attempted for small atoms and molecules. However, the results of such atomic calculations, some of which are discussed in this section, allow us to draw some general conclusions which should be applicable to larger systems.

Oscillator strengths of boron and carbon. The most difficult step in the theoretical study of electronic transitions is the calculation of oscillator strengths, also referred to as f values. There are three equivalent forms in which the oscillator strength f_{ab} between two electronic states ψ_a and ψ_b is commonly expressed (389), the "length" form

$$f_{ab} = \tfrac{2}{3}(E_b - E_a) \left| \int \psi_a^* \left\{ \sum_{i=1}^{n} r_i \right\} \psi_b \, d\tau \right|^2 \qquad (II,6)$$

the "velocity" form

$$f_{ab} = \tfrac{2}{3}(E_b - E_a)^{-1} \left| \int \psi_a^* \left\{ -i \sum_{i=1}^{n} \nabla_i \right\} \psi_b \, d\tau \right|^2 \qquad (II,7)$$

and the "acceleration" form

$$f_{ab} = \tfrac{2}{3}(E_b - E_a)^{-3} \left| \int \psi_a^* \left\{ -\sum_{i=1}^{n} \nabla_i V \right\} \psi_b \, d\tau \right|^2 \qquad (II,8)$$

where E_b and E_a are the energies associated with ψ_b and ψ_a and V is the potential energy for the n-electron system. The three forms are equivalent only in the sense that they will yield the same oscillator strength if the _exact_ electronic wave functions ψ_a and ψ_b are used in (II,6), (II,7), and (II,8). Experience has shown that the "acceleration" form (II,8) is of little practical value, since the wave functions ψ_a and ψ_b must be _extremely_ accurate before a reasonable f value is obtained (345). Since the energy difference ($E_b - E_a$) is frequently known from experiment, the crux of the oscillator strength problem is in the calculation of the integrals or <u>transition moments</u> on the right hand side of (II,6) through (II,8).

Until very recently, the calculation of transition probabilities has been based on an approximation to the SCF approximation. As an example consider the form of the RHF wave functions for the ground and first excited states of the sodium atom

$$\psi_{RHF}(^2S\ Na) = 1s\alpha\ 1s\beta\ 2s\alpha\ 2s\beta\ 2p_{-1}\alpha\ 2p_{-1}\beta\ 2p_0\alpha\ 2p_0\beta\ 2p_{+1}\alpha\ 2p_{+1}\beta\ 3s\alpha$$

(II,9)

$$\psi_{RHF}(^2P\ Na) = 1s\alpha\ 1s\beta\ 2s\alpha\ 2s\beta\ 2p_{-1}\alpha\ 2p_{-1}\beta\ 2p_0\alpha\ 2p_0\beta\ 2p_{+1}\alpha\ 2p_{+1}\beta\ 3p_0\alpha$$

(II,10)

We can of course very easily carry out RHF calculations on these two states of sodium. However, the RHF orbitals for ^2S Na will <u>not</u> be exactly the same as those for ^2P Na, since the two calculations are carried out independently. However, the traditional approach to the calculation of transition probabilities (83) assumes that the 1s, 2s, and 2p orbitals of the two wave functions are identical. Using this approximation, the calculation of the f value becomes very simple, the expression for the length form reducing to

$$f(^2S, ^2P) = \frac{2}{3}[E(^2P) - E(^2S)]\left|\int \chi_{3s}^*(1)\, r_1\, \chi_{3p}(1)\, dv(1)\right|^2 \qquad (II,11)$$

using the notation developed in chapter I. The oscillator strength in (II,11) is seen to be given by a single one-electron integral.

If we wish to carry out a rigorous calculation within the RHF formulation, the effects of nonorthogonality of the 1s, 2s, and 2p orbitals between the ^{2}S and ^{2}P states of sodium must be explicitly considered. And, as mentioned in chapter I, the calculation of matrix elements between Slater determinants constructed of nonorthogonal orbitals is a difficult task (363). However, for single determinant wave functions only a single such matrix element need be evaluated. Table II-4 gives a comparison by Weiss (390) between rigorously evaluated near RHF f values and experiment for several first-row atoms and ions.

Table II-4 answers the first theoretical question we wish to raise concerning oscillator strengths, namely, how reliable is the RHF approximation. One must conclude that RHF atomic oscillator strengths usually differ from experiment by a factor of 2 to 3. There are exceptions of course: the nitrogen ^{4}S ($2s^2\ 2p^3$) to ^{4}P ($2s\ 2p^4$) oscillator strength differs from one experimental value by a factor of 6, while the oxygen f value given in Table II-4 is very close agreement with the experimental value. It is worth pointing out that the <u>experimental</u> determination of oscillator strength is also very difficult, as discrepancies in Table II-4 between different experimental values for the same transition would seem to indicate.

Table II-4. Near Hartree-Fock oscillator strengths compared with experiment. The theoretical boron and carbon results are those of Weiss (390), while those for nitrogen and oxygen are from the calculations of Kelly (201). Each state is described by its spectroscopic term symbol and orbital occupancy excluding the 1s electrons.

Atoms	Transition	f values Near Hartree-Fock	Experiment
B	$^2P(2s^2\ 2p) \rightarrow\ ^2D(2s\ 2p^2)$	0.143	0.048
C	$^3P(2s^2\ 2p^2) \rightarrow\ ^3D(2s\ 2p^3)$	0.286	0.091, 0.076
	$\rightarrow\ ^3P(2s\ 2p^3)$	0.202	0.039
	$\rightarrow\ ^3P(2s^2\ 2p\ 3s)$	0.075	0.17, 0.13
C^+	$^2P(2s^2\ 2p) \rightarrow\ ^2D(2s\ 2p^2)$	0.274	0.114
	$^2P(2s\ 2p^2) \rightarrow\ ^2D(2p^3)$	0.228	0.136
N	$^4S(2s^2\ 2p^3) \rightarrow\ ^4P(2s\ 2p^4)$	0.491	0.137, 0.080
	$\rightarrow\ ^4P(2s^2\ 2p^2\ 3s)$	0.099	0.350, 0.259
	$^2D(2s^2\ 2p^3) \rightarrow\ ^2D(2s^2\ 2p^2\ 3s)$	0.062	0.110, 0.095
	$\rightarrow\ ^2P(2s^2\ 2p^2\ 3s)$	0.048	0.111, 0.078
	$^2P(2s^2\ 2p^3) \rightarrow\ ^2P(2s^2\ 2p^2\ 3s)$	0.049	0.093, 0.064
N^+	$^3P(2s^2\ 2p^2) \rightarrow\ ^3D(2s\ 2p^3)$	0.240	0.109
	$\rightarrow\ ^3P(2s\ 2p^3)$	0.172	0.131
	$\rightarrow\ ^3S(2s\ 2p^3)$	0.323	0.189
	$\rightarrow\ ^3P(2s^2\ 2p\ 3s)$	0.089	0.067
O	$^3P(2p^4) \rightarrow\ ^3S(2p^3\ 3s)$	0.030	0.035

It is clear that in order to make reliable theoretical predictions of oscillator strengths, one must go beyond the RHF approximation. We know that by including all of the correlation energy in wave functions ψ_a and ψ_b in (II,6) we will obtain the correct oscillator strength. Perhaps the best theoretical question to ask, then, is how <u>little</u> of the correlation energy can we describe and yet predict reasonable f values. Excellent work addressed to the above question has been carried out by Weiss (390,391) for B, C, and several isoelectronic ions.

A reasonable approximation adopted by Weiss (390) is to ignore the correlation energy associated with the 1s core orbitals. In terms of CI, this means that all the configurations included in the wave function have the 1s orbital doubly-occupied. So for the ground state of carbon, which has orbital occupancy $1s^2\ 2s^2\ 2p^2$, only correlation effects involving the 2s and 2p orbitals are considered. The theoretical key to Weiss's approach is the pseudonatural orbital method (see section IC for a discussion) of Edmiston and Krauss (112). Weiss first calculates the pseudonatural orbitals for the 2s2p pair. The resulting 2s2p PSNO's are ordered by occupation number according to their ability to describe the 2s 2p correlation.

An important aspect of Weiss's work (390) is that he also calculates the PSNO's for the $2s^2$ and $2p^2$ pairs and finds that the three independently determined sets of orbitals are very similar. This means that any one of the three sets of PSNO's can be used to describe <u>all</u> the correlation involving 2s and 2p orbitals in a nearly optimum manner. With the orbitals ordered by occupation number, Weiss is then able to use only the most important PSNO's in a small (less than 40 configurations) CI which picks up over 90% of the L-shell correlation energy.

Using wave functions for several electronic states obtained as discussed above, Weiss evaluates the length (II,6) and velocity (II,7) formulas to predict the oscillator strengths of particular interest. Since the orbitals used to construct the CI wave function for one electronic state are not orthogonal to those for any other electronic state, the nonorthogonality problem can be quite difficult. Weiss's results (390, 391) are compared with the near RHF results and experiment in Table II-5.

The agreement between Weiss's oscillator strengths (obtained by CI) and experiment is excellent, particularly when it is realized that the experimental f values may be subject to uncertainties of 20% or more. For many cases, e.g., all three transitions shown in Table II-5 for C^+, the improvement provided by CI over the RHF values is indeed striking. It can also be seen in a number of cases that the agreement between length and velocity f values is much better for the CI results than the RHF. In conclusion, it

Table II-5. Oscillator strengths of B, C, and C^+ predicted by configuration interaction (390,391). Only those C^+ values are included for which experimental values are available.

Atom	Transition	Near RHF Length	Near RHF Velocity	CI Length	CI Velocity	Experiment
B	$^2P(2s^22p) \rightarrow {}^2S(2s^23s)$	0.052	0.063	0.067	0.074	0.055
	$\rightarrow {}^2D(2s2p^2)$	0.339	0.336	0.067	0.087	0.059, 0.048
	$\rightarrow {}^2D(2s^23d)$	0.109	0.092	0.197	0.189	0.175
	$^2S(2s^23s) \rightarrow {}^2P(2s^23p)$	1.269	0.995	1.199	0.978	--
	$^2P(2s^23p) \rightarrow {}^2D(2s^23d)$	1.036	0.691	0.768	0.765	--
	$^4P(2s2p^2) \rightarrow {}^4S(2p^3)$	0.266	0.146	0.213	0.225	--
C	$^3P(2s^22p^2) \rightarrow {}^3D(2s2p^3)$	0.286	0.332	0.102	0.117	0.091, 0.076
	$\rightarrow {}^3P(2s2p^3)$	0.202	0.171	0.097	0.105	0.039
	$\rightarrow {}^3P(2s^22p3s)$	0.075	0.094	0.108	0.123	--
	$^1D(2s^22p^2) \rightarrow {}^1P(2s^22p3s)$	0.079	0.084	0.092	0.108	<0.101
	$^1S(2s^22p^2) \rightarrow {}^1P(2s^22p3s)$	0.097	0.098	0.081	0.090	--
C^+	$^2P(2s^22p) \rightarrow {}^2D(2s2p^2)$	0.274	0.256	0.121	0.124	0.114
	$^2S(2s2p^2) \rightarrow {}^2P(2s^23p)$	0.0	0.001	0.127	0.137	0.133
	$^2P(2s2p^2) \rightarrow {}^2D(2p^3)$	0.228	0.627	0.101	0.092	0.136

seems quite reasonable to assume that the predicted f values for which no experimental values are available are in fact quite reliable. With this in mind, Weiss has predicted f values for the five electron ions from C^+ to P^{+10}.

In order to predict the oscillator strengths in Table II-5, as pointed out above it was necessary for Weiss to compute accurate correlated wave functions for several electronic states of B and C. The total energies of these wave functions are used in Table II-6 to predict the positions or term values of these electronic states. The near RHF relative positions of the states of B and C are only in fair to good agreement with experiment. However, the CI term values are in excellent agreement with experiment. Of particular interest to chemists are the splittings between the three lowest or valence states of carbon (^3P, ^1D, and ^1S), all of which arise from the orbital occupancy $1s^2 2s^2 2p^2$.

From the theoretical standpoint, the work of Weiss (390,391) is important because it shows in a quantitative way that two correlation sensitive properties, oscillator strengths and term values, may be described in a very satisfactory way <u>without</u> con-

Table II-6. Energies in eV of the excited states of boron and carbon with respect to the ground electronic states (390,391).

Orbital Occupancy	Spectroscopic Designation		Near RHF	CI	Experiment
$2s^2 2p$	^2P	B	0.00	0.00	0.00
$2s 2p^2$	^4P	B	2.13	3.52	3.57
$2s^2 3s$	^2S	B	4.82	4.95	4.96
$2s 2p^2$	^2D	B	5.91	5.99	5.93
$2s^2 3p$	^2P	B	5.79	6.00	6.03
$2s^2 3d$	^2D	B	6.40	6.75	6.79
$2p^3$	^4S	B	10.91	12.03	12.03
$2s^2 2p^2$	^3P	C	0.00	0.00	0.00
	^1D	C	1.56	1.30	1.26
	^1S	C	3.78	2.71	2.68
$2s^2 2p 3s$	^3P	C	7.24	7.52	7.48
	^1P	C	7.40	7.68	7.68
$2s 2p^3$	^3D	C	8.01	7.99	7.94
	^3P	C	9.55	9.56	9.33

sideration of electron correlation involving inner shell orbitals. From a more general perspective, these calculations are of interest because they describe a working procedure for the reliable prediction of two fundamental properties of atoms and molecules.

Accurate wave functions for Be, C, and Ne. The most sophisticated and rapidly convergent configuration interaction wave functions reported in the literature for first-row atoms are those obtained by Bunge and coworkers (61-63). These calculations introduce a number of new theoretical ideas and shed considerable light on the nature of electron correlation in many-electron systems.

The Bunge calculations make use of natural orbital concepts, but in a somewhat different manner than either the pseudonatural orbital or iterative natural orbital (27) methods. The calculation on the ground state of beryllium illustrates the approach taken (62). The first step is a self-consistent-field calculation, carried out using a large basis set of Slater functions. Then a calculation is carried out including the RHF configuration $1s\alpha\ 1s\beta\ 2s\alpha\ 2s\beta$ plus all configurations whose orbital occupancies may be specified $x_i x_j\ 2s^2$ and $1s^2\ x_i x_j$ where x_i and x_j are all orbitals in the basis set except the SCF orbitals 1s and 2s. After the expansion coefficients c of Eq. (I,52) are obtained for this CI wave function, the density matrix is set up and diagonalized, yielding a set of natural orbitals.

The above procedure corresponds to obtaining a set of orbitals to describe both the $1s^2$ and the $2s^2$ correlation. Bunge, in fact finds that the natural orbitals with highest occupation number are quite similar to either (a) one of the $1s^2$ PSNO's or (b) one of the $2s^2$ PSNO's. The advantage in finding these natural orbitals from a single calculation is that the optimum orbitals for correlating $1s^2$ and $2s^2$ are determined in the presence of each other.

Using the natural orbitals thus obtained, Bunge proceeds to carry out a CI calculation designed to account for as much of the correlation energy as possible. An exhaustive search is carried out, using perturbation theory as well as numerous trial variational calculations, to determine which configurations should be included in the final wave function. Eventually a 180 configuration wave function, constructed from 1492 Slater determinants, is decided upon (62). The variational energy of this wave function is -14.6642 hartrees, which may be compared to the estimated RHF energy -14.5730 and the estimated nonrelativistic exact energy -14.6664 hartrees. It may be seen that 98% of the correlation energy of Be has been obtained (354).

All singly-excited configurations have zero H matrix elements with the SCF configuration for the closed shell ^1S ground state of beryllium. Therefore these configurations can only contribute to the final wave function by way of interaction with doubly- and triply-excited configurations. Bunge finds for Be that single excitations lower the energy by only -0.00058 hartrees, or 0.6%

of the calculated correlation energy. Triple excitations are found to lower the energy by 0.00025 hartrees and quadruple excitations by 0.00352 hartrees. The latter two numbers correspond to 0.3% and 3.8% of the calculated correlation energy. Perhaps the most important aspect of the Bunge calculation is the unequivocal demonstration of the relative unimportance (4.1% of the correlation energy) of configurations differing by more than two orbitals from the SCF wave function.

If triple and quadruple excitations had turned out to be important, the quantitative calculation of the correlation energy would be extremely difficult. This is because of the very large number of triply- and quadruply-excited configurations. For example, using a double zeta basis set for the water molecule, there are 224 single and double excitations, 1334 triple excitations, and 5221 quadruple excitations if the $1a_1$ inner shell MO is always held doubly-occupied (170).

The double excitations for ^1S Be are of three types: (a) the two 1s SCF orbitals are replaced, (b) the two 2s SCF orbitals are replaced, and (c) one of the 1s and one of the 2s orbitals are replaced. Double excitations of type (c) are the most difficult to deal with since an orbital occupancy of the type $1s2s\ x_i x_j$ will give rise to two linearly independent ^1S configurations. Table II-7 shows the most important configurations of each type in the Be wave function of Bunge (62).

After the SCF, the most important configuration is $1s^2 2p^2$ or $2s^2 \to 2p^2$ in the excitation notation. The importance of this configuration is well known and it has been called the <u>degeneracy</u> effect (388) since the 2s and 2p orbitals are degenerate in the (admittedly crude) hydrogenic approximation. The 1s2s expansion converges the slowest since the natural orbitals were determined for the $1s^2$ and $2s^2$ pairs. Were it possible to rigorously divide the correlation energy into contributions from the $1s^2$, $2s^2$, and 1s2s pairs, Bunge estimates that these contributions would be 0.0426, 0.0455, and 0.0053 hartrees. This situation is reminiscent of the LiH results of Bender and Davidson (27) in that two doubly-occupied SCF orbitals are located in different regions of space. <r> for the 1s orbital is 0.41 bohrs, while <r> for the 2s orbital is 2.65 bohrs, and as a result the intershell 1s2s correlation energy is found to be small.

The exact restricted Hartree-Fock wave function for Be may be obtained using a basis set consisting only of s orbitals. However, Table II-7 shows clearly that p and d orbitals must be included to obtain a significant fraction of the correlation energy. Bunge has also estimated the effect of f and g orbitals on the correlation energy of Be and these estimates are seen in Table II-8. As one goes across the first row of the periodic table to larger atoms, it has been shown by a variety of calculations (61,63,198,277,381) that the effects of higher spherical harmonics become more important.

Table II-7. Important configurations in the 180 configuration wave function of Bunge (62) for the ground state of Be. The orbitals beyond 1s and 2s are natural orbitals, not to be confused with those orbitals which describe the excited electronic states of Be. The energy criterion is defined by Bunge (62) and gives an idea of the contribution of each configuration to the correlation energy.

Excitation	Coefficient	Energy Criterion
$1s^2 2s^2$	0.9531	--
$1s^2 \to 3p^2$	0.0285	0.0203
$1s^2 \to 4s^2$	0.0168	0.0079
$1s^2 \to 3s4s$	0.0139	0.0039
$1s^2 \to 2p3p$	0.0085	0.0011
$1s^2 \to 3s^2$	0.0065	0.0006
$1s^2 \to 4d^2$	0.0064	0.0026
$1s^2 \to 5p^2$	0.0051	0.0019
$2s^2 \to 2p^2$	0.2933	0.0409
$2s^2 \to 3s^2$	0.0396	0.0023
$2s^2 \to 3d^2$	0.0166	0.0014
$1s2s \to 2p3p$	0.0093	0.0009
	0.0058	0.0004
$1s2s \to 3p4p$	0.0077	0.0007
$1s2s \to 2p^2$	0.0057	0.0002
$1s2s \to 4s^2$	0.0055	0.0007

Table II-8. Estimated contributions of orbitals with different ℓ values to the correlation energy of Be.

	^1S Be
E (s limit)	−14.5920
% Correlation Energy	20.3
E (sp limit)	−14.6608
% Correlation Energy	94.0
E (spd limit)	−14.66453
% Correlation Energy	98.0
E (spdf limit)	−14.66570
% Correlation Energy	99.3
E (spdfg limit)	−14.66598
% Correlation Energy	99.6
Exact	−14.66639

On the basis of accurate calculations such as that discussed above, Bunge and coworkers (61, 63) have estimated the effects of g, h, i,.... for the ground states of carbon and neon to be 3% and 8% of the total correlation energy. Since it is very difficult to include even g orbitals in atomic (not to mention molecular) calculations, Bunge's estimates show clearly that the quantitative calculation of correlation energies for atoms larger than neon may be nearly intractable within an orbital framework. We will return to the question of the importance of different ℓ values in our discussion of pair correlation energies.

Bunge's calculations on carbon (61) and neon (63) yield total energies -37.8338 and -128.8868 hartrees. The 234 configuration (4836 determinants) ^{3}P C wave function accounts for about 93% of the estimated correlation energy. For neon about 89% of the correlation energy is recovered using a 231 configuration (5343 distinct Slater determinants) wave function. It is expected that both of these wave functions are nearly as rapidly convergent (lowest energy using the fewest configurations) as is possible for a CI wave function constructed from one-electron functions.

As was the case for beryllium, the CI wave functions of Bunge and coworkers for C and Ne show that configurations differing by 3 or 4 orbitals from the SCF orbital occupancy are relatively unimportant. For carbon, triple excitations correspond to 0.6% of the calculated correlation energy and quadruple excitations are 0.5%. The actual contributions to the exact carbon wave function are not likely to be more than twice the above percentages. For neon the evaluation of the importance of triple and quadruple excitations is particularly difficult due to the enormous number of such configurations. Nevertheless, Bunge and Peixoto (63) carried out an exhaustive search of triple and quadruple excitations and estimate that such configurations contribute only about 1% of the total correlation energy. This work seems to show conclusively that triple and quadruple excitations will not contribute more than a few percent of the correlation energy when the SCF configuration dominates the wave function. For carbon the coefficient of the SCF configuration in the final normalized wave function is 0.9727 while for neon it is 0.9836 (61, 63).

A final point to be made concerns Bunge's treatment of configurations arising from orbital occupancies with several unpaired electrons. For closed shell atoms this is not a problem, since an orbital occupancy differing by one or two orbitals from the SCF can yield no more than two configurations of ^1S symmetry. However, for the open-shell ^3P ground state of carbon, consider the excitation $1s2s \to 3s4s$, which corresponds to orbital occupancy $1s2s2p^23s4s$. There are 27 linear independent ^3P configurations which can be constructed from this orbital occupancy. However, these ^3P configurations can be expressed in such a way that only two of the 27 have nonvanishing H matrix elements with the SCF configuration (I,36). It is found (61) that inclusion of the chosen two configurations in the wave function gives essentially all the correlation energy obtained by including all 27. Bunge refers to this procedure as the <u>partitioning of degenerate spaces</u> and has used projection operators (235) in an elegant manner to carry out this partitioning in a completely general way.

Hyperfine structure of first-row atoms. Hyperfine structure (hfs) results from the fact that the nucleus is not a point charge, but has a structure of its own. This nuclear structure can include a magnetic-dipole moment, an electric-quadrupole moment, and, occasionally, higher electric and magnetic moments. The nuclear moments are coupled with the electronic charge distribution to produce a hyperfine structure. An understanding of hfs is essential to the study of such varied topics as electron spin resonance (ESR) in aromatic molecules, the Knight shift in metals, and angular correlations of nuclear gamma rays. In light of the enormous importance of hfs in chemistry and physics, it is not surprising that a great deal of theoretical effort has been expended in order to understand the hfs of the simplest systems, the first-row atoms.

An open-shell atom containing a nucleus with $I \geq \frac{1}{2}$ has a magnetic hfs due to the nonvanishing magnetic dipole moment of the nucleus. It can be shown (374) that the magnetic-dipole hfs can be expressed in terms of three electronic expectation values, one of which is the spin density at the nucleus $|\psi(0)|^2$ or Fermi contact interaction, discussed in section IIB. For an atom with an unfilled 2p orbital outside closed shells, the second expectation value (the _orbital_ contribution to the hfs) appears on the left hand side of the following equation:

$$\int \psi_e^* \left\{ \sum_{i=1}^{n} \frac{\ell_{zi}}{r_i^3} \right\} \psi_e \, d\tau = k_1 \int \chi_{2p}^*(r) \left(\frac{1}{r^3}\right) \chi_{2p}(r) r^2 dr \qquad (II,12)$$

where k_1 is a well-defined vector coupling coefficient (374) and $\chi_{2p}(r)$ gives the radial dependence of the 2p atomic orbital. Eq. (II,12) is <u>not</u> valid in general, but only holds when the expectation value on the left is calculated from a single configuration wave function. Using single configuration wave functions, the third expectation value (or <u>spin-dipolar</u>) contribution to the hfs is

$$\int \psi_e^* \left\{ \sum_{i=1}^{n} \frac{(3\cos^2\theta_i - 1)}{r_i^3} s_{zi} \right\} \psi_e \, d\tau = k_2 \langle r^{-3} \rangle \tag{II,13}$$

in which the integral on the right-hand side of (II,12) is abbreviated $\langle r^{-3} \rangle$.

If the nucleus has a spin of 1 or greater, the nonzero electric-quadrupole moment of the nucleus will interact with the <u>electric field gradient</u>

$$\int \psi_e^* \left\{ \sum_{i=1}^{n} \frac{(3\cos^2\theta_i - 1)}{r_i^3} \right\} \psi_e \, d\tau = k_3 \langle r^{-3} \rangle \tag{II,14}$$

to yield an electric-quadrupole contribution to the hfs.

Thus we see that in the restricted Hartree-Fock approximation, the hyperfine structure of the first-row atoms is given in terms of two parameters, the spin density $|\psi(0)|^2$ and $\langle r^{-3} \rangle$, which only involves the radial part of the 2p RHF orbital. Unfortunately, for more sophisticated wave functions the simple relations (II,12)-(II,14) are not valid and <u>four</u> independent parameters ($|\psi(0)|^2$ and the three expectation values on the left hand sides of the above three equations are in fact required to correctly describe the atomic hfs. However, to keep the analogy with the RHF theory, we may write each of the expectation values as the same constant (k_1, k_2, or k_3) times an adjustable parameter $\langle r_\ell^{-3} \rangle$, $\langle r_s^{-3} \rangle$, or $\langle r_q^{-3} \rangle$

$$\int \psi_e^* \left\{ \sum_{i=1}^n \frac{\ell_{zi}}{r_i^3} \right\} \psi_e \, d\tau = k_1 \langle r_\ell^{-3} \rangle \tag{II,15}$$

$$\int \psi_e^* \left\{ \sum_{i=1}^n \frac{(3\cos^2\theta_i - 1)}{r_i^3} s_{zi} \right\} \psi_e \, d\tau = k_2 \langle r_s^{-3} \rangle \tag{II,16}$$

$$\int \psi_e^* \left\{ \sum_{i=1}^n \frac{(3\cos^2\theta_i - 1)}{r_i^3} \right\} \psi_e \, d\tau = k_3 \langle r_q^{-3} \rangle \tag{II,17}$$

The values of the empirical parameters $\langle r_\ell^{-3} \rangle$, $\langle r_s^{-3} \rangle$, and $\langle r_q^{-3} \rangle$ may be obtained either experimentally or theoretically by determining one of the expectation values and dividing by the appropriate vector coupling coefficient k_1, k_2, or k_3. For RHF wave functions with an unfilled 2p orbital outside closed shells

$$\langle r_\ell^{-3} \rangle = \langle r_s^{-3} \rangle = \langle r_q^{-3} \rangle = \langle r^{-3} \rangle = \int \chi_{2p}^*(r) \left(\frac{1}{r^3}\right) \chi_{2p}(r) r^2 \, dr \tag{II,18}$$

Among the types of variational wave functions used to investigate atomic hfs are many of "extended" SCF wave function discussed in chapter I, section B. In this section we discuss the results from SUHF, SEHF, and SOHF wave functions. In addition two limited types of CI wave functions, termed <u>polarization</u> (335) and <u>first-order</u> (336) wave functions, have been extensively used for the study of atomic hfs. The polarization wave function includes the RHF configuration plus all singly excited configurations, i.e., all configurations of the proper symmetry arising from orbital occupancies differing from the RHF by one orbital. This type of wave function takes advantage of the now well-known fact (279), that single excitations, particularly 1s → ns and nd and 2s → ns and nd, play an important role in accurate determinations of the hfs parameters. These configurations describe the effects of "core polarization" so frequently discussed in the literature (127,138).

The first-order wave function (336) includes the RHF configuration, plus all degeneracy effect configurations (76) and all single excitations with respect to both. For the ground state of carbon these degeneracy effect configurations are $1s^2 \rightarrow 2p^2$, $1s2s \rightarrow 2p^2$, and $2s^2 \rightarrow 2p^2$. In more general terms the first-order wave function includes all configurations in which no more than a single electron is promoted to an orbital beyond the valence shell. This breakdown of correlation effects is for the most part consistent with the ideas of Silverstone and Sinanoglu (352). The first-order wave function includes all configurations which have <u>no counterpart</u> in true closed-shell systems such as ^1S Ne. The name first-order arose because for open-shell systems this type of wave function obeys a theorem analogous to Brillouin's theorem for closed-shell systems. The Be atom in its ground state is, unlike neon, not a true closed shell atom because the unoccupied 2p orbital is nearly degenerate with the doubly-occupied 2s.

As a concrete example, Table II-9 gives the configurations included in the polarization and first-order wave functions for ^3P C. Even with a very large basis set, the number of configurations in polarization and first-order wave functions is modest; the largest CI being the ^3P carbon first-order wave function, which includes 183 configurations (336).

Table II-10 gives the hfs predictions for B, C, N, O, and F. As pointed out in section IIB, the most serious failing of the RHF model as regards atomic hyperfine structure is its inability to predict a nonzero spin density at the nucleus for atoms such as those in Table II-10. From Table II-10 it can be seen that the relation (II,18) is also a serious flaw in the RHF model. In oxygen and fluorine, for which experimental values of both $\langle r_\ell^{-3} \rangle$ and $\langle r_s^{-3} \rangle$ are known, the two values differ significantly both from the RHF value and from each other. As discussed in section IIB, the SUHF spin densities *are* nonzero but about twice the experimental values for N, O, and F. However, many of the inherent problems of the RHF method with respect to the $\langle r^{-3} \rangle$ values carry over to the SUHF. All three values $\langle r_\ell^{-3} \rangle$, $\langle r_s^{-3} \rangle$, and $\langle r_q^{-3} \rangle$ are identical for B and C, while for O and F, $\langle r_\ell^{-3} \rangle$ and $\langle r_q^{-3} \rangle$ are identical.

Table II-9. Types of configurations included in the polarization and first-order wave functions for the ground state of the carbon atom. ns refers to all s orbitals in the basis set except 1s and 2s. Similarly np means all p orbitals besides 2p in the basis, while nd refers to 3d,4d,5d,...and nf to all functions with $\ell=3$. The polarization includes only the first six types of orbital occupancies. The first-order wave function include all the orbital occupancies.

Excitation notation	Orbital Occupancy	Number of ^3P Configurations per Orbital Occupancy
Hartree-Fock	$1s^2 2s^2 2p^2$	1
$1s \to ns$	$1s\, 2s^2 2p^2\, ns$	2
$1s \to nd$	$1s\, 2s^2 2p^2\, nd$	3
$2s \to ns$	$1s^2\, 2s\, 2p^2\, ns$	2
$2s \to nd$	$1s^2\, 2s\, 2p^2\, nd$	3
$2p \to np$	$1s^2 2s^2 2p\, np$	1
~~		
$1s^2 \to 2p^2$	$2s^2 2p^4$	1
$1s^2 \to 2p\, np$	$2s^2 2p^3\, np$	3
$1s^2 \to 2p\, nf$	$2s^2 2p^3\, nf$	1
$1s\, 2s \to 2p^2$	$1s\, 2s\, 2p^4$	2
$1s\, 2s \to 2p\, np$	$1s\, 2s\, 2p^3\, np$	9
$1s\, 2s \to 2p\, nf$	$1s\, 2s\, 2p^3\, nf$	3
$2s^2 \to 2p^2$	$1s^2\, 2p^4$	1
$2s^2 \to 2p\, np$	$1s^2\, 2p^3\, np$	3
$2s^2 \to 2p\, nf$	$1s^2\, 2p^3\, nf$	1

Table II-10. Predicted hyperfine structure parameters in atomic units for the ground states of the first-row atoms. The different theoretical methods are described in the text.

Calculation Reference	RHF 335	SUHF 13,136	SEHF 136	SOHF 193,194	Polarization 335	First-Order 336	Many-Body Calculations 110,199,279	Experiment
Boron $\|\psi(0)\|^2$	0.0	0.0171	0.0362	0.0022	0.0073	0.0041	0.0038	--
$\langle r_\ell^{-3}\rangle$	0.7755	0.7817	0.7943	0.7900	0.7674	0.7572	0.7789	--
$\langle r_s^{-3}\rangle$	0.7755	0.7817	0.7924	0.7894	0.8301	0.8167	0.8368	--
$\langle r_q^{-3}\rangle$	0.7755	0.7817	0.7943	0.7900	0.7436	0.6833	0.7054	--
Carbon $\|\psi(0)\|^2$	0.0	0.0772	0.0733	0.0423	0.0277	0.0228		--
$\langle r_\ell^{-3}\rangle$	1.692	1.709	1.723	1.721	1.679	1.663		--
$\langle r_s^{-3}\rangle$	1.692	1.709	1.720	1.720	1.782	1.769		--
$\langle r_q^{-3}\rangle$	1.692	1.709	1.723	1.721	1.637	1.537		--
Nitrogen $\|\psi(0)\|^2$	0.0	0.1874	0.1579	0.1200	0.0730	0.0714	0.0976	0.0972
Oxygen $\|\psi(0)\|^2$	0.0	0.1954	0.2137		0.0610	0.0628	0.1202	0.1138
$\langle r_\ell^{-3}\rangle$	4.973	4.370	4.279		4.613	4.570	4.547	4.58
$\langle r_s^{-3}\rangle$	4.973	4.746	4.647		5.125	5.100	5.126	5.19
$\langle r_q^{-3}\rangle$	4.973	4.370	4.279		4.334	4.307	4.205	--
Fluorine $\|\psi(0)\|^2$	0.0	0.1333	0.2454		0.0470	0.0496		0.0717
$\langle r_\ell^{-3}\rangle$	7.545	7.033	6.899		7.276	7.234		7.35
$\langle r_s^{-3}\rangle$	7.545	7.611	7.407		7.963	7.950		8.14
$\langle r_q^{-3}\rangle$	7.545	7.033	6.899		6.880	6.852		--

The SEHF spin density for nitrogen represents an improvement over the SUHF value, but the oxygen and fluorine values of $|\psi(0)|^2$ are in poorer agreement with experiment than the SUHF values. The SEHF values of $<r^{-3}>$ suffer from the fact that $<r_\ell^{-3}>$ and $<r_q^{-3}>$ are identical for B, C, O, and F. However, they represent an improvement over the SUHF values since for B and C the values of $<r_s^{-3}>$ are slightly different from $<r_\ell^{-3}>$ and $<r_q^{-3}>$. Only for B, C, and N are calculations within the SOHF framework available (193,194). This is because the nonorthogonality problem discussed in section IB is currently intractable for systems with more than 7 electrons. The only comparison which can be made between SOHF and experiment is for the spin density of nitrogen, which is predicted to be 123% of experiment. In light of the difficulty involved in the calculation of $|\psi(0)|^2$, this agreement must be considered rather good. The SOHF approximation is weak in that it predicts $<r_\ell^{-3}> = <r_q^{-3}>$ for B and C.

The polarization and first-order results are seen to be in much more consistent agreement with experiment than the methods discussed above. For each atom the calculated values $<r_\ell^{-3}>$, $<r_s^{-3}>$, and $<r_q^{-3}>$ are distinctly different, as they must be. Of the two types of CI, the polarization wave function results are in closer agreement with experiment. The spin densities of N, O, and F are consistently low, being 75, 54, and 66% of experiment. For ^3P O, the polarization values of $<r_\ell^{-3}>$ and $<r_s^{-3}>$

differ from experiment by 0.7% and 1.2%. For ^2P F the deviations are slightly greater, 1.0% and 2.2%.

Thus it would appear that the polarization wave function is capable of predictions of $\langle r_\ell^{-3} \rangle$, $\langle r_s^{-3} \rangle$, and $\langle r_q^{-3} \rangle$ reliable to within ~2% of the exact values. This observation has been used to predict nuclear electric quadrupole moments for B^{10}, B^{11}, C^{11}, and O^{17}. This is done using the fact that the quadrupole coupling constant is just eqQ, where e is the electronic charge, q the electric field gradient, and Q the nuclear electric-quadrupole moment. Q is deduced from the experimental quadrupole coupling constant and the theoretical electric field gradient, which is proportional to $\langle r_q^{-3} \rangle$. Quadrupole moments obtained in this way (335) should be the most accurate values known to date.

Also included in Table II-10 are the results of Nesbet (279) for boron, Dutta et.al. (110) for nitrogen, and Kelly (199) for oxygen. These calculations use many-body theory to evaluate correlation effects and are very difficult to carry out. However, the results are in excellent agreement with experiment for N and O. In fact the agreement with the experimental spin density for nitrogen is probably fortuitously good. For oxygen Kelly's spin density is excellent, 6% greater than experiment. The values of $\langle r_\ell^{-3} \rangle$ and $\langle r_s^{-3} \rangle$ are in the same excellent agreement with experiment as are the polarization function values. For $\langle r_q^{-3} \rangle$, where no experimental value is available, the many body value is 3% less than that obtained from the polarization wave function. The agreement between the Nesbet's boron results and the polarization values is also good, particularly for $<r_\ell^{-3}>$ and $<r_s^{-3}>$. The values of $<r_q^{-3}>$ differ by about 5%. Although the boron polarization spin density is nearly twice as large as Nesbet's, the absolute difference is very small and Nesbet gives a second value, 0.0076 atomic units, which must be considered about equally reliable (279). Both Kelly and Nesbet use their calculated $<r_q^{-3}>$ values to predict nuclear electric-quadrupole moments.

With an eye to the possibility of applying the above methods to the hfs of molecules, let us compare the difficulty of the calculations summarized in Table II-10. Each RHF or SUHF calculation may be carried out on a computer such as the IBM 360/65 in a few seconds. The SEHF, polarization, and first-order calculations are somewhat more difficult but all require less than 10 minutes on the above computer. The computation time used in the SOHF calculations goes up astronomically with the number of electrons and is likely to be unreasonable for atoms larger than neon. Nesbet's boron calculation is difficult and requires several hours of machine time. The method of Kelly, also used by Dutta, et.al., has not been fully programmed due to its intricacy and the nitrogen calculations referred to (110) required the expenditure of more than a year's work by a competent graduate student.

D. PAIR CORRELATION APPROXIMATIONS

The calculations of Bunge (61 - 63) described in the previous section are very difficult. If it weren't necessary to calculate all of the correlation energy at once, the problem would be much simpler. The essential motivation behind the idea of <u>pair correlations</u> is the goal of dividing the correlation energy into many small pieces, each of which may be computed separately.

One of two related methods is usually employed in the theoretical evaluation of pair correlations. The first method is the many body perturbation theory (MBPT) of Brueckner and Goldstone (137), extended to atomic systems by Kelly (200). Kelly (198,199,200) and Das and coworkers (110) have shown MBPT to be a very powerful method for the treatment of a variety of problems involving electron correlation. However, MBPT makes extensive use of Feynman diagrams, which most chemists are not familiar with, and for this reason we will not discuss MBPT herein. The second method, usually called pair correlation theory and associated with the names Sinanoglu (358) and Nesbet (278) fits in nicely with our development of configuration interaction. The rather complicated and intensely debated relationship between MBPT and pair correlation theory has been discussed in detail by Freed (126).

<u>Spinorbital pairs - Be, Mg, and Ar</u>. The Hartree-Fock approximation leaves out the instantaneous repulsions between pairs of

electrons. Therefore it is reasonable to assume that most of the correlation energy can be accounted for by consideration of a set of functions describing in more detail the interactions between each pair of electrons. A particularly happy situation would arise if the pair correlation functions describing the different pairs of electrons were independent and their contributions to the correlation energy were additive.

In pair correlation approximations for closed-shell systems one assumes that the correlation energy E_c is given by (355)

$$E_c \approx \sum_{ij} e(i,j) \qquad (II,19)$$

In its usual form the pair correlation approximation summation in (II,19) goes over all the occupied spinorbitals ϕ_i in the RHF wave function. Each pair correlation energy $e(i,j)$ is calculated independently. It is possible to make formal arguments as to why (II,19) should be approximately correct (356). Nonetheless, the pair correlation approximation is essentially intuitive, arising from our confidence in the orbital theory of electronic structure. As to its merits as a quantitative theory, these can only be determined by detailed calculations which make comparisons with accurate variational calculations and with experiment.

As a concrete example, consider the ^1S ground state of the beryllium atom. The RHF wave function is of the form

$$\psi_{RHF} = 1s\alpha \; 1s\beta \; 2s\alpha \; 2s\beta \qquad (II,20)$$

For Be there are six pair correlation energies of the usual kind: $e(1s\alpha,1s\beta)$, $e(1s\alpha,2s\alpha)$, $e(1s\alpha,2s\beta)$, $e(1s\beta,2s\alpha)$, $e(1s\beta,2s\beta)$, $e(2s\alpha,2s\beta)$. These pair correlation energies have been accurately evaluated by Nesbet (275) and his results are seen in Table II-11. It may be noted, as one might expect, that $e(1s\alpha,2s\alpha) = e(1s\beta,2s\beta)$ and $e(1s\alpha,2s\beta) = e(1s\beta,2s\alpha)$. The results of Table II-11 show many features in common with Bunge's Be calculation. In particular, the $1s^2$ and $2s^2$ pair correlations are large and the contributions to the 1s2s correlation are small. Table II-11 also upholds a cherished belief among theoretical chemists--that "electrons with the same spin (function) avoid each other". The pair correlation energy due to a 1s and 2s orbitals with α spin is 2.5 times less than that obtained when one of electrons being correlated is $1s\alpha$ and the other $2s\beta$.

We have not yet mentioned how the spinorbital pair correlation energies were obtained. It turns out that Nesbet calculated the $e(i,j)$ using a CI approach in terms of Slater determinants (275). First a large basis set of s, p, d, and f orbitals is chosen and an SCF calculation carried out. Then to calculate, for example, $e(1s\alpha,1s\alpha)$, a CI is carried out including the SCF determinant plus all determinants of the type $2s\alpha\ 2s\beta\ xy$, where x and y are all the spinorbitals besides 1s and 2s in the basis set. $e(1s\alpha,1s\alpha)$ is then just the difference between the energy of the above CI and the SCF energy. Put in another way, the $1s^2$ pair correlation energy

Table II-11. Spinorbital pair correlation energies of Nesbet (275) for the ground state of Be.

Pair Correlation	Energy (hartrees)
$e(1s\alpha,1s\beta)$	0.04183
$e(1s\alpha,2s\alpha)$	0.00081
$e(1s\alpha,2s\beta)$	0.00212
$e(1s\beta,2s\alpha)$	0.00212
$e(1s\beta,2s\beta)$	0.00081
$e(2s\alpha,2s\beta)$	0.04535

$e(1s\alpha,1s\alpha)$ is just the energy lowering (below the SCF) due to Slater determinants which may be designated $1s\alpha$ $1s\beta$ → xy. Similarly $e(1s\alpha,2s\alpha)$ is obtained by performing a CI including the SCF determinant plus all determinants $1s\beta$ $2s\beta$ xy which result from the replacement of the $1s\alpha$ and $2s\alpha$ spinorbitals from the SCF determinant. $e(1s\alpha,2s\alpha)$ is just the difference between the above CI energy and the SCF energy.

The usefulness of the pair correlation idea depends on how well the approximate relation (II,19) holds. The sum of the six spinorbital pair correlation energies in Table II-11 is 0.09304 hartrees. For comparison the "experimental" correlation energy of Be is 0.094 hartrees (377). The sum of the pairs is therefore 99% of the true correlation energy. Thus it appears that for beryllium the spinorbital pair correlation approximation is an excellent approximation. It should be pointed out that Nesbet's basis set is unavoidably less than complete, and a comparable calculation using a complete set of functions might yield 101 or 102% of the true correlation energy. In a variational calculation, of course, we can never obtain more than 100% of the correlation energy, since the calculated energy is a rigorous upper bound to the exact nonrelativistic energy.

More recently Nesbet has reported pair correlation energies for the second row atoms (280). Particularly relevant to the present discussion are the results obtained for the closed-shell systems Mg and Ar. Table II-12 gives the calculated pair correlation energies in abbreviated form. For reference the magnesium RHF orbital occupancy is $1s^2 2s^2 2p^6 3s^2$ and that for argon $1s^2 2s^2 2p^6 3s^2 3p^6$. The largest contribution in Table II-12 to the correlation energy of Mg is from the $3s^2$ pair. The $3s^2$ pair correlation energy in Mg is primarily due to the $3s^2 \rightarrow 3p^2$ "degeneracy" configuration. Since the 3p SCF orbital is fully occupied for Ar, there is no such degeneracy effect and Table II-12 shows that for Ar the $3s^2$ pair correlation energy is small. By far the largest contribution to the correlation energy of argon comes from the (3p,3p) spinorbital pairs, e.g., $(3p_{-1}\alpha, 3p_{+1}\beta)$. The next largest effect is from the intershell (3s, 3p) pairs, of which there are 12 in Nesbet's formulation. The calculation of pair correlation energies such

Table II-12. Summary of the pair correlation energies of Nesbet (280) for the electronic ground states of Mg and Ar. Each number tabulated is the sum of all spinorbital pair correlations associated with the two radial orbitals. For example, e(2s,3s) is the sum of the four energies e(2sα,3sα), e(2sα,3sβ), e(2sβ,3sα), and e(2sβ,3sβ). All energies are in hartrees.

	^1S Mg	^1S Ar
e(2s,3s)	0.00163	0.00293
e(2p,3s)	0.01893	0.01589
e(2s,3p)	--	0.00800
e(2p,3p)	--	0.04980
e(3s,3s)	0.03375	0.01026
e(3s,3p)	--	0.05858
e(3p,3p)	--	0.20415

as those seen in Table II-12 represents a very constructive first step in the evaluation of electron correlation in molecules containing second row atoms.

Symmetry-adapted pairs for neon and the question of additivity. There is at least one theoretical objection to the idea of spinorbital pair correlations, in addition to the fact that the sum of the pairs is not an upper bound to the true correlation energy.

This objection is that in many of the spinorbital pair correlation CI calculations discussed above, the wave function obtained is not an exact eigenfunction of the symmetry operators $\underset{\sim}{L}^2$ and $\underset{\sim}{S}^2$. For example, in the (1sα,2sβ) Be calculation, the determinants 1sβ 2sα 3sα 4sβ and 1sβ 2sα 3sβ 4sα would be included, corresponding to the excitations 1sα 2sβ → 3sα 4sβ and 1sα 2sβ → 3sβ 4sα. In the (1sβ,2sα) beryllium calculation would be included the two determinants 1sα 2sβ 3sα 4sβ and 1sα 2sβ 3sβ 4sα. Unfortunately, the simplest ^{1}S <u>configuration</u> which can be constructed from the orbital occupancy 1s2s3s4s includes all four of the above Slater determinants. Since the (1sα,2sβ) and (1sβ,2sα) calculations are completely independent, neither of the two wave functions will be ^{1}S L-S eigenfunctions.

The most general way of guaranteeing that each pair correlation wave function will be an exact eigenfunction of all the symmetry operators of the point group is to include in the same calculation all configurations arising from each particular orbital occupancy. For the Be atom, the (1sα,2sα), (1sα,2sβ), (1sβ,2sα), (1sβ,2sβ) calculations will no longer be independent but incorporated into a single pair correlation, the (1s,2s) <u>symmetry-adapted pair correlation</u> (332,381). Although there are four times as many Slater determinants in the symmetry-adapted (1s,2s) calculation, the size of the eigenvalue problem will not be four times as large as that required to evaluate e(1sα,2sα). This is because the symmetry-adapted pair calculations are carried out in terms of configurations, which are symmetrized linear combinations of determinants. For the 1s2s3s4s problem, only two linearly independent ^{1}S eigenfunctions may be constructed from the six acceptable Slater determinants.

It is of course possible that the question of symmetry-adapted versus spinorbital pairs is not of practical importance. For Be, the summation of the spinorbital pair correlation energies was in close agreement with the experimental correlation energy. To test the differences between the two pair correlation schemes, a direct comparison, using the same basis set and the same Slater determinants, is needed. Such a comparative study has been made for Ne by Viers, et.al. (381) and is seen in Table II-13. All 434 configurations which arise during the six symmetry-adapted pair calculations have also been used in a variational calculation which yields an energy -128.8767 hartrees, or 0.3296 hartrees correlation energy. We know from the work of Bunge and Peixoto (63) that triple and quadruple excitations are only about 1% of the neon correlation energy. Therefore the variational calculation of Viers, et.al. (381) which includes all single and double excitations should yield a total energy very close to the limit of the basis, i.e., that which would be obtained by a full CI.

Table II-13 shows that summation of the spinorbital pairs yields 113% of the variationally calculated correlation energy.

Table II-13. Direct comparison of spinorbital and symmetry-adapted pair correlations for the neon atom (381). As in Table II-12, each number given under the column "spinorbital pairs" is the sum of all spinorbital pair correlation energies associated with the two radial orbitals.

Pair correlations	Symmetry-adapted pairs	Spinorbital pairs
(1s,1s)	0.03897	0.03897
(1s,2s)	0.00500	0.00502
(1s,2p)	0.01902	0.01935
(2s,2s)	0.01133	0.01133
(2s,2p)	0.07562	0.08265
(2p,2p)	0.18562	0.21646
Sum	0.33555	0.37377

The sum of symmetry-adapted pairs is, however, very reasonable, being 101.8% of the variational result. Thus these calculations predict, extrapolating to a complete basis set, that for neon, the sum of the spinorbital pair correlation energies will be about 110% of the true correlation energy, while the sum of the symmetry-adapted pairs should be within 1% of the exact correlation energy. Table II-13 shows that the nonadittivity among the spinorbital pairs is almost entirely due to the (2s,2p) and (2p,2p) contributions. The spinorbital contribution under (2s,2p) in Table II-13 is the sum of twelve individual pair correlation energies, $e(2s\alpha, 2p_{-1}\alpha)$, $e(2s\alpha, 2p_{-1}\beta)$, $e(2s\alpha, 2p_{0}\alpha)$, etc. The sum of these twelve contributions is 9.3% greater than the symmetry-adapted pair correlation energy $e(2s,2p)$, obtained from a single calculation including all Slater determinants from the twelve spinorbital calculations. The (2p,2p) correlation is 16.6% greater from the spinorbital than the symmetry-adapted calculation.

On the basis of the above results and related work by Barr and Davidson (20), Nesbet, et.al. carried out exhaustive calculations (277) of the spinorbital pair correlation energies of neon. They used a basis set of s,p,d,f,g,h, and i orbitals! For the L shell (2s and 2p) pairs these results (277) are seen in Table II-14. The sum of the spinorbital pair correlation energies from this calculation is -0.4103 hartrees or 106.6% of Bunge's estimate of the neon correlation energy. However, even this spdfghi basis set is incomplete and in addition Bunge estimates (63) that j,k,.... orbitals are responsible for 2% of the correlation energy. Therefore the results of this exhaustive calculation are in harmony with the earlier estimate (based on the symmetry-adapted pair results) that spinorbital pairs yield ∼110% of correlation energy for ^1S Ne.

The fact that the spinorbital pair correlation energies may overestimate the correlation energy of Ne by 10% is only a serious problem if a truly quantitative theory is required.

Table II-14. Valence shell spinorbital pair correlation energies for neon (277). Only unique pairs are indicated and the "weight" given is the number of equivalent pairs of each kind.

Spinorbital pair	Weight	Correlation energy (hartrees)
$(2s\alpha, 2s\beta)$	1	0.01173
$(2s\beta, 2p_{-1}\beta)$	4	0.00349
$(2s\beta, 2p_0\beta)$	2	0.00350
$(2s\beta, 2p_{-1}\alpha)$	4	0.01159
$(2s\beta, 2p_0\alpha)$	2	0.01163
$(2p_{-1}\beta, 2p_0\beta)$	4	0.01145
$(2p_{-1}\beta, 2p_{+1}\beta)$	2	0.01147
$(2p_{-1}\beta, 2p_{-1}\alpha)$	2	0.01867
$(2p_{-1}\beta, 2p_0\alpha)$	4	0.01509
$(2p_{-1}\beta, 2p_{+1}\alpha)$	2	0.02426
$(2p_0\beta, 2p_0\alpha)$	1	0.02814

The truth of the matter is that in a qualitative sense the correlation energy for first-row atoms is nearly separable into many small independently calculable quantities. In this sense, the correlation energy has a fairly well-defined structure. We will return to the question of additivity in our chapter III discussion of diatomic molecules.

Treatment of open shells--nitrogen. As pointed out in the section IIC discussion of first-order wave functions, there are many correlation effects in open-shell systems which have no counterpart in, for example, the Ne atom. These effects are quantitatively evaluated for the ^4S ground state of nitrogen in Table II-15. These correlation effects are essentially of two types: (a) polarization effects such as $2s \to 3d$, which as Table II-15 shows, is extremely important for nitrogen; (b) degeneracy effects such as $2s^2 \to 2p^2$, which is not possible for nitrogen since the $1s^2 2p^5$ orbital occupancy does not give rise to a ^4S configuration. For the ground states of B, C, N, O, and F these correlation effects account for approximately 47%, 40%, 30%, 20%, and 10% of the total correlation energy (337). There is an increasing amount of numerical evidence that the systematics of electron correlation in open-shell atoms are much clearer after taking into account these polarization and degeneracy effects (337,353,357).

In many cases, spinorbital pair correlation energies may be calculated in a manner similar to that adopted for closed-shell systems. However, since Brillouin's theorem does not hold, the effects of single excitations must somehow be included. Nesbet (276) has handled this problem by defining single particle contributions to the correlation energy. For nitrogen, the $2s\beta$ single particle contribution to the correlation energy is the

Table II-15. Correction effects in the nitrogen atom with no counterpart for true closed shell systems (300). Each energy lowering given is that obtained by carrying out a CI including the SCF wave functions plus all configurations of the type given on the same line.

Type of excitation	Energy lowering relative to SCF
$1s \to ns$ (n>2)	0.00026
$1s \to nd$ (n>2)	0.00064
$2s \to ns$	0.00452
$2s \to nd$	0.04629
$2p \to np$ (n>2)	0.00092
$1s^2 \to 2pnp$	0.00092
$1s\,2s \to 2pnp$	0.00129
$2s^2 \to 2pnp$	0.00318

lowering of the SCF energy due to a CI including all singly-excited determinants $2s\beta \to x$. This quantity, which may be called $e(2s)$ is found to be 0.00176 hartrees for nitrogen. As pointed out previously in a different situation, such a singly-excited CI wave function will not be an exact eigenfunction of $\underset{\sim}{L}^2$ and $\underset{\sim}{S}^2$. Given this definition of a single particle contribution to the correlation energy, Nesbet evaluates the pair correlation energies as before and sums the one particle and pair correlation energies to give estimates of the total correlation energy of B, C, N, O and F (276).

A perhaps more logical approach to pair correlations would be to realize the unique character of polarization and degeneracy effects and include them in a first-order wave function before proceeding to evaluate the correlation effects expected to be similar to those observed for closed shell systems. This approach has been followed by Platas and Schaefer in a study of the nitrogen atom (300). The $2p^2$ symmetry-adapted pair correlation energy is then calculated as the lowering of the first-order energy due to addition of all configurations $2p^2 \rightarrow xy$. The total energy is then expected to be closely approximated (to within 1% of the correlation energy) as the first-order wave function energy plus the sum of the symmetry-adapted pair correlation energies.

An interesting feature of the symmetry-adapted pair correlation energies is that they may be quite accurately divided into contributions due to different classes of configurations (300, 381). Such a breakdown is seen for nitrogen in Table II-16. For the $2p^2$ pair the sum of the three contributions is 0.02491 hartrees. The lowering of the first-order energy obtained by including all three classes of configurations in a variational calculation is 0.02404 hartrees. The overestimation due to summation is only 3.6%. The data of Table II-16 may also be used to discuss the contributions of different spherical harmonics (that is, s,p,d,f,...

Table II-16. Breakdown of the symmetry-adapted pair correlation energies for ^4S N (300). ns=3s,4s,...; np=3p,4p...; nd=3d,4d,...; nf=4f,5f,... .

Class of configurations	Energy lowering (hartrees)
e(1s,1s)	
$1s^2 \to ns\ ms$	0.01242
$1s^2 \to np\ mp$	0.02241
$1s^2 \to nd\ md$	0.00340
$1s^2 \to nf\ mf$	0.00072
$1s^2 \to ns\ md$	0.00001
e(1s,2s)	
$1s2s \to ns\ ms$	0.00156
$1s2s \to np\ mp$	0.00255
$1s2s \to nd\ md$	0.00054
$1s2s \to nf\ mf$	0.00009
e(1s,2p)	
$1s2p \to ns\ mp$	0.00281
$1s2p \to np\ md$	0.00573
$1s2p \to nd\ mf$	0.00150
e(2s,2s)	
$2s^2 \to ns\ ms$	0.00375
$2s^2 \to np\ mp$	0.00181

cont.

Class of configurations	Energy lowering (hartrees)
$2s^2 \to nd\ md$	0.00601
$2s^2 \to nf\ mf$	0.00120
$2s^2 \to ns\ md$	0.00016
$2s^2 \to np\ mf$	0.00002
$e(2s,2p)$	
$2s2p \to ns\ mp$	0.02039
$2s2p \to np\ md$	0.00659
$2s2p \to nd\ mf$	0.00776
$e(2p,2p)$	
$2p^2 \to np\ mp$	0.01236
$2p^2 \to nd\ md$	0.01098
$2p^2 \to nf\ mf$	0.00157

orbitals) to the correlation energy. A glance at Table II-16 shows that when f orbitals are included, the symmetry-adapted pairs have nearly converged with respect to ℓ-value except for the 2s2p pair. Configurations of the type $2s2p \to nf\ mg$ are probably quite important. A final point worth mentioning is that the sort of breakdown seen in Table II-16 gives the same sort of detailed information concerning the structure of the correlation energy as does the many body perturbation theory (200).

III
DIATOMIC MOLECULES

Prior to 1965, nearly all nonempirical electronic structure calculations reported in the literature were for diatomic molecules. The primary cause of this situation was the difficulty encountered in calculating two-electron integrals involving functions centered on three or four different nuclei. In a diatomic molecule calculation, the most difficult integrals are the two-center integrals, involving functions centered on both nuclei. Using Slater functions, these integrals may be evaluated very rapidly using sophisticated semi-analytic methods developed over the past twenty years (328,383). Due to the ease of integral calculation and other simplifications introduced by the high symmetry, it is still true today that our most detailed information concerning the electronic structure of molecules has come from ab initio calculations on diatomic molecules.

Some of the earliest calculations on diatomic molecules were carried out using Kotani's tables of two-center integrals (213). With the introduction of fully automatic computer codes, much larger calculations became feasible. In 1957 at the University of Texas, a minimum basis set valence bond calculation was carried out on lithium hydride including 20 VB structures or configurations (259).

The first SCF calculation using a larger basis was that of Nesbet on hydrogen fluoride (271). And the first systematic study of molecular electronic structure was carried out by Ransil with minimum basis sets for molecules from LiH through F_2 (310).

Another significant achievement was the development at the University of Chicago of computer programs (178,384) for the calculation of SCF wave functions approaching rather closely the Hartree-Fock limit for molecules such as N_2. Since 1966, most of the important advances in the calculation of diatomic molecule wave functions have centered about the treatment of electron correlation by way of configuration interaction (27,29,96,342). Concurrently, SCF procedures have been perfected to the point where near Hartree-Fock wave functions have been reported for molecules as large as KrF^+ and KrF (229).

In this chapter we discuss some of the more recent electronic structure calculations on diatomic molecules in order to a) give a clear idea of current capabilities, b) point out some of the strengths and weaknesses of different theoretical methods, and c) demonstrate that a number of problems of real chemical interest have already been studied by theoretical chemists.

A. NEAR HARTREE-FOCK RESULTS

Spectroscopic constants and potential curves for N_2 and N_2^+. In the work of Cade, Sales, and Wahl (65) on the nitrogen molecule, we see the first near Hartree-Fock (here we mean within 0.001 hartrees of the true HF energy) wave function for a molecule larger than one of the simple hydrides. These results are so close to the correct Hartree-Fock results that they allow us to make some conclusions concerning the chemical usefulness of the RHF approximation. The ground state of N_2 and the three lowest states of N_2^+ were studied, and the orbital occupancies for these four states are

$$1\sigma_g^2 \, 1\sigma_u^2 \, 2\sigma_g^2 \, 2\sigma_u^2 \, 3\sigma_g^2 \, 1\pi_u^4 \qquad X \; {}^1\Sigma_g^+ \quad N_2 \qquad \text{(III,1)}$$

$$1\sigma_g^2 \, 1\sigma_u^2 \, 2\sigma_g^2 \, 2\sigma_u^2 \, 3\sigma_g \, 1\pi_u^4 \qquad X \; {}^2\Sigma_g^+ \quad N_2^+ \qquad \text{(III,2)}$$

$$1\sigma_g^2 \, 1\sigma_u^2 \, 2\sigma_g^2 \, 2\sigma_u^2 \, 3\sigma_g^2 \, 1\pi_u^3 \qquad A \; {}^2\Pi_u \quad N_2^+ \qquad \text{(III,3)}$$

$$1\sigma_g^2 \, 1\sigma_u^2 \, 2\sigma_g^2 \, 2\sigma_u \, 3\sigma_g^2 \, 1\pi_u^4 \qquad B \; {}^2\Sigma_u^+ \quad N_2^+ \qquad \text{(III,4)}$$

In (III,1) through (III,4) both the term symbol (e.g., ${}^2\Pi_u$) and the accepted spectroscopic designation (e.g., A) are indicated for each of the four states. Relative to neutral N_2, the three states of N_2^+ may be thought of as those resulting from the removal of an electron from each of the three highest occupied molecular orbitals $3\sigma_g$, $1\pi_u$, and $2\sigma_u$.

In the process of obtaining their wave functions for N_2 and N_2^+, Cade et al. (65) established a straightforward (but very expensive) procedure for the computation of very-close-to Hartree-Fock wave functions. The first step is to choose an extended basis of Slater functions optimized for each atom in the molecule. For N_2 the chosen basis set was intermediate in size between the double zeta basis seen in Table I-12 and the extended sets in Table I-13. The N_2 energy obtained with this basis was -108.8967 hartrees. Secondly, a rather large number of polarization functions was added to the basis. To describe the $2\sigma_g$ and $3\sigma_g$ orbitals three different 3d Slater functions and one

4f function were added. The energy at this stage was -108.9897 hartrees. The third stage in the procedure involves the optimization of all orbital exponents ζ by repeated SCF calculations. While this is a rather standard procedure for atoms, calculations of this type on a molecule even as small as N_2 are extremely time consuming. The final energy obtained for N_2 is -108.9928 hartrees, and is likely to be within 0.001 hartrees of the exact RHF energy.

A number of properties calculated from the wave functions of Cade, Sale, and Wahl are seen in Table III-1. As is generally found within the Hartree-Fock framework, the calculated dissociation energy 5.27 eV is much less than experiment, 9.91 eV. For exact RHF wave functions the virial ratio $-V/T$ should be exactly 2, and the closeness of the calculated value is a necessary condition for the SCF wave function to be considered adequate. Koopmans' theorem holds for N_2 and the fact that the $2\sigma_u$, $3\sigma_g$, and $1\pi_u$ orbital energies are all small in absolute value implies, as is well known, that there should be three low-lying electronic states of N_2^+.

Following the procedure outlined above, calculations were carried out for about 15 different internuclear separations for each of the four electronic states. Potential energy curves have been drawn from the calculated energies and are compared with the experimental Rydberg-Klein-Rees potential curves in Figure III-1. This figure illustrates rather dramatically two of the inherent limitations of the RHF approximation. First, the RHF potential curve for N_2 does not dissociate to two RHF nitrogen atom wave functions. For large internuclear separations, the SCF wave function for the $X\ ^1\Sigma_g^+$ state of N_2 is in fact a complicated mixture of ground and excited state wave functions for

Table III-1. Near Hartree Fock properties for the $X\ ^1\Sigma_g^+$ ground state of N_2 at its equilibrium internuclear separation. All properties are in atomic units. CM refers to the fact that a property has been calculated with respect to the center of mass of the molecule.

Total Energy	-108.9928
Dissociation Energy (eV)	5.27
$-V/T$	2.0019
Orbital Energies	
$1\sigma_g$	-15.6820
$1\sigma_u$	-15.6783
$2\sigma_g$	-1.4736
$2\sigma_u$	-0.7780
$3\sigma_g$	-0.6350
$1\pi_u$	-0.6154
$\langle X^2 \rangle_{CM}$	7.7007
$\langle Z^2 \rangle_{CM}$	23.6172
Quadrupole Moment	-0.9473
$\langle 1/r_N \rangle$	21.6580
Electric Field Gradient	1.3653

atomic species from N^{+3} to N^{-3}. Similarly the three N_2^+ RHF curves do not display the correct dissociation behavior, and for large values of R (the internuclear separation) all three theoretical N_2^+ curves rise much too sharply.

An even more obvious failing of the RHF approximation, seen in Figure III-1, is that the experimentally known order of the two lowest electronic states of N_2^+ is <u>reversed</u>. That is, the near RHF calculations predict the $A\ ^2\Pi_u$ state to be the ground state of N_2^+ whereas, in fact, the $X\ ^2\Sigma_g^+$ state is the ground state. In retrospect this result is not too surprising since the experimental separation between the X and A states of N_2^+ is only ~1.15 eV. Since the correlation energy of the $X\ ^2\Sigma^+$ state is nearly 2 eV greater than that of the $A\ ^2\Pi_u$ state, the order of the states is predicted incorrectly. More generally, it can be stated that the unreliability of RHF electronic state position predictions is now a well-known phenomenon. Although energy differences are usually predicted to ~2 eV accuracy within the RHF approximation, it is frequently the case that this degree of accuracy is inadequate for the resolution of problems of chemical interest.

Spectroscopic constants obtained by a Dunham analysis of the calculated SCF energies for N_2 and N_2^+ are seen in Table III-2. The agreement with experiment is typical of near Hartree-Fock results and several trends are seen. In all four cases the SCF bond distance R_e is less than experiment, the deviations ranging from 0.061 to 0.097 bohrs. Other such near RHF calcula-

---

Figure III-1. Near Hartree-Fock (a) and experimental (b) potential energy curves for N_2 and N_2^+. The relative position of the energy scales has been adjusted so that the minima of the two ground state curves coincide. The absolute energies of the experimental curve are much lower since the theoretical curve excludes electron correlation and relativistic effects. ▶

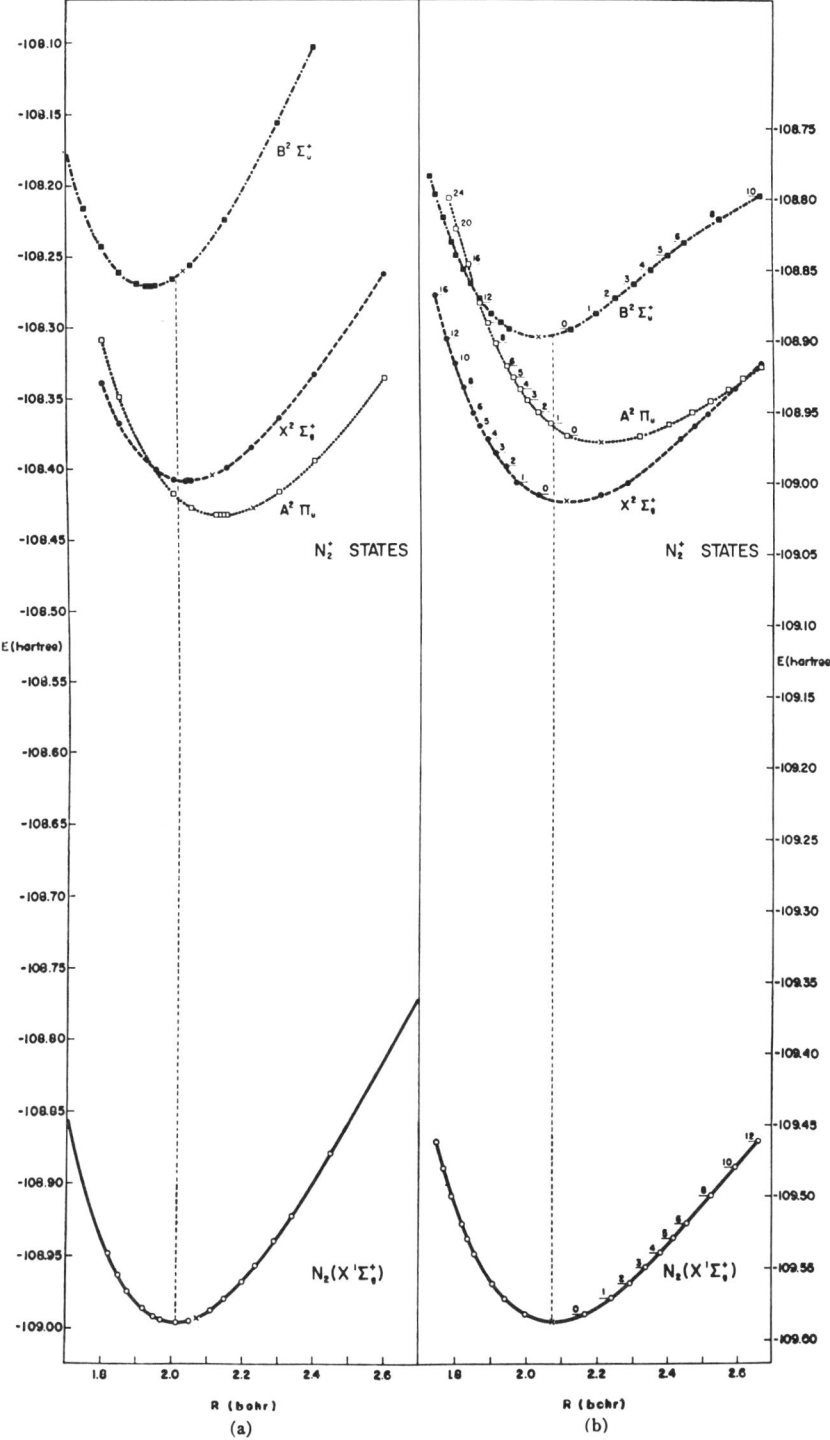

Table III-2. Spectroscopic constants for N_2 and N_2^+ obtained from near Hartree-Fock energies (65). Unless indicated, all values are in wave numbers (cm^{-1}). Experimental values are given in parentheses.

	R_e (bohrs)	ω_e	B_e	$\omega_e x_e$	α_e
N_2 (X $^1\Sigma_g^+$)	2.013 (2.074)	2730 (2358)	2.121 (1.999)	8.38 (14.19)	0.0135 (0.0178)
N_2^+ (X $^2\Sigma_g^+$)	2.041 (2.113)	2571 (2207)	2.065 (1.932)	9.81 (16.14)	0.0148 (0.020)
N_2^+ (A $^2\Pi_u$)	2.134 (2.222)	2313 (1903)	1.887 (1.740)	6.08 (14.91)	0.0155 (0.018)
N_2^+ (B $^2\Sigma_u^+$)	1.935 (2.032)	3102 (2420)	2.296 (2.073)	19.88 (23.19)	0.0128 (0.020)

tions have shown the same trend, and it appears safe to say that a general characteristic of the RHF approximation is the prediction of bond distances as much as 0.05 Å too short. The vibration frequencies ω_e (which are related to the force constants for these molecules) are all greater than experiment, being 116%, 116%, 122% and 128% of the correct values. Prediction of vibration frequencies 20% greater than experiment is another general feature of RHF calculations. The anharmonicity corrections $\omega_e x_e$ are so sensitive to the shape of the potential curves that reliable values are not obtained from the SCF results, and the calculated values are of the order of half the experimental values. The values of α_e, the correction constant for the rotational levels, are between 64 and 86% of experiment. Thus, one is likely to conclude on the basis of these N_2 and N_2^+ results that spectroscopic constants are not very reliably predicted within the Hartree-Fock approximation.

Molecular properties of the alkali halides. The alkali halides have a particularly important place in chemistry since they are the simplest known "ionic" molecules. An important theoretical study of LiCl, NaF, NaCl, and KF using near Hartree-Fock wave functions has been carried out by Matcha (241). Accurate SCF calculations were earlier carried out on the simplest alkali halide, LiF, by McLean and Yoshimine (251). In these calculations a variety of molecular properties were reported including both spectroscopic constants and one-electron properties. In addition, calculated electric field gradients were used to predict a number of nuclear quadrupole moments. In contrast to N_2 and N_2^+ results seen above, for the alkali halides the RHF approximation is found to give quite reliable results for a number of properties.

The method used for choosing basis sets for LiF through KF is related to that used by Cade, Sales, and Wahl for N_2 (65) but requires far less optimization of orbital exponents. An atom-optimized extended basis set is chosen and polarization functions are added. For LiCl, for example, these polarization functions include one 3d and one 4f function centered on both the Li and Cl atoms. However, exponent optimization is carried out only for the polarization functions. The LiF wave functions obtained in this way are expected to yield SCF energies with 0.0005 hartree of the HF limit, while for the larger molecules the results are of necessity somewhat less accurate. However for NaCl, with 20 electrons, the calculated energies should still be within 0.01 hartrees of the HF energies.

Table III-3 summarizes the calculated one electron properties for the alkali halides. It is perhaps worth pointing out that the diamagnetic shielding

Table III-3. Properties of the alkali halides AX as determined from near Hartree-Fock wave functions (241,251) at the equilibrium internuclear separations. A refers to the alkali atom and X to the halogen atom. Experimental values are in parentheses. Values are in atomic units unless specified.

	LiF	LiCl	NaF	NaCl	KF
Energy	-106.9894	-467.0547	-261.3785	-621.4574	-698.6850
-V/2T		0.99999	1.00002	0.99999	0.99995
$\langle 1/r_A \rangle$		10.153	37.927	39.192	77.142
$\langle 1/r_X \rangle$		65.3263	29.854	67.022	31.482
$\langle r_A \rangle$		72.26	44.05	89.33	56.14
$\langle r_X \rangle$		26.29	46.73	63.99	85.54
$\langle r_A^2 \rangle$		292.89	151.13	397.59	200.27
$\langle r_X^2 \rangle$		66.33	153.67	244.59	340.60
Dipole Moment (debyes)	6.295(6.28)	7.256(7.128)	8.367(8.206)	9.183(9.002)	8.720(8.6)
Field Gradient at A	-0.036	-0.021	-0.378	-0.250	-0.600
Field Gradient at X	-0.276	0.123	-0.082	0.200	-0.349

σ^d at each nucleus is directly proportional to the expectation value $<1/r>$ where r is the distance from the nucleus in question. The consistently excellent virial ratios $-V/2T$ (the exact value is 1.0) are an indication that the wave functions are rather close to their RHF limits.

The most striking result seen in Table III-3 is the excellent agreement between SCF and experimental dipole moments. The largest deviation between calculated and theoretical values is for sodium chloride and is still only 0.181 debyes. In fact it must be pointed out that in general RHF dipole moments are not expected to be in as close agreement with experiment as is the case for the alkali halides. Matcha also has calculated dipole moments as a function of internuclear separation and expressed the results as a function of vibrational quantum number. For LiCl the near RHF wave functions yield

$$\mu_\nu = 7.2182 + 0.0753 \, (\nu + \tfrac{1}{2}) + 0.0002 \, (\nu + \tfrac{1}{2})^2 \qquad \text{(III,5)}$$

whereas the experimentally determined expression is

$$\mu_\nu = 7.0853 + 0.0864 \, (\nu + \tfrac{1}{2}) + 0.0006 \, (\nu + \tfrac{1}{2})^2 \qquad \text{(III,6)}$$

Similar calculations were reported (241) for NaF, NaCl, and KF, and in all three cases the agreement with experiment is significantly better than that for LiCl for the coefficient of the $(\nu + \tfrac{1}{2})$ term.

Since the calculated properties as a function of internuclear separation appear to be quite reliable Figure III-2 shows the dipole moment and field gradients of LiCl as a function of R. The dipole moment has an essentially linear dependence on internuclear separation and the dipole moment derivative

$\partial\mu/\partial R$ is 4.356 debyes per Å, in good agreement with the experimental value, 4.6. The electric field gradients at both Li and at Cl increase monotonically with R, but Figure III-2 shows this increase to be very gradual for the field gradient at the lithium nucleus. Matcha reports similar results for NaF, NaCl, and KF; namely, both field gradients increase monotonically with R but the R dependence is much stronger for the field gradient at the halogen nucleus.

From theoretical field gradients q and experimentally determined quadrupole coupling constants eqQ, one may deduce the values of nuclear electric quadrupole moments. As pointed out in section IIC, the most reliable nuclear quadrupole moments currently available are for B^{10}, B^{11}, and O^{17}, obtained from atomic beam measurements and atomic structure calculations including electron correlation. Since the alkali halide calculations discussed here do not include electron correlation, one intuitively feels that the calculated field gradients are probably reliable to within at best 10%. From the LiCl and NaCl calculations the quadrupole moments of Cl^{35} are predicted to be -0.106 and -0.117 x 10^{-24} cm^2 and for Cl^{37} -0.084 and -0.095 x 10^{-24} cm^2. The differences in the predicted values is one indication of the expected reliability of the Q values. From the NaF and NaCl calculations the quadrupole moment of Na^{23} is predicted to be 0.0951 and 0.0950 x 10^{-24} cm^2. Finally the KF calculations have been used to deduce $Q(K^{39})$ = 0.0563 x 10^{-24} cm^2 and $Q(K^{41})$ = 0.0685 x 10^{-24} cm^2. The above predicted quadrupole moments for Na^{23}, Cl^{35}, Cl^{37}, K^{39}, and K^{41} may be subject to uncertainties as large as 20%. Nevertheless they are the most reliable values available for these nuclei and this aspect of Matcha's work in itself justifies the cost of the calculations.

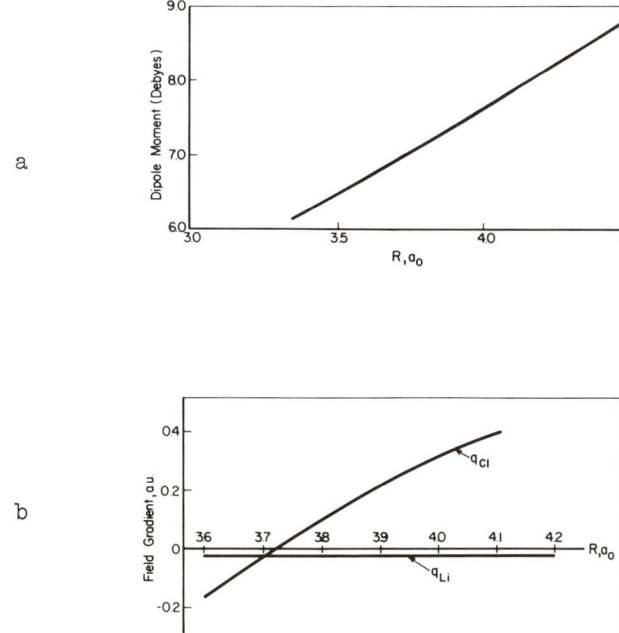

Figure III-2. Near Hartree-Fock dipole moments (figure 2a) and field gradients (figure 2b) as a function of internuclear separation in bohrs for lithium chloride (241).

Table III-4 shows the predicted spectroscopic constants for the alkali halides. As usual, dissociation energies D_e are obtained by subtraction of the molecular SCF energy from the sum of the two atomic SCF energies. Except for the D_e values, the calculated spectroscopic constants are in good agreement with experiment. This good agreement would not have been expected from the N_2 and N_2^+ results seen in Table III-2. The bond distances are within 0.006, 0.011, 0.025 and 0.084 bohrs of experiment and it is probable that a smaller bond distance would have been obtained for KF had a larger basis set been used. As was the case for N_2 and N_2^+, the vibrational frequencies are all greater than experiment. However, the actual agreement is much closer for the alkali halides and in no case is the SCF ω_e more than 5% greater than the experimental value. Even the anharmonicity corrections $\omega_e x_e$ are rather accurately predicted, being 111%, 115%, 126%, and 102% of experiment for LiCl, NaF, NaCl, and KF.

The results of Table III-4 lead to the question: why are the alkali halide spectroscopic constants so much more accurately predicted by near RHF wave functions than are the same spectroscopic constants for N_2 and the three lowest states of N_2^+? It appears that at least part of the answer to this question lies with the fact that the alkali halide SCF wave functions dissociate in a physically realistic manner. For LiCl, for example, at infinite internuclear separation, the molecular RHF wave function becomes the product of an RHF wave function for Li^+ and an RHF wave function for Cl^-. In reality, of course, there is an avoided crossing between the ionic $Li^+ + Cl^-$ and covalent $Li + Cl$ potential curves and LiCl actually dissociates to the somewhat lower limit, $Li + Cl$. However, near Re, the ionic description dominates the true wave function and thus the RHF potential curve has essentially the correct shape in this region. The N_2 and N_2^+ RHF wave functions, on the contrary, dissociate to highly unrealistic combinations of ionic wave functions such as $N^{+3} + N^{-3}$.

We must conclude, then, that properties computed from RHF wave functions are likely to be more reliable for certain classes of molecules than for others. In addition to the alkali halides, it has been shown that spectroscopic constants for the hydrides from LiH through HCl are quite accurately

Table III-4. Spectroscopic constants for the alkali halides obtained from near Hartree-Fock wave functions. Unless indicated, constants are in cm^{-1}. Experimental values are given in parentheses.

	^{7}Li^{35}Cl	^{23}Na^{19}F	^{23}Na^{35}Cl	^{39}K^{19}F
D_e (eV)	3.83 (4.84)	3.05 (4.95)	3.18 (4.22)	3.05 (5.09)
R_e (bohrs)	3.825 (3.819)	3.628 (3.639)	4.485 (4.460)	4.188 (4.104)
ω_e	672 (641)	558 (536)	378 (365)	449 (426)
B_e	0.7042 (0.7065)	0.439 (0.437)	0.2155 (0.2181)	0.2686 (0.2799)
$\omega_e x_e$	4.67 (4.2)	4.39 (3.83)	2.59 (2.05)	2.48 (2.43)
α_e	0.00444	0.00468 (0.00457)	0.00180 (0.00162)	0.00214 (0.00233)
force constant (millidynes/Å)	1.55 (1.41)	1.91 (1.76)	1.17	1.51 (1.31)

predicted from near RHF wave functions (66). However, the near RHF spectroscopic constants for F$_2$ (384) and O$_2$ (342) are (like those of N$_2$ and N$_2^+$) in rather poor agreement with experiment. For the lowest $^1\Sigma^+$ states of the alkaline earth oxides BeO, MgO, CaO, and SrO, spectroscopic constants obtained within the Hartree-Fock formulation are in no better than fair agreement with experiment (401). In general then, we conclude that near

RHF spectroscopic constants should be considered at best qualitatively reliable except for certain classes of molecules such as the alkali halides and the hydrides, for which good agreement with experiment has been documented.

<u>Paramagnetism predicted for BH</u>. Magnetic properties of molecules are of considerable interest to chemists. In particular, the magnetic susceptibility of a molecule determines its diamagnetism or paramagnetism, and the nuclear magnetic shielding tensor determines the magnitude of NMR chemical shifts. It is frequently stated that closed-shell molecules are diamagnetic and that paramagnetism is only possible for open-shell systems. However, this does not appear to be rigorously true (153), and based on rigorous quantum mechanical calculations, Stevens, Lipscomb, and Hegstrom have predicted the ground $^1\Sigma^+$ state of BH (366) to be paramagnetic.

In Van Vleck's theory of susceptibilities (376) the magnetic susceptibility is given as

$$\chi = \frac{\mu^2}{3kT} + \chi^p + \chi^d \qquad (III,7)$$

in which μ is the permanent molecular magnetic dipole moment, χ^p is the temperature-independent paramagnetic susceptibility and χ^d is the diamagnetic susceptibility, also temperature-independent. For open-shell molecules, μ is nonzero and the first-term in (III,7) dominates the expression, yielding a temperature-dependent paramagnetism. However, for closed-shell molecules, μ is identically zero and the magnetic susceptibility is determined by χ^p and χ^d. From perturbation theory, the Van Vleck χ^p

expression (376) for the electronic ground state of a molecule described by wave function ψ_o is

$$\chi^p = \frac{2}{3} \sum_{j=1}^{n} (E_j - E_o)^{-1} \left| \int \psi_o^* \left[\sum_{i=1}^{n} m_i^o \right] \psi_j \, d\tau \right|^2 \qquad (III,8)$$

The j summation in (III,8) goes over all the excited states of the molecule and the operator m_i^o represents the instantaneous magnetic dipole moment of the ith electron. The diamagnetic susceptibility χ^d is given by a simple expectation value

$$\chi^d = -\frac{e^2}{6mc^2} \int \psi_o^* \left[\sum_{i=1}^{n} r_i^2 \right] \psi_o \, d\tau \qquad (III,9)$$

χ^p is of necessity positive as is χ^d negative, and it is usually thought that χ^p cannot be larger in absolute magnitude than χ^d for closed-shell systems (142). This would imply that the total magnetic susceptibility χ is negative and thus that all closed-shell systems are diamagnetic.

There is one obvious difficulty which arises in the quantum mechanical calculation of the paramagnetic susceptibility χ^p using Eq. (III,8). The wave functions ψ_j for <u>all</u> the excited electronic states of the molecule are required. Needless to say, it is not practical to obtain all these wave functions, and several alternatives have been suggested to overcome this problem. Stevens, Lipscomb, et al. (222,366) use the <u>perturbed or coupled Hartree-Fock method</u>, in which the normal SCF wave function is used as a zero-order wave function. The perturbation, which in this case is the strength of the static external magnetic field, is then applied to obtain a first-order wave function, which yields χ^p.

The effect of the magnetic field in coupled SCF calculations is to mix functions of higher angular momentum with the unperturbed SCF orbitals. In particular, for diatomic molecules, orbitals are mixed with values of λ which differ by 1 from the ordinary SCF orbitals. For BH, the ground state orbital occupancy is $1\sigma^2\ 2\sigma^2\ 3\sigma^2$ and the effect of the perturbation is to allow these orbitals to take on a certain amount of π character. Thus, while the exact RHF wave function for BH may be obtained using only basis functions of σ ($\lambda=0$) symmetry, π functions must be added to the basis set if one hopes to calculate χ^p from coupled Hartree-Fock theory.

A second difficulty arising in the calculation of magnetic properties is that of <u>gauge invariance</u> (98). The simplest type of gauge transformation is a change in origin of the molecular coordinate system. It is clearly desirable that calculated magnetic properties be independent of coordinate origin, or as is usually stated, gauge invariant. If exact wave functions are used, the calculated value of χ will be invariant to changes in the coordinate system. In finite basis set calculations, however, the magnetic properties are found to depend on the gauge origin. Actually, this gauge invariance is in one sense helpful, because magnetic properties found to vary substantially with changes in the coordinate system are not likely to be reliable.

Table III-5 shows the magnetic properties of BH calculated by Hegstrom and Lipscomb (366b). Earlier calculations of the same type by

Stevens and Lipscomb (366a) did not use a sufficiently large basis to provide satisfactory gauge invariance. The basis set used for the calculations with origin at H in Table III-5 used three pπ and two dπ Slater functions centered on B and three pπ functions centered on hydrogen. Thus it is seen that rather large basis sets are required to guarantee reasonable gauge invariance. It must be pointed out that the individual contributions χ^p and χ^d will not be gauge invariant, but only the total magnetic susceptibility χ. This is not surprising in light of the fact that χ^d is proportional to $\langle r^2 \rangle$, and the calculated value of $\langle r^2 \rangle$ will surely be different with respect to the B atom as origin than for a coordinate system with the H atom at the origin.

The total magnetic susceptibility χ from Table III-5 is 18.75 ppm (parts per million) with the gauge origin at B and 18.52 ppm with the coordinate system center at H. The difference, 0.23 ppm, is only a small fraction of the total susceptibility. The paramagnetic contribution clearly dominates χ and it is certainly reasonable to believe that BH is paramagnetic. Similar calculations for CO and N_2 (222) yield χ^p values in good agreement with experiment. An experimental confirmation of this closed-shell paramagnetism would be very significant.

Some explanation of the other magnetic properties shown in Table III-5 is required. The magnetic shielding σ at each nucleus, like the magnetic susceptibility, is a sum of paramagnetic (σ^p) and diamagnetic (σ^d) contributions. The total shielding σ should be gauge invariant and the shielding at the boron nucleus $\sigma(B)$ and the proton $\sigma(H)$ are seen to

Table III-5. Predicted magnetic properties of BH from coupled SCF calculations (366). Magnetic susceptibilities and shieldings are in ppm, while spin rotation constants C are in kilocycles per second.

	Gauge origin at B	Gauge origin at H
χ^p	36.77	56.30
χ^d	-18.02	-37.78
χ	18.75	18.52
$\sigma^p(B)$	-471.3	-468.6
$\sigma^d(B)$	209.6	203.3
$\sigma(B)$	-261.7	-265.3
$C(B)$	492.9	496.7
$\sigma^p(H)$	4.83	-32.62
$\sigma^d(H)$	19.63	58.07
$\sigma(H)$	24.46	25.44
$C(H)$	-14.92	-18.17

be nearly invariant. The fact that the magnetic shielding at boron is negative is referred to as <u>antishielding.</u> For BH, this is due to the negative paramagnetic contribution to the shielding being larger in absolute value than the positive diamagnetic part. Also seen in Table III-5 are the spin rotation constants $C(B)$ and $C(H)$ at the B and H nuclei. The spin rotation constant at a nucleus is experimentally measurable and related to σ^p at the nucleus in question. The exact spin-rotation constants are gauge invariant, although for BH, C(H) varies by about 20% between the two gauges in Table III-5.

In the coupled Hartree-Fock framework, each magnetic property is a sum of contributions from the different occupied molecular orbitals. Table III-6 shows such a breakdown in terms of MO's from the calculations of Hegstrom and Lipscomb (366b). The contributions labeled 1σ, 2σ and 3σ in Table III-6 include of course the contributions from the mixing of π orbitals due to the magnetic field. With either gauge origin, Table III-6 shows that the paramagnetism of BH may be primarily ascribed to the large contribution to χ^p from the lone-pair or nonbonding 3σ orbital. The magnetic shielding at H, on the other hand, is dominated by the contribution from the 2σ orbital, which is essentially the bonding orbital for BH.

Similar calculations for the aluminum hydride molecule (which is isoelectronic in valence structure to BH) are seen in Table III-7. These calculations, by Laws, Stevens, and Lipscomb (222) predict that AlH is slightly diamagnetic and, also unlike BH, the Al nucleus does not exhibit antishielding. However, in a more detailed treatment, the total magnetic susceptibility of a diatomic molecule may be divided into parallel and perpendicular components, and for some molecules both components have been measured. For AlH, the calculations predict the parallel component to be diamagnetic by 21.87 ppm but the perpendicular component to be paramagnetic by 8.88 ppm.

Table III-6. Orbital contributions to some magnetic properties of BH. Units are as in Table III-5.

	Gauge origin at B		
	1σ	2σ	3σ
χ^p	0.00	4.50	32.27
χ^d	-0.23	-8.29	-9.51
χ	-0.23	-3.79	22.76
$\sigma^p(H)$	0.51	12.99	-8.61
$\sigma^d(H)$	0.00	10.87	8.76
$\sigma(H)$	0.51	23.86	0.15

	Gauge origin at H		
	1σ	2σ	3σ
χ^p	8.31	1.48	46.52
χ^d	-8.82	-6.68	-22.29
χ	-0.51	-5.20	24.23
$\sigma^p(H)$	-14.74	-3.61	-14.24
$\sigma^d(H)$	15.25	28.18	14.64
$\sigma(H)$	0.51	24.57	0.40

Table III-7. Theoretical magnetic properties for AlH (222). Except for the spin-rotation constants, the gauge origin is at the center of electronic charge.

χ^p	32.56 ppm
χ^d	-33.93
χ	- 1.37
$\sigma^p(Al)$	-572.63 ppm
$\sigma^d(Al)$	795.35
$\sigma(Al)$	222.71
$\sigma^p(H)$	0.32
$\sigma^d(H)$	25.81
$\sigma(H)$	26.13
C (Al)	260.46 kilocycles/second
C (H)	- 9.34

The feature about BH which differs from AlH is that the lowest unoccupied orbital (1π for BH) is of different symmetry from any of the occupied SCF orbitals. For AlH, 2π is the lowest unoccupied orbital. The strong mixing of the BH 1π orbital with the 3σ orbital in the presence of the magnetic field may, in a simple picture, be considered responsible for the paramagnetism.

Application of an external magnetic field induces electron currents in a molecule. Laws, et al. (222) have calculated these currents, and they are seen for AlH in Figure III-3. Plots such as that of Figure III-3 give a useful qualitative picture of the perturbation effects induced by a magnetic field.

<u>Transition probabilities for N_2^+, C_2^-, CN, CO^+, BO, BF^+, and BeF.</u> Considering the importance of molecular transition probabilities in chemistry, it is perhaps surprising how few theoretical calculations of oscillator strengths have been carried out. It is not particularly surprising that little work has been done beyond the Hartree-Fock approximation, since the calculation of oscillator strengths might involve the computation of large numbers of matrix elements involving nonorthogonal orbitals. However, if single configuration SCF wave functions are used, only one such matrix element is required. From the atomic results discussion in Section IIC, we might expect near RHF oscillator strengths for molecules to be reliable to within a factor of about 5. Although this accuracy might at first glance seem terrible, for some purposes order of magnitude predictions of f values can be very useful.

The first molecular calculations of electronic transition probabilities using near RHF wave functions were those of Huo (180) involving the four lowest excited states of NH and CH. Huo's calculated oscillator strengths, as expected, are only in order-of-magnitude agreement with available experimental data. Recently Henneker and Popkie (156) have undertaken a very thorough study of electronic transition probabilities in diatomic

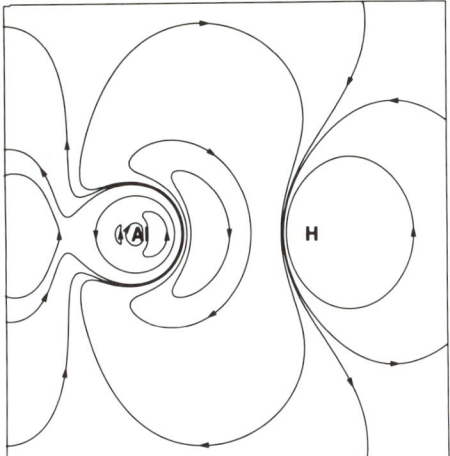

Figure III-3. Direction of the current density induced in aluminum hydride by a magnetic field perpendicular to the molecular axis. Paramagnetic currents circulate clockwise.

---

molecules using near Hartree-Fock wave functions. One of the distinguishing features of the work is that contributions to the transition probability from both the vibrational and electronic wave functions are considered in detail. In the present subsection, we discuss the results of Popkie and Henneker for the 13 electron isoelectronic series containing CN, BO, and BeF (304).

The electronic states of N_2^+ and C_2^- considered by Popkie and Henneker are the X $^2\Sigma_g^+$, A $^2\Pi_u$, and B $^2\Sigma_u^+$ states, with orbital occupancies given by Eqs. (III,2)-(III,4). For the heteronuclear molecules CN, CO^+, BO, BF^+, and BeF, the states considered may be designated

$$1\sigma^2\ 2\sigma^2\ 3\sigma^2\ 4\sigma^2\ 5\sigma\ 1\pi^4 \qquad X\ ^2\Sigma^+ \qquad \text{(III,10)}$$

$$1\sigma^2\ 2\sigma^2\ 3\sigma^2\ 4\sigma^2\ 5\sigma^2\ 1\pi^3 \qquad A\ ^2\Pi_i \qquad \text{(III,11)}$$

$$1\sigma^2\ 2\sigma^2\ 3\sigma^2\ 4\sigma^2\ 1\pi^4\ 2\pi \qquad ^2\Pi_r \qquad \text{(III,12)}$$

Only for CN has the third state been observed, and for CN it is designated H $^2\Pi_r$. The wave functions used in this work are of very-close-to (within 0.001 hartree in total energy) RHF quality and were obtained by Cade at the University of Chicago using the programs developed by Wahl (384) and Huo (178).

As pointed out in section IIC, all the early calculations of transition probabilities were carried out using the <u>active electron approximation</u>. In this approximation the nonorthogonality of the orbitals from the two single-configuration SCF wave functions is ignored and the oscillator strength is given as a single one-electron integral. For the "length" form (II,6) of the oscillator strength between the ground (III,2) and first-excited (III,3) states of N_2^+ or C_2^-, the active electron expression is

$$f(X\ ^2\Sigma_g^+,\ A\ ^2\Pi_u) \approx \frac{2}{3}\left[E(^2\Pi_u) - E(^2\Sigma_g^+)\right]\left|\int 1\pi_u^*(1)\ r_1\ 3\sigma_g(1)\ dv(1)\right|^2. \qquad \text{(III,13)}$$

In this expression the $1\pi_u$ orbital is taken for the X $^2\Sigma_g^+$ wave function and the $3\sigma_g$ orbital from the A $^2\Pi$ SCF function. In a rigorous calculation, of course, the fact that (except as required by symmetry) no one of the X $^2\Sigma^+$ state RHF orbitals is truly orthogonal to the orbitals obtained in the completely independent A $^2\Pi_u$ calculation must be considered, and the result is a rather complicated expression involving overlap integrals and exchange terms.

A second frequently used approximate scheme for the calculation of oscillator strengths is the <u>virtual orbital approximation</u>. In this approximation Eq. (III,13) is also used but the excited state A $^2\Pi_u$ SCF wave function need not be calculated at all, since both the $1\pi_u$ and $3\sigma_g$ orbitals are taken from the ground state wave function. For N_2^+, Table III-8 shows a comparison of the two approximate schemes with rigorous calculations. In all cases the transition moments $\langle x \rangle$ and $-i \langle p_x \rangle$ predicted using the virtual orbital approximation are in closer agreement with the correct transition moments than are the active electron values. This is a bit surprising since the active electron approximation allows a more accurate description of the A $^2\Pi_u$ state. In light of our expectation that RHF oscillator strengths are at best reliable to within a factor of 5, both the active electron and virtual orbital approximations give reasonable results.

For the $X\ ^2\Sigma_g^+ - A\ ^2\Pi_u$ and $X\ ^2\Sigma_g^+ - B\ ^2\Sigma_u^+$ transitions of N_2^+, Popkie and Henneker have made a careful theoretical analysis of experimental data to obtain "experimental" values of the transition moments M_x as a function of internuclear separation

$$M_x = \int \psi_a^* \left\{ -i \sum_i p_{xi} \right\} \psi_b \, d\tau \qquad (III,14)$$

Such an analysis requires the determination of the contribution to the observed oscillator strength due to the vibrational wave functions for the two states. For exact wave functions one can show (156) within the Born-Oppenheimer approximation that

$$M_x = \left[E_b - E_a \right] \int \psi_a^* \left\{ \sum_i x_i \right\} \psi_b \, d\tau \qquad (III,15)$$

Table III-8. Comparison of active electron and virtual orbital approximations for the calculation of transition moment integrals for the X $^2\Sigma_g^+$ - A $^2\Pi_u$ transition in N_2^+ (304). Values are in atomic units.

Internuclear Separation (bohrs)	Active Electron Approximation		Virtual Orbital Approximation		Rigorous Calculation [a]	
	$<x>$	$-i<p_x>$	$<x>$	$-i<p_x>$	$<x>$	$-i<p_x>$
1.8	0.2810	0.0277	0.2380	0.0331	0.2526	0.0486
2.0	0.2651	0.0187	0.2183	0.0266	0.2331	0.0413
2.3	0.2298	0.0107	0.1797	0.0221	0.1931	0.0346
2.6	0.1860	0.0079	0.1370	0.0212	0.1477	0.0310

[a] $<x>$ and $-i<p_x>$ evaluated with no approximations from near Hartree-Fock wave functions for the two states.

Figure III-4 compares calculated and experimental transition moments M_z as a function of R.

Figure III-4 shows three theoretically obtained transition moments $M_z(R)$

a) $-i\langle p_z \rangle$ is the z component of the expression (III,14) and differs from experiment by as much as a factor of 3.

b) $\Delta E_{HF} \langle z \rangle$ is obtained using the calculated RHF energy differences between the X $^2\Sigma_g^+$ and B $^2\Sigma_u^+$ potential curves. These results are seen to be in the poorest agreement with experiment of the three theoretical curves.

c) $\Delta E_{RKR} \langle z \rangle$ is obtained using the experimental energy differences from the RKR potential curves. This is an intelligent approach to take, since

Figure III-4. Theoretical and experimental transition moments in atomic units for the N_2^+ first negative system as a function of internuclear separation in bohrs. Also shown are the near Hartree-Fock and experimental Rydberg-Klein-Rees potential energy curves. Energies $U(R)$ are in hartrees.

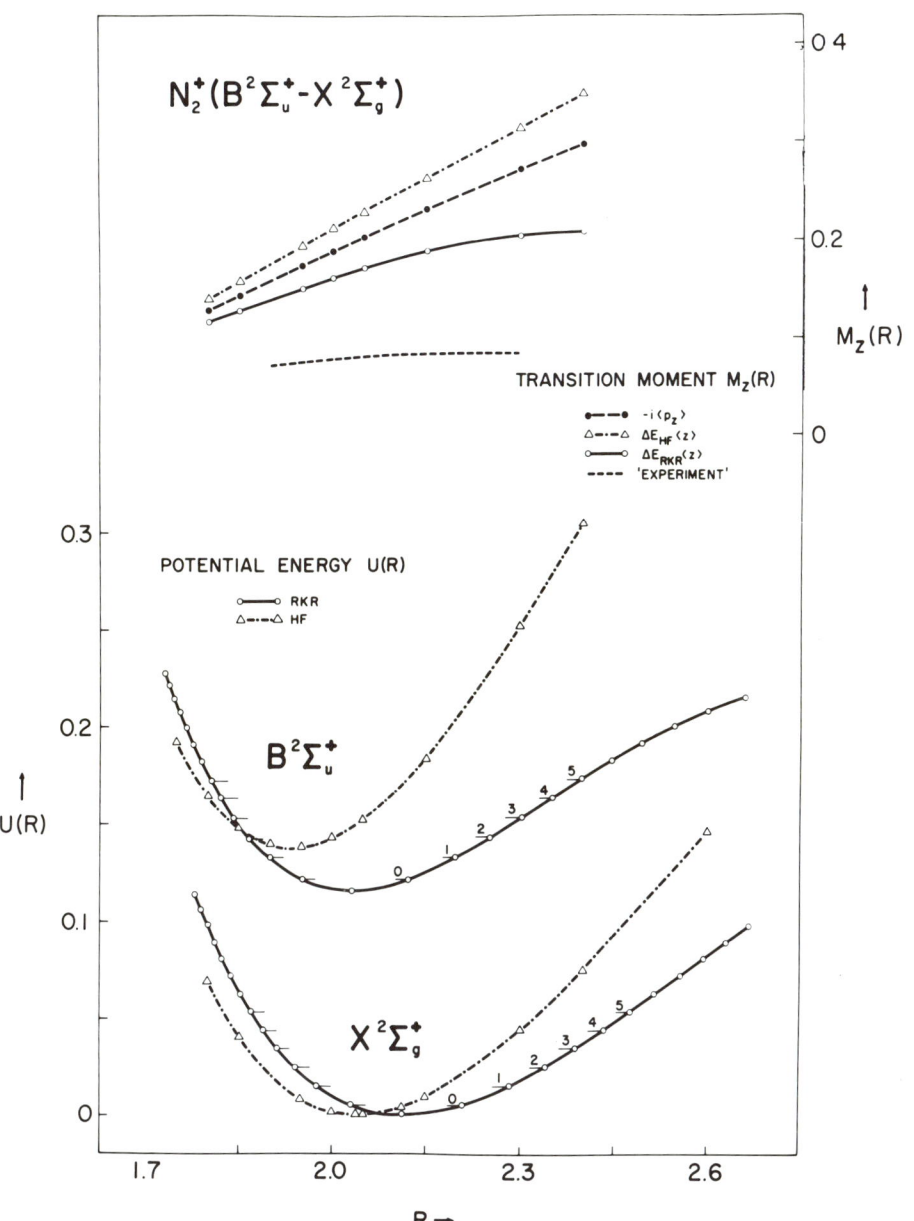

experimental energy differences are usually available for transitions of interest, whereas the transition moments or oscillator strengths are rarely known. The $\Delta E_{RKR} \langle z \rangle$ results give the best agreement with experiment, differing by less than a factor of 2 at R_e.

If the Hartree-Fock approximation were exact, of course, all three transition moments a), b), and c) would be identical and equal to the experimental values. For the Meinel system of N_2^+, corresponding to the $X\ ^2\Sigma_g^+ - A\ ^2\Pi_u$ transition, Popkie and Henneker (304) find $\Delta E_{RKR} \langle x \rangle$ to give much better agreement with experiment than forms a) and b) above.

Table III-9 summarizes Popkie and Henneker's results for the 13-electron isoelectronic series. The worst agreement with experiment is for the $B\ ^2\Sigma_u^+ - X\ ^2\Sigma_g^+$ transition of N_2^+, where the calculated oscillator strength is about 5 times greater than experiment. However, there is considerable reason for optimism in Table III-9. The calculated oscillator strengths for the $A\ ^2\Pi_i - X\ ^2\Sigma^+$ transitions of CN and CO^+ differ from experiment by a factor of about three. The extreme difficulties involved in the experimental measurement of oscillator strengths may be seen both in the small number of experimental values in Table III-9 and in the large difference between the two experimental values for the Meinel N_2^+ system. The most reliable theoretical oscillator strengths are probably those for the heteronuclear $^2\Pi_r - X\ ^2\Sigma^+$ systems. For all five molecules CN, CO^+, BO, BF^+, and BeF, the transition moments $-i\langle p_x \rangle$ and $\Delta E \langle x \rangle$ (which must be identical for exact wave functions) are quite similar. In conclusion, it seems that near RHF oscillator strength calculations such as those of Popkie and Henneker, while not of high absolute accuracy, are

Table III-9. Near Hartree-Fock transition moments and electronic oscillator strengths for some 13-electron systems. Theoretical values refer to a single internuclear separation near R_e.

Transition	$-i \langle p_x \rangle$	$\Delta E_{calc} \langle x \rangle$	f	f(experiment)
$(A\ ^2\Pi_u - X\ ^2\Sigma_g^+)$				
C_2^-	0.0195	-0.0087	0.0166	
N_2^+	0.0408	-0.0029	0.0252	0.0029, 0.0064
$(B\ ^2\Sigma_u^+ - X\ ^2\Sigma_g^+)$				
C_2^-	0.1795	0.2301	0.224	
N_2^+	0.1925	0.2164	0.193	0.038
$(A\ ^2\Pi_i - X\ ^2\Sigma^+)$				
CN	0.0231	-0.0003	0.0161	0.0058
CO$^+$	0.0292	0.0176	0.0177	0.0056
BO	0.0419	0.0250	0.0294	
BF$^+$	0.0529	0.0421	0.0231	
BeF	0.0701	0.0583	0.0380	
$(^2\Pi_r - X\ ^2\Sigma^+)$				
CN	0.0462	0.0465	0.0158	
CO$^+$	0.1231	0.1215	0.105	
BO	0.1333	0.1687	0.190	
BF$^+$	0.1411	0.1434	0.220	
BeF	0.1424	0.1674	0.371	

nevertheless of very significant value in the study of electronic transition probabilities.

Exchange splitting of the inner shell ionization potentials of NO. Koopmans' theorem was discussed in Section IB and it was pointed out that for open-shell molecules, it is usually not possible to identify orbital energies with ionization potentials. As an example consider the NO molecule, with $^2\Pi$ ground state orbital occupancy $1\sigma^2\ 2\sigma^2\ 3\sigma^2\ 4\sigma^2\ 5\sigma^2\ 1\pi^4\ 2\pi$. If one of the innermost 1σ electrons is ejected, the resulting orbital occupancy is

$$1\sigma\ 2\sigma^2\ 3\sigma^2\ 4\sigma^2\ 5\sigma^2\ 1\pi^4\ 2\pi \qquad (III,16)$$

However, the simplest consideration of symmetry (157) shows that the above orbital occupancy, with 1σ and 2π orbitals unpaired, gives rise to two electronic states, $^3\Pi$ and $^1\Pi$.

Thus we see that for the NO radical, there are two distinct 1σ hole states. The splitting, or energy difference, between the $^3\Pi$ and $^1\Pi$ states is referred to as an exchange-induced splitting. This terminology is a result of the fact that, if single configuration wave functions are used to describe both states, the theoretical energy difference is given by a single exchange integral $(1\sigma\ 2\pi\ |\ 2\pi\ 1\sigma)$ in the notation of (I,20). A fundamental breakthrough with respect to this problem was made by Siegbahn and coworkers (350), who were able to experimentally measure the exchange-induced splittings for the core electrons of both NO and O_2. More recently, Davis and Shirley have determined the NO splittings to a precision of a few hundredths of an electron volt (99). Near Hartree-Fock calculations of the exchange splittings in NO^+

have been carried out by Bagus and Schaefer (14) and the present subsection is a discussion of these calculations.

An interesting aspect of the calculations of Bagus and Schaefer (14) is that they present an alternative scheme to that of Cade, Sales, and Wahl (65) for obtaining very close to RHF wave functions. Recall that the latter authors used an exhaustively optimized basis set of five s, three p, three d, and one f Slater functions on the two N atoms in N_2. More than two hundred separate calculations were carried out by Cade, Sales, and Wahl in order to optimize the orbital exponents ζ of the Slater functions. A much larger basis set, seen for the O atom in Table III-10, was used by Bagus and Schaefer, but <u>no</u> exponent optimization was performed. That is, only a single SCF calculation was carried out. The basis set in Table III-10 may be labeled (7s 6p 3d 2f) and is nearly twice as large as that of Cade, et al. (65). Since the total computation time is proportional to the fourth power of the number of basis functions, a single calculation with the Bagus-Schaefer basis set might take sixteen ($2^4 = 16$) times longer than a single calculation with the Cade, et al. set. The total energies obtained from both calculations are very close to the true RHF energy, and a comparative study for O_2 shows the larger set of Table III-10 to give an energy 0.0008 hartrees lower than that obtained with an exhaustively optimized smaller set. In terms of computation time expended, it may be seen that the single calculation with the large set is a more economical way of closely approaching the RHF limit than are the 200 calculations required to optimize the smaller set.

Although the 1σ orbital energy for NO does not correspond to either the

Table III-10. Basis set of Slater functions centered on oxygen for hole state calculations on NO. A comparable basis set centered on the nitrogen atom was used.

Function	Orbital Exponent ζ
1s	12.418
1s	6.995
3s	8.681
2s	3.900
2s	2.922
2s	1.818
3s	1.600
2p	8.450
2p	4.700
2p	3.744
2p	2.121
2p	1.318
3p	1.300
3d	4.000
3d	3.000
3d	2.000
4f	4.000
4f	2.000

$^3\Pi$ or $^1\Pi$ NO$^+$ energy, one can carry out calculations for the $^3\Pi$ and $^1\Pi$ states which are completely analogous to the Koopmans' theorem results for closed shells. Such calculations involve a <u>frozen orbital approximation</u>, in which single configuration wave functions are constructed from the neutral NO ground state SCF orbitals and the corresponding energies evaluated. The ionization potentials are then simply the differences between the NO ground state energy and the NO$^+$ energies. In addition Schaefer and Bagus (14) carried out direct SCF calculations on the appropriate $^3\Pi$ and $^1\Pi$ states of NO$^+$. The latter calculations, as discussed in section IIA, are expected to give the more reliable results since the neutral molecule and positive ion states are treated on an equal footing.

The results of the above two types of calculations are compared with experiment in Table III-11. The oxygen 1s hole states referred to in Table III-11 result from ejection of the 1σ electron, while the nitrogen 1s hole states arise from the removal of an electron from the 2σ orbital. The first point to be made is that the frozen orbital or Koopmans' type results yield ionization potentials about 20 eV greater than experiment. This is of course due to the fact that the NO$^+$ energies used to determine the ionization potentials are not variational. The direct hole state calculations, however (as was seen in the Ne and Ar results of Bagus (10) discussed in section IIA), yield ionization potentials no more than 1.5 eV from experiment.

For the 1σ or 1s oxygen hole states, the frozen orbital splitting is 0.73 eV, the direct calculation splitting 0.48 eV, and the more accurate

Table III-11. Summary of near Hartree-Fock calculations on the 1σ and 2σ hole states of the NO radical. I.P. is an abbreviation for ionization potential.

State of NO$^+$	Frozen Orbital Approximation		Direct Hole State Calculation		Experiment
	E(hartrees)	I.P. (eV)	E(hartrees)	I.P. (eV)	
Oxygen 1s					
$^3\Pi$	-108.6097	562.93	-109.3771	542.05	543.3[a], x[b]
$^1\Pi$	-108.5829	563.66	-109.3594	542.53	544.0[a], x + 0.530±0.021[b]
Nitrogen 1s					
$^3\Pi$	-113.5847	427.56	-114.1872	411.17	410.3[a], x[b]
$^1\Pi$	-113.5396	428.79	-114.1375	412.52	411.8[a], x + 1.412±0.016[b]

[a] Reference 350.
[b] Reference 99. Only the splittings are measured in this work.

experimental value 0.53 ± 0.02 eV. For the 2σ or nitrogen 1s hole, the same three splittings are 1.23 eV, 1.35 eV, and 1.412 ± 0.016 eV. We see that

a) the frozen orbital calculations are in qualitative agreement with experiment, and

b) the direct hole state calculations agree nearly quantitatively with experiment.

The name "exchange-induced" splittings is only meaningful in the context of single configuration wave functions, and the results of Table III-11 do indeed show that near Hartree-Fock wave functions provide an accurate description of the phenomena in question for NO.

Comparable calculations have been carried out (15) on the 1s hole states of the O_2 molecule and yield a qualitatively important result. Since the $1\sigma_g$ and $1\sigma_u$ orbital energies are nearly identical, the ionization potentials resulting from the ejection of $1\sigma_g$ and $1\sigma_u$ electrons are for practical purposes equal. If a $1\sigma_g$ electron is removed the resulting O_2^+ orbital occupancy is

$$1\sigma_g \ 1\sigma_u^2 \ 2\sigma_g^2 \ 2\sigma_u^2 \ 3\sigma_g^2 \ 1\pi_u^4 \ 1\pi_g^2 \qquad (III,17)$$

From the unpaired $1\sigma_g \ 1\pi_g^2$ electrons, four electronic states arise, of $^4\Sigma_g^-$, $^2\Sigma_g^-$, $^2\Delta_g$, and $^2\Sigma_g^+$ symmetry. For our purpose only the $^4\Sigma_g^-$ state need be considered and this state has been designated $^4\Sigma^-$ by Siegbahn et al. (350) since a state of equal energy but $^4\Sigma_u^-$ symmetry arises from removal of a $1\sigma_u$ orbital from O_2.

If one uses the frozen orbital approximation, i.e., evaluates the energy of (III,17) using the ground state O_2 SCF orbitals, the $^4\Sigma^-$ ionization potential is 563.5 eV. A direct $^4\Sigma^-$ calculation using orbital occupancy (III,17) yields ionization potential 554.4 eV, while the experimental value is 543.1 eV (350). In light of the NO results seen in Table III-11, the difference of 11.3 eV between the direct SCF calculation and experiment is on the surface difficult to understand.

The problem was found (15) to lie in the symmetry restriction implicit in (III,17), namely that each molecular orbital has either g or u inversion symmetry. If inversion symmetry is abandoned, the O_2^+ orbital occupancy becomes

$$1\sigma\ 2\sigma^2\ 3\sigma^2\ 4\sigma^2\ 5\sigma^2\ 1\pi^4\ 2\pi^2 \qquad (III,18)$$

A direct hole state SCF calculation of the $^4\Sigma^-$ state of O_2^+ arising from (III,18) yields ionization potential 542.0 eV, which differs by only 1.1 eV from experiment. The above calculations (15) provide the first theoretical evidence for the importance of a _localized_ description of hole states of molecules containing equivalent atoms. A well-correlated wavefunction can retain g or u symmetry while describing a hole as being localized, but an RHF wavefunction cannot because of its simple form. An analysis of the near RHF wave function of form (III,18) shows that the singly-occupied 1σ orbital is essentially an 1s orbital on oxygen atom A, while the doubly-occupied 2σ orbital is localized on atom B.

B. MULTICONFIGURATION RESULTS

Most of our current understanding of electron correlation in molecules comes from _ab initio_ calculations on diatomic molecules. This section is devoted to a discussion of some of the more accurate and interesting of these calculations.

Pair correlations in the first-row hydrides. In light of the nearly additive nature of atomic pair correlation energies (see section IID), it is not surprising that as theoretical methods for the evaluation of electron correlation in diatomic molecules became available, the topic of pair correlations was addressed almost immediately. The pioneering work in this area was done by Bender and Davidson on the series LiH through HF (28,29,97). The work of Bender and Davidson is of particular importance, since they compared each variationally calculated correlation energy with that obtained by summing the independently computed pair correlation energies via Eq. (II,19).

Since the goal of the Bender-Davidson study (29) was to obtain variationally at least 70% of the correlation energy for all the hydrides through HF, large basis sets were used. For HF, for example, a (5s 6p 4d 2f) set of Slater functions was centered on fluorine and a (2s 2p 1d 1f) set centered on hydrogen. The SCF energies obtained with these extended basis sets appear to be within 0.005 hartrees of the true Hartree-Fock energies in all cases. Very large CI calculations were carried out by Bender and Davidson, who included all single and double excitations of significant import. The use of the iterative natural orbital procedure (27) was extremely helpful in this regard since it guarantees that configurations constructed from those natural orbitals with smallest occupation numbers will be quite unimportant.

Table III-12 shows the calculated pair correlation energies for the larger hydrides. To help understand the notation used in Table III-12, it is useful to write the SCF wave function for hydrogen fluoride in the form

$$\psi_{SCF} = 1\sigma\alpha \; 1\sigma\beta \; 2\sigma\alpha \; 2\sigma\beta \; 3\sigma\alpha \; 3\sigma\beta \; 1\pi_{+}\alpha \; 1\pi_{-}\alpha \; 1\pi_{+}\beta \; 1\pi_{-}\beta \qquad (III,19)$$

Then the form of the SCF wave function for the $^2\Pi$ ground state of OH is that found by dropping the last spin orbital in (III,19). In an analogous manner, the SCF wave functions for $^3\Sigma^-$ NH and $^2\Pi$ CH are obtained by deleting the last two and three spinorbitals of (III,19). The pair correlation energies of Table III-12 of course refer to the energy lowering obtained by replacing two orbitals of (III,19) with higher orbitals.

The pair correlation energies of Bender and Davidson are intermediate in complexity between the spinorbital and symmetry-adapted pairs discussed in section IID. As such they may be called <u>spin-adapted pair correlation energies</u>. Bender and Davidson compute only a single 1σ 2σ pair correlation energy whereas there are four such spinorbital pair correlation energies, namely $e(1\sigma\alpha, 2\sigma\alpha)$, $e(1\sigma\alpha, 2\sigma\beta)$, $e(1\sigma\beta, 2\sigma\alpha)$, and $e(1\sigma\beta, 2\sigma\beta)$. Thus for the hydrides through BH, for which the SCF orbitals are all σ orbitals, the spin-adapted pair correlation energies are the same as the symmetry-adapted pair correlation energies discussed in section IID. The difference between spin-adapted and symmetry-adapted formulations only becomes apparent when π orbitals (corresponding to a doubly degenerate irreducible representation) become occupied in the SCF wave function.

For hydrogen fluoride there is only a single 1π 1π symmetry-adapted pair correlation energy. $e(1\pi, 1\pi)$ is obtained by carrying out a CI including the SCF configuration plus all $^1\Sigma^+$ configurations arising from orbital occupancies of the type

$$1\sigma^2 \ 2\sigma^2 \ 3\sigma^2 \ 1\pi^2 \ xy \qquad \text{(III,20)}$$

where x and y are two orbitals not occupied in the SCF approximation. However, there are three 1π 1π pair correlation energies $e(1\pi_+, 1\pi_-)$, $e(1\pi_+, 1\pi_+)$, and

Table III-12. Pair correlation energies in hartrees for CH, NH, OH, and FH (29). See text for a discussion.

	CH	NH	OH	FH
Contributions of single excitations	-0.0052	-0.0132	-0.0073	-0.0018
$1\sigma - 1\sigma$	-0.0377	-0.0378	-0.0369	-0.0364
$1\sigma - 2\sigma$	-0.0025	-0.0046	-0.0044	-0.0050
$2\sigma - 2\sigma$	-0.0315	-0.0120	-0.0126	-0.0107
$1\sigma - 3\sigma$	-0.0050	-0.0032	-0.0035	-0.0045
$2\sigma - 3\sigma$	-0.0278	-0.0299	-0.0255	-0.0252
$3\sigma - 3\sigma$	-0.0282	-0.0300	-0.0329	-0.0336
$1\sigma - 1\pi_+$	-0.0025	-0.0025	0.0051	-0.0052
$2\sigma - 1\pi_+$	-0.0177	-0.0195	-0.0345	-0.0258
$3\sigma - 1\pi_+$	-0.0239	-0.0178	-0.0384	-0.0461
$1\sigma - 1\pi_-$		-0.0025	-0.0027	-0.0052
$2\sigma - 1\pi_-$		-0.0195	-0.0131	-0.0258
$3\sigma - 1\pi_-$		-0.0178	0.0198	-0.0461
$1\pi_+ - 1\pi_-$		-0.0098	-0.0311	0.0618
$1\pi_+ - 1\pi_+$			-0.0166	-0.0166
$1\pi_- - 1\pi_-$				-0.0166

$e(1\pi_-, 1\pi_-)$ in the spin-adapted formulation. The first of these spin-adapted pair correlation energies is obtained by carrying out a CI calculation including all singlet (S = 0) functions arising from spatial products of the type

$$1\sigma^2\ 2\sigma^2\ 3\sigma^2\ 1\pi_+\ 1\pi_-\ x\ y \qquad (III,21)$$

Such functions (spin-adapted linear combinations of determinants) will not be true $^1\Sigma^+$ configurations but will also have components of $^1\Sigma^-$ symmetry. $e(1\pi_+,1\pi_-)$ is obtained as the difference between the SCF energy and the CI energy found including the SCF configuration plus all singlet functions of the type (III,21). In a similar manner Bender and Davidson evaluate $e(1\pi_+,1\pi_+)$ via a CI including singlet spin functions of the form

$$1\sigma^2\ 2\sigma^2\ 3\sigma^2\ 1\pi_-^2\ x\ y \qquad (III,22)$$

Finally $e(1\pi_-,1\pi_-)$ is found from spin-adapted functions arising from

$$1\sigma^2\ 2\sigma^2\ 3\sigma^2\ 1\pi_+^2\ x\ y \qquad (III,23)$$

It should be noted that Table III-12 includes a contribution to the correlation energy from single excitations. For the open-shell hydrides this is understandable, but for hydrogen fluoride, Brillouin's theorem (section IB) states that singly-excited configurations should not lower the SCF energy. The reason for the small 0.0018 hartrees energy lowering for HF is that the 1σ, 2σ, 3σ, and 1π orbitals referred to in Table III-12 are not the SCF orbitals but rather the natural orbitals. However, differences between the first four natural orbitals and the SCF orbitals are (as expected) small for the hydrides and do not affect any of the qualitative conclusions of Bender and Davidson (29).

As was the case for the first-row atoms, the electron correlation involving the core electrons (1σ for the hydrides) is relatively constant, $e(1\sigma,1\sigma)$ being seen to vary little in Table III-12 from CH to FH. Most of the correlation is due to the valence electrons, which occupy the 2σ, 3σ, and 1π orbitals in the single configuration approximation. $e(2\sigma,2\sigma)$ is much larger for CH than for NH, OH, or FH, since only for CH is the very important degeneracy configuration $2\sigma^2 \to 1\pi^2$ allowed by symmetry and the Pauli exclusion principle.

Each variationally obtained correlation energy E_c is compared with the sum of the pair correlation energies $\sum_{i \leq j} e(i,j)$ in Table III-13. This table also shows, for example, that 2401 $^3\Pi$ configurations were included in the OH variation calculation. Each of the same 2401 configurations was used during one of the calculations required to obtain the OH pair correlation energies seen in Table III-12.

The same qualitative result seen in Table III-13 was found from pair correlation calculations on first-row atoms. Namely, as one goes to larger molecules (or atoms) the pair correlation approximation becomes progressively worse. For hydrogen fluoride, the variational calculation provides 74% of the correlation energy, while the sum of the pair correlation energies gives 95% of the correlation. Since the basis set used by Bender and Davidson is probably capable of yielding (via full CI) no more than 80% of the correlation energy, the pair correlation approximation is seriously in error. Although the use of symmetry-adapted pairs [e.g., Eq.(III,20] would probably lower the sum of the pairs by ~5% (certainly no more than 10%), the basic idea of summing pair correlation energies appears less realistic for molecules than for atoms.

Table III-13. Validity of the pair correlation energy approximation for the ground states of the first-row hydrides (29).

	LiH	BeH	BH	CH	NH	OH	FH
Ground state symmetry	$^1\Sigma^+$	$^2\Sigma^+$	$^1\Sigma^+$	$^2\Pi$	$^3\Sigma^-$	$^2\Pi$	$^1\Sigma^+$
Bond distance R_e (bohrs)	3.015	2.538	2.336	2.124	1.9614	1.8342	1.7328
Number of configurations	939	1039	1123	1667	3379	2401	1517
Total variational energy	-8.0600	-15.2324	-25.2621	-38.4399	-54.1620	-75.6422	-100.3564
Variationally calculated correlation energy	0.0731	0.0801	0.1332	0.1629	0.1871	0.2258	0.2913
Percentage of estimated correlation energy	87	--	85	81	74	70	74
Sum of pair correlation energies	0.0740	0.0808	0.1484	0.1768	0.2069	0.2771	0.3646
Percentage of estimated correlation energy	89	--	95	90	87	89	95

Additional quantitative information concerning the additivity of pair correlation energies in molecules is given by the calculations of Siu and Davidson (359) on carbon monoxide. Using the Bender-Davidson approach they obtain a 2484 configuration wave function for CO yielding a total energy -113.14557 hartrees, which corresponds to 70% of the estimated correlation energy. However, the sum of the spin-adapted pair correlation energies corresponds to 93% of the correlation energy. It therefore appears that for CO, the sum of the exact pair correlation energies (obtained using a complete basis set) is more than 120% of the correlation energy. This certainly raises serious questions concerning the usefulness of the pair correlation approximation (II,19) in predictions of molecular correlation energies.

An interesting and related question raised by Bender and Davidson (97) concerns the effect on calculated pair energies of a unitary transformation of the occupied SCF orbitals. It is well known (317) that for closed-shell systems, the SCF wave function is invariant and the SCF energy unchanged by a unitary transformation of the doubly-occupied SCF orbitals. The simplest hydride to consider is LiH with orbital occupancy $1\sigma^2\ 2\sigma^2$, and a unitary transformation

$$1\sigma' = [1\sigma + \alpha\ 2\sigma](1 + \alpha^2)^{-\frac{1}{2}}$$
$$2\sigma' = [-\alpha\ 1\sigma + 2\sigma](1 + \alpha^2)^{-\frac{1}{2}}$$

(III,24)

is such that the two new orbitals remain normalized and orthogonal. Bender and Davidson found that by continuously varying the parameter α, which determines the unitary transformation, a wide variety of pair correlation energies e(1σ',1σ'), e(1σ',2σ'), and e(2σ',2σ') could be obtained. The sum of the pair correlation energies is closest to the variationally calculated correlation energy if α is close to 0, i.e., if the canonical or ordinary SCF orbitals are used.

Table III-14 shows the results of similar calculations for BH. Since we now know in general that summing pair correlation energies overestimates the true correlation energy, it is perhaps a good idea to look for pair correlation schemes which do not overestimate the correlation too severely. Table III-14 shows that the use of localized (see section VE) orbitals allows the sum of the e(i,j) to be less than that obtained with the ordinary or canonical SCF orbitals. By intentionally maximizing the sum of the pairs as a function of unitary transformation, a very much larger correlation energy, 0.14890 hartrees, may also be obtained. Table III-14 thus suggests that the use of localized orbitals may lessen the nonadditivity of pair correlation energies. Kutzelnigg and coworkers (2) have in fact calculated localized pair correlation energies for both diatomic and polyatomic molecules, and some of their results will be discussed in section VB.

The separated pair approximation: LiH, BH, NH. Ever since the 1916 manifesto of G. N. Lewis, chemists have found it useful to think of electronic structure in terms of electron pairs, imagined to be localized in space, separable, and transferable in many cases from molecule to molecule. Hurley, Lennard-Jones, and Pople (181) put the electron pair concept in a mathematical form in the separated pair approximation (SPA). It should be pointed out at the onset that the separated pair approximation is very different from pair correlation

Table III-14. Pair correlation energies in hartrees for BH as a function of the unitary transformation chosen to define the occupied orbitals 1σ, 2σ, and 3σ (97).

	Ordinary SCF Orbitals	Localized Orbitals	Maximum sum of pairs
$e(1\sigma,1\sigma)$	0.03728	0.03870	0.03728
$e(1\sigma,2\sigma)$	0.00326	0.00223	0.00267
$e(2\sigma,2\sigma)$	0.02950	0.03507	0.03512
$e(1\sigma,3\sigma)$	0.00367	0.00327	0.00427
$e(2\sigma,3\sigma)$	0.02930	0.01828	0.02086
$e(3\sigma,3\sigma)$	0.03831	0.04169	0.04870
Sum of pairs	0.14132	0.13924	0.14890

theory. First of all, the SPA makes use of a single variational wave function, while pair correlation theory involves the summation of independently computed pair energies. Secondly, certain types of electron correlation are intentionally excluded in the separated pair approximation, while pair correlation theory attempts to evaluate essentially all correlation effects.

As a concrete example consider the LiH molecule, for which the separated pair wave function is of the form (255)

$$A(4) \; [\Lambda_C \; (\alpha\beta - \beta\alpha)/\sqrt{2}] \; [\Lambda_B \; (\alpha\beta - \beta\alpha)/\sqrt{2}] \qquad (III,25)$$

where $A(4)$ is the antisymmetrizer for four electrons. Λ_C is a spatial function describing the two core electrons and Λ_B similarly describes the two bonding electrons. As the spin functions in (III,25) show, the core and bonding electron pairs are independently singlet coupled. The separated pair wave function of (III,25) is equivalent to a CI wave function including orbital occupancies

$$
\begin{array}{cc}
1\sigma^2 & 2\sigma^2 \\
1\sigma^2 & wx \\
yz & 2\sigma^2
\end{array}
\qquad (\text{III},26)
$$

What is clearly not included in a CI of type (III,26) is the intershell 1σ 2σ correlation energy. And, as Bender and Davidson (97) have shown, no choice of 1σ and 2σ orbitals can make this intershell correlation energy vanish. So we see that the separated pair approximation ignores the electron correlation between different localized electron pairs.

An approximation incorporated by Hurley, Lennard-Jones, and Pople (181) to make the separated pair formulation tractable is that of <u>strong orthogonality</u>. Strong orthogonality requires that each electron pair function Λ_i (1,2) be orthogonal to the other (n/2 - 1) electron pair functions or geminals. For the LiH case this amounts to

$$
\int \Lambda_C^*(1,2)\, \Lambda_B(1,3)\, d\nu(1) = 0 \qquad (\text{III},27)
$$

In the CI picture, represented by orbital occupancies (III,26), strong orthogonality requires all orbitals w and x (describing correlation in the bonding pair) to be orthogonal to all orbitals y and z (which describe the inner shell correlation.

Our most quantitative information concerning the validity of the SPA

Table III-15. Energies in hartrees for LiH, BH, and NH within the separated pair approximation (255, 351).

	$^1\Sigma^+$ LiH	$^1\Sigma^+$ BH	$^3\Sigma^-$ NH
Estimated RHF energy	-7.9873	-25.1314	-54.9784
Separated pair energy	-8.0542	-25.2053	-55.0335
Correlation energy obtained	0.0669	0.0739	0.0551
Estimated correlation energy	0.083	0.155	0.249
Percent correlation energy obtained via separated pair	81	48	22

derives from the calculations of Ruedenberg, Silver, and Mehler on LiH, BH, and NH (255, 351). This work was carried out using extended basis sets of Slater functions. For NH, a N(5s 4p 1d) and H(2s 1p) set was optimized by Ruedenberg et al. (351) by repeated separated pair calculations on the molecule. We may estimate that a full CI calculation using this basis set would yield ~60% of the correlation energy of NH. The computational method used by Ruedenberg and coworkers for finding the best (in the variational sense) separated pair wave function within a given basis set is rather complicated and the reader is referred to their work (255) for a discussion.

Table III-15 shows the calculated SPA total energies for LiH, BH, and NH in their ground electronic states. The most readily seen point in Table III-15 is that the percentage of correlation energy obtained goes down drastically from LiH to BH to NH. This is partly due to the fact that the NH basis set is less complete than that for LiH. However, the fact that for NH the SPA gives only about 1/3 of the correlation energy that would be obtained by full CI (with the same basis) is due to the model itself. This failure of the separated pair method is due both to the strong orthogonality constraint and the form of the wave function itself.

For BH (orbital occupancy $1\sigma^2\ 2\sigma^2\ 3\sigma^2$) strong orthogonality demands that the set of orbitals used to correlate the $2\sigma^2$ pair must be orthogonal to and distinct from those orbitals used to correlate the $3\sigma^2$ pair. And, in fact the first natural orbital which correlates $2\sigma^2$ is quite different from the first natural orbital correlating the $3\sigma^2$ pair. However, the higher orbitals optimally suited to correlating the $2\sigma^2$ and $3\sigma^2$ pairs are not mutually orthogonal. This is one of the two reasons the separated pair approximation will yield at most about 60% of the correlation energy of BH.

The second reason the SPA cannot account for all of the correlation energy is that it is not possible to sufficiently localize the electron pairs that the interpair correlation energies vanish. This might have been suspected from the pair correlation energy results of Table III-14 for BH and Table III-12 for NH. For BH, the core $1\sigma^2$ electron pair is nearly independent of the bonding $2\sigma^2$ pair and the $3\sigma^2$ lone pair. However, there is no way to make $e(2\sigma,3\sigma)$, the correlation between the bonding and lone pair orbitals, be very small. For NH, the SPA is worse, since pair correlations between three ($2\sigma^2$, $3\sigma^2$, and triplet-coupled $1\pi^2$) pairs are all relatively large. The success of the separated pair approach for lithium hydride is of course due to the fact (not present in larger molecules) that there are only two electron pairs, and they are located in very distinctly different regions of space.

In fairly evaluating the above shortcomings of the SPA, it should be remembered that the primary goal of theoretical chemistry is not necessarily the calculation of 100% or even 50% of the correlation energy of a molecule. More important are the reliable prediction of observables and the development of reliable models for the interpretation of experimental results. With respect to the latter goal, it is important to point out that Ruedenberg, et al. (255, 351) find the two-electron functions Λ describing the core electron pairs of LiH, BH, and NH to be nearly identical to those obtained from independent calculations on the atoms Li, B, and N. Along the same lines, Levy, et al. (226) have reported separated pair calculations which suggest that inner shell, lone pair, and bonding pairs may be transferable between water and hydrogen peroxide. In addition, some very recent work using the SPA suggests that it may be qualitatively useful in predicting reaction coordinates for chemical reactions (177). It seems fair to conclude that the SPA may be of considerable value as an interpretive tool.

Table III-16 gives additional quantitative information on the SPA results for NH. The calculated dissociation energy D_e, 2.65 eV, is only marginally better than the near RHF value and remains 1.15 eV less than experiment. The best agreement with experiment is obtained for the SPA predicted bond distance, which is only 0.0005 bohr too long. In addition, the SPA predicts the LiH bond distance to be 1% too large and the BH R_e to be 0.5% too small (255). The most serious

Table III-16. Properties of the ground electronic state of NH within the separated pair approximation (255,351). Unless indicated, properties are in atomic units.

	Near Hartree-Fock	Separated Pair	Experiment
D_e (eV)	2.10	2.65	~3.8
R_e (bohrs)	1.923	1.9619	1.9614
ω_e (cm^{-1})	3556	4910	3126
B_e (cm^{-1})	17.319	16.625	16.668
α_e (cm^{-1})	0.572	0.466	0.646
$\omega_e x_e$ (cm^{-1})	78.3	66.8	78.5
dipole moment (debyes)	-1.627	-1.675	
quadrupole moment		0.738	
$< r_N^2 >$		2.149	
$< r_H^2 >$		5.357	
$< 1/r_N >$		2.357	
$< 1/r_H >$		0.576	

that of Nesbet (273) and the calculated dipole moment was 0.361 debyes, C^+O^- in sign. However, the experimental dipole moment was thought to be of negative sign, C^-O^+, and of magnitude 0.112 debye. In light of a) previous good agreement between SCF dipole moments and experiment and b) the knowledge of every chemist weakness of SPA spectroscopic constants for NH is that the predicted vibrational frequency ω_e is nearly 70% greater than experiment. For LiH, the calculated ω_e is only 5.5% greater than experiment but for BH the difference is 23.7%. These results seem to indicate that the SPA is not capable of yielding reliable force constants.

A final interesting point concerning the SPA is that it allows a rigorous breakdown of one-electron properties in terms of contributions from the different electron pairs. The analysis of Silver, et al. (351) shows for example that the dipole moment of NH is determined almost completely by the nonbonding lone pair. The contribution from the core electron pair is -0.003 debye, while the contributions from the lone pair, bonding pair, and triplet-coupled pairs are -1.850, -0.102, and +0.280 debyes.

Accurate dipole moment calculations: CO. In section A of this chapter, we saw that near Hartree-Fock wave functions for the alkali halides yield dipole moments in close agreement with experiment. Unfortunately this is not always the case, especially if the dipole moment is small, i.e., less than 1 debye. In such cases the nuclear and electronic contributions, Eq. (I,29), are frequently of opposite sign and nearly equal magnitude. The carbon monoxide molecule provides the classic example of a near Hartree-Fock dipole moment being in serious disagreement with experiment. Here we discuss a reliable method for the theoretical prediction of dipole moments, with special reference to the CO problem.

From our discussion of basis sets, we know that the smallest basis likely to yield dipole moments of close to RHF accuracy is the double-zeta-plus-polarization set. For carbon monoxide the first SCF calculation with such a basis was that the oxygen atom is more electronegative than carbon, Nesbet investigated the validity of the experimental determination of the sign of the CO dipole moment. He came to the conclusion that the C^-O^+ sign determination was not definitive, since it rested upon the sign of the molecular g factor, for which either a positive or a negative value was feasible. Next, an SCF calculation by Huo (178) with a larger, fully optimized basis set gave an SCF dipole moment 0.274, with sign C^+O^- again. Using Huo's wave function, McLean and Yoshimine (246) have computed $<r^2>$, which may be combined with the molecular g value to predict the diamagnetic susceptibility. McLean and Yoshimine found that only if a negative g factor (corresponding to C^-O^+) is used does one obtain reasonable agreement with the experimental diamagnetic susceptibility. It thus seems fairly clear that the sign of the CO dipole moment is C^-O^+, as originally thought. The true RHF dipole should be very close to 0.280 debyes, the value obtained by McLean and Yoshimine (250) using a very large basis. Thus it must be concluded that the Hartree-Fock dipole moment differs from experiment by 0.4 debyes and has the incorrect sign.

The neglect of electron correlation is responsible for the incorrect sign of the RHF dipole moment. However, a straightforward CI will not necessarily yield reliable dipole moments. For example Bender and Davidson (28) calculated a wave function for hydrogen fluoride which accounted for 50% of the correlation energy but yielded a dipole moment -1.6485 debyes, compared to the estimated RHF value -1.65 debyes, and experiment -1.74 debyes. The reason that the CI dipole moment is not improved over the RHF value is, according to Bender and Davidson, that the basis set used was not sufficient to approach the RHF limit.

That is, a dipole moment error (due to basis set) present at the SCF level will not be removed by ordinary CI. From similar calculations the same conclusion may be drawn, and it appears to be true in general that the quantitative prediction of one-electron properties requires calculated SCF properties to approach their RHF values before the inclusion of electron correlation.

The first satisfactory quantum mechanical calculation of the dipole moment of CO was that of Grimaldi, Lecourt, and Moser (140). The rather surprising result of their calculation was that <u>single excitations</u> are primarily responsible for the correlation contribution to the CO dipole moment. This result was unexpected since singly-excited configurations make only a small contribution to the correlation energy and interact with the SCF configuration only indirectly, through the double excitations. The results of Grimaldi, et al. are summarized in Table III-17.

Using the extended basis set of Huo (178), Grimaldi, et al. selected the most important doubly-excited configuration using perturbation theory. However, as Table III-17 shows clearly, the inclusion of even the 200 most important configurations still results in a dipole moment of the wrong sign. But, by adding to the CI <u>all</u> single excitations (62 in this case), μ takes on the correct sign and is within 0.035 debyes of experiment. Note that the energy lowering due to the single excitations is only 0.002 hartrees. Related calculations have been carried out on the first-row hydrides by Bender and Davidson (29). The basis sets and configurations used were described in the discussion of pair correlations earlier in this section. As now seems wise in the calculation of molecular properties, <u>all</u> configurations differing from the SCF by one orbital were included. Table III-18 shows that for the four experimentally known dipole moments, the CI values are significantly improved over the SCF results and within 3% of experiment. It is also interesting to note that in all cases the inclusion of electron correlation lowers the magnitude of the dipole moments. The predicted BeH, BH, and NH dipole moments may be considered very reliable.

Green (139) has further refined the method of Grimaldi, Lecourt, and Moser (140) and in addition applied it to two open-shell excited states of experimental interest, the a $^3\Pi$ state of CO and the A $^1\Pi$ state of CS. For the open-shell molecules the iterative natural orbital method (27) of Bender and Davidson was used to decrease the number of important configurations. Green's results are seen in Table III-19. Taken as a whole, the results of Grimaldi, et al., Bender and Davidson, and Green clearly outline a practical approach to the nearly quantitative prediction of dipole moments.

Potential energy curves for the low-lying states of CN. Spectroscopists have long been interested in the electronic spectra of diatomic molecules, particularly species such as CN, N_2, NO, and O_2. However, the spectrum of O_2, for example, is very complicated due to the fact that many different electronic states contribute. In order to theoretically predict the electronic spectrum of O_2 it is necessary to calculate potential energy curves for all the low-lying electronic states. One reliable approach to this problem might be that of Bender and Davidson (29), namely large CI (several thousand configurations) using extended basis sets and the iterative natural orbital method. However, this approach would currently (1971) be unreasonably expensive due to the large number of electronic states (at least 30) and internuclear separations (at least 5 for

Table III-17. Effects of electron correlation on the dipole moment of carbon monoxide (140).

Wave function	Energy (hartrees)	μ (debyes)
Self-Consistent-Field (SCF)	-112.7879	0.274
SCF + 10 double excitations	-112.8783	0.213
SCF + 20 double excitations	-112.9147	0.186
SCF + 60 double excitations	-112.9738	0.175
SCF + 100 double excitations	-112.9996	0.175
SCF + 138 double excitations	-113.0159	0.173
SCF + 200 double excitations	-113.0338	0.183
SCF + 62 singles + 138 doubles	-113.0179	-0.077
Experiment		-0.112

Table III-18. Dipole moments in debyes of the first-row hydrides calculated by large configuration interaction (29).

	Near Hartree-Fock	CI	Experiment
LiH	6.002	5.853	5.82
BeH	0.282	0.248	
BH	-1.733	-1.470	
CH	-1.570	-1.427	-1.40
NH	-1.627	-1.587	
OH	-1.780	-1.633	-1.66
FH	-1.942	-1.816	-1.82

Table III-19. Limited CI calculations of dipole moments for larger molecules (139). In each calculation, the most important double excitations plus <u>all</u> (or nearly all) single excitations are included. Positive entry means + - polarity for molecule as written.

Molecule	State	μ(SCF)	μ(CI)	Experiment
CS	X $^1\Sigma^+$	-1.56	-2.03	-1.97
CS	A $^1\Pi$	--	-0.63	-0.63
NaLi	X $^1\Sigma^+$	0.94	0.99	--
CO	a $^3\Pi$	2.46	1.55	1.38

each state) which must be considered in order to predict the O_2 spectrum. Less time consuming would be a series of SCF calculations, but we know that SCF potential energy curves do not display the correct dissociation behavior and that predicted SCF electronic energy separations are not reliable. What is needed is a relatively simple method which yields qualitatively correct potential energy curves.

A surprisingly effective method of predicting potential energy curves involves full valence shell CI using a minimum basis set. This approach is opposite in philosophy from calculations discussed earlier in the book which used extended basis sets and limited CI. Although minimum basis sets must in general be used with skepticism, this appears to be one area in which they yield chemically useful results. The first comprehensive, minimum basis, full valence CI study was that of Schaefer and Harris for the O_2 molecule (333). Potential curves for 62 electronic states were obtained and except for the c $^1\Sigma_u^-$ state, the order of electronic states agreed with experiment. However the experimental position of the c $^1\Sigma_u^-$ state is not beyond question, and Herzberg (158) has suggested a renumbering of the c state vibrational levels which would place the c $^1\Sigma_u^-$ state in the position predicted by the calculations of Schaefer and Harris (333). Some care must be used in choosing molecules for which to carry out minimum basis, full valence CI calculations. For example, Harris and Michels (147) have shown that such calculations give poor results for the OH radical, but that good results are obtained if a 2p function is added to the 1s function centered on hydrogen.

In addition to O_2, qualitatively useful minimum basis, full valence CI calculations have been carried out for the electronic states of NH, N_2, CO, CN, and SiO. The results for CN (155) are of particular interest due to the prediction of a number of low-lying electronic states which have not been observed experimentally. It may be useful to recall that the CN ground state orbital occupancy is given by Eq. (III,10). Even using a minimum basis set (1s, 2s, and 2p Slater functions on each atom), a relatively large number of configurations must be included in full valence CI calculations for CN. This is because configurations differing by as many as <u>seven</u> orbitals from (III,10) must be included. The orbital occupancy $1\sigma^2\ 2\sigma^2\ 3\sigma^2\ 5\sigma^2\ 6\sigma\ 2\pi^4$, for example, must be included in the $^2\Sigma^+$ calculations. In this way 320 configurations comprise the full valence CI for $^2\Sigma^+$ CN. The symmetry requiring the largest number of configurations is $^2\Pi$, with 486 configurations necessary. A very important result of the use of full CI in this way is that each molecular wave function dissociates smoothly to the atomic wave functions required by the Wigner-Witmer rules (157).

The CN calculations were undertaken due to the astrophysical and chemical importance of this radical. More specifically, Heil and Schaefer (155) recognized that the simplest molecular orbital arguments predict the existence of several low-lying quartet (S = 3/2) states of CN. Since the electronic transitions between these quartet states and the eight known doublet (S = 1/2) states of CN are forbidden (due to $\Delta S = 1$), it is not too surprising that no quartet states have been observed. Spectroscopists have assigned four doublet states to the following two orbital occupancies:

$$E\ ^2\Sigma^+,\ F\ ^2\Delta,\ J\ ^2\Delta \qquad 1\sigma^2\ 2\sigma^2\ 3\sigma^2\ 4\sigma^2\ 5\sigma\ 1\pi^3\ 2\pi \qquad (III,28)$$

$$H\ ^2\Pi \qquad 1\sigma^2\ 2\sigma^2\ 3\sigma^2\ 4\sigma^2\ 5\sigma^2\ 1\pi^2\ 2\pi \qquad (III,29)$$

The simplest consideration of symmetry (157) shows that the first of these orbital occupancies gives rise to $^4\Sigma^+$, $^4\Sigma^-$, and $^4\Delta$ states, and that the second yields a $^4\Pi$ state. And in fact we would expect from Hund's first rule that the predicted quartet states lie <u>below</u> the experimentally known doublets. Since the H $^2\Pi$ and E $^2\Sigma^+$ states are the fourth and fifth excited states of CN, one might expect the quartet states to be quite low-lying.

Figure III-5 shows the potential energy curves for the eight lowest electronic states of CN, as theoretically predicted by Heil and Schaefer (155). The quantitative calculations bear out the simple MO arguments and the $^4\Sigma^+$ state is seen to lie only slightly above the B $^2\Sigma^+$ state, extremely well-known from the "violet" B - X system. The calculations thus predict the unknown $^4\Sigma^+$, $^4\Pi$, $^4\Delta$, and $^4\Sigma^-$ states to be the third, fourth, fifth, and seventh lowest excited electronic states of CN.

More precise information regarding these calculations is seen in Table III-20. Due to the relative simplicity of the calculations, one should not expect the spectroscopic constants to be quantitatively accurate. However, like the O_2 results (333) mentioned above, the CN predictions are surprisingly good. The ordering of the eight known doublet states is correct with the exception of the H $^2\Pi$ state. The H $^2\Pi$ state is calculated to lie 0.07 eV below the F $^2\Delta$ state while in fact it lies 0.16 eV above the F $^2\Delta$ state. The good agreement with experiment for the ordering of states gives us some confidence that the seven experimentally unknown states of CN seen in Table III-20 are in reasonable positions. As is usually the case, the predicted bond distances are all too large, the deviations from experiment ranging from 0.043 to 0.104 Å. Vibrational frequencies ω_e differ from experiment by no more than 18% (the B $^2\Sigma^+$ state).

Accurate potential energy curves for the ground states of F_2 and O_2. Calculations of the type described above for CN provide useful and (usually) qualitatively correct predictions of chemical interest. However, such calculations are not usually adequate for quantitative applications. For example, the dissociation energy D_e obtained for O_2 from a minimum basis, full valence CI is 3.81 eV. Although this represents a significant improvement over the near RHF value, 1.43 eV, it is nevertheless 1.4 eV less than the experimental value, 5.21 eV. To obtain dissociation energies and potential energy curves of higher accuracy it is absolutely necessary to use larger than minimum basis sets. However, it is also true that straightforward CI (including the SCF con-

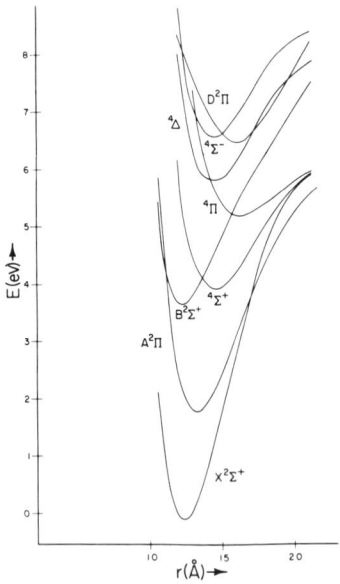

Figure III-5. Predicted potential energy curves for CN.

figuration plus all single and double excitations therefrom) with a large basis will not give reasonable potential curves. The problem is that as most diatomic molecules are pulled apart, the SCF approximation becomes very poor. That is, there are additional configurations (besides the SCF) which enter the wave function with large coefficients. In general, then, a reasonable potential curve is obtained by brute force CI only if it includes all single and double excitations with respect to all configurations which enter the wave function with coefficient greater than, say, 0.3, at any internuclear separation. Very few such calculations have been carried out to date, and one logically looks for simpler approaches to the quantitative calculation of potential curves.

Table III-20. Theoretical and experimental (values in parentheses) spectroscopic constants for CN.

	T_e (eV)	D_e (eV)	r_e (Å)	ω_e (cm^{-1})
X $^2\Sigma^+$	0.0 (0.0)	6.18 (7.75±0.2)	1.236 (1.172)	1939 (2069)
A $^2\Pi$	1.88 (1.15)	4.30	1.324 (1.233)	1621 (1814)
B $^2\Sigma^+$	3.77 (3.19)	4.34	1.226 (1.151)	1765 (2164)
$^4\Sigma^+$	4.02	2.16	1.448	1249
$^4\Pi$	5.33	0.85	1.629	873
$^4\Delta$	5.91	2.19	1.441	1274
D $^2\Pi$	6.59 (6.76)	3.00	1.590 (1.499)	1042 (1005)
$^4\Sigma^-$	6.67	1.42	1.446	1286
$^2\Phi$	7.74	1.86	1.621	951
H $^2\Pi$	7.81 (7.56)	1.79	1.414 (1.310)	1651
E $^2\Sigma^+$	7.86 (7.33)	1.75	1.364 (1.320)	1717 (1681)
F $^2\Delta$	7.88 (7.40)	1.72	1.461 (1.378)	1254
$^4\Sigma^+$	7.90	1.26	1.749	
$^2\Sigma^-$	7.99	1.61	1.453	1226
J $^2\Delta$	8.43 (8.09)	1.17	1.457 (1.414)	1212 (1122)

Such an approach has been developed by Wahl and coworkers, the method of optimized valence configurations (OVC). This method was developed with particular emphasis on F_2, but has since been applied to a number of other molecules. For F_2 one of the earliest extended basis SCF calculations, by Wahl(384), showed the Hartree-Fock energy of F_2 to lie more than 1 eV above the energy of two F atoms in the RHF approximation. Since F_2 is a well-known molecule with dissociation energy ~1.6 eV, the RHF prediction of instability is clearly unsatisfactory. The first breakthrough on this problem came with the calculation by Das and Wahl (94) of two-configuration SCF wave functions for F_2

$$\psi = a\ 1\sigma_g^2\ 1\sigma_u^2\ 2\sigma_g^2\ 2\sigma_u^2\ 3\sigma_g^2\ 1\pi_u^4\ 1\pi_g^4$$
$$+ b\ 1\sigma_g^2\ 1\sigma_u^2\ 2\sigma_g^2\ 2\sigma_u^2\ 3\sigma_u^2\ 1\pi_u^4\ 1\pi_g^4 \qquad (III,30)$$

The first of the two $^1\Sigma_g^+$ configurations in (III,30) is the SCF, but the second configuration must be included to allow the wave function to dissociate to two SCF fluorine atom wave functions. The fact that F_2 cannot dissociate properly using the SCF configuration alone is due to the RHF restriction that all orbitals be doubly-occupied. And a reasonable description of F_2 at large internuclear separation necessitates the $2p\sigma$ orbitals of fluorine atoms A and B to be singly occupied. The two configuration SCF treatment of F_2 provided correct dissociation to SCF atoms and a bound F_2 molecule with predicted dissociation energy 0.54 eV (94).

The OVC method is not restricted to two configurations, and typically as many as ten configurations expected to be important are included. The O in OVC refers to the fact that the multiconfiguration SCF method (see section IC) is used to optimally determine the orbitals used in the small CI. The V refers to the fact that chemical intuition is used and configurations are limited to those in which only the valence SCF orbitals are replaced. For F_2 it required several years for Das and Wahl to determine which configurations should be included beyond the two in (III,30) to yield a highly accurate potential curve. Table III-21 shows the four additional configurations chosen. The final six-configuration extended basis set MCSCF wave functions yield a dissociation energy of 1.71 eV (96). Although there has been some recent controversy concerning the experimental value of D_e for F_2, it now appears that the correct value is close to 1.65 eV. Thus the final OVC potential curve for F_2 may be nearly as reliable as the experimental curve.

The only serious objection which has been raised concerning the OVC method concerns the rather arbitrary selection of configurations. The importance of the so-called "split-shell" configurations 5 and 6 in Table III-21 at first came as a surprise even to Das and Wahl. However, based on the experience of several calculations, Das and Wahl (96) have developed a rather orderly two-step procedure for the selection of OVC configurations: a) include all configurations necessary to insure that the molecule dissociates into SCF atoms in their appropriate electronic states, and b) add all configurations whose contributions are small at large internuclear separations but large near R_e.

Perhaps the most important aspect of the OVC calculations of Wahl and co-workers (385) is that reliable potential curves are obtained although only a small percentage (~10% for F_2) of the molecular correlation energy is recovered. More precisely, the OVC method appears to pick up just that part of the correlation

Table III-21. Configurations included in the OVC wave function of Das and Wahl (96) for F_2.

	Excitation notation	Orbital Occupancy
1.	--	$1\sigma_g^2\ 1\sigma_u^2\ 2\sigma_g^2\ 2\sigma_u^2\ 3\sigma_g^2\ 1\pi_u^4\ 1\pi_g^4$
2.	$3\sigma_g^2 \rightarrow 3\sigma_u^2$	$1\sigma_g^2\ 1\sigma_u^2\ 2\sigma_g^2\ 2\sigma_u^2\ 3\sigma_u^2\ 1\pi_u^4\ 1\pi_g^4$
3.	$1\pi_u^2 \rightarrow 3\sigma_u^2$	$1\sigma_g^2\ 1\sigma_u^2\ 2\sigma_g^2\ 2\sigma_u^2\ 3\sigma_g^2\ 3\sigma_u^2\ 1\pi_u^2\ 1\pi_g^4$
4.	$1\pi_g^2 \rightarrow 3\sigma_u^2$	$1\sigma_g^2\ 1\sigma_u^2\ 2\sigma_g^2\ 2\sigma_u^2\ 3\sigma_g^2\ 3\sigma_u^2\ 1\pi_u^4\ 1\pi_g^2$
5.	$3\sigma_g\ 1\pi_u \rightarrow 3\sigma_u\ 2\pi_g$	$1\sigma_g^2\ 1\sigma_u^2\ 2\sigma_g^2\ 2\sigma_u^2\ 3\sigma_g\ 3\sigma_u\ 1\pi_u^3\ 1\pi_g^4\ 2\pi_g$
6.	$3\sigma_g\ 1\pi_g \rightarrow 3\sigma_u\ 2\pi_u$	$1\sigma_g^2\ 1\sigma_u^2\ 2\sigma_g^2\ 2\sigma_u^2\ 3\sigma_g\ 3\sigma_u\ 1\pi_u^4\ 1\pi_g^3\ 2\pi_u$

energy which varies with internuclear separation. A related idea, that of calculating the structure sensitive part of the correlation energy, finds a mathematical form in the first-order wave function, discussed in sections IIC and IID with respect to atomic hyperfine structure and pair correlations in open-shell systems. First-order wave functions have been calculated for a number of diatomic molecules and found to yield quite reliable dissociation energies and potential curves.

A study of the oxygen molecule (342) illustrates the first-order approach. In these O_2 calculations a double zeta plus polarization basis set of Slater functions was used. This basis yields an O_2 SCF energy 0.0193 hartrees or

0.53 eV above a near Hartree-Fock calculation and is about the smallest that can be used if reliable dissociation energies are required. As pointed out earlier, the first-order wave function in principle includes all configurations in which no more than a single electron occupies an orbital beyond the valence shell. To keep the number of configurations for O_2 reasonably small, Schaefer (342) has imposed additional restrictions: a) only configurations differing by one or two orbitals from the SCF configuration are included, b) excitations out of the core $1\sigma_g$ and $1\sigma_u$ orbitals are excluded, and c) configurations in which the semi-valence-like $2\sigma_g$ and $2\sigma_u$ orbitals (corresponding to the 2s oxygen atom orbitals) are replaced cannot include even one orbital beyond the valence shell.

The structure of the first-order wave function is most easily understood after it is recognized that the O_2 SCF orbital occupancy is $1\sigma_g^2 1\sigma_u^2 2\sigma_g^2 2\sigma_u^2 3\sigma_g^2 1\pi_u^4 1\pi_g^2$ and thus the $1\pi_g$ and $3\sigma_u$ orbitals are the valence orbitals not fully occupied in the RHF approximation. In fact, the correlation energy obtained via the first-order wave function for O_2 is just that correlation energy associated with the unoccupied valence orbitals $1\pi_g$ and $3\sigma_u$. Table III-22 shows the orbital occupancies thus included. Note that the last orbital occupancy is a quadruple excitation which must be included in order to describe two SCF oxygen atoms at infinite separations. With the basis set mentioned above, the first-order wave function for the $X\ ^3\Sigma_g^-$ ground state of O_2 consists of 128 configurations.

In order for the first-order wave function to correspond to the desired physical model, it is necessary that the $1\sigma_g$, $1\sigma_u$, $2\sigma_g$, $2\sigma_u$, $3\sigma_g$, $3\sigma_u$, $1\pi_u$ and $1\pi_g$ orbitals be optimally determined. Although the MCSCF method could in principle be applied to the 128 configuration wave function for O_2, in practice this is not possible. Instead the iterative natural orbital method (27) was used, with 5 iterations being required for convergence.

The results of the oxygen calculations are summarized in Figure III-6 and Table III-23. The first-order wave function dissociates to two oxygen atom wave functions of energy only 0.001 hartrees below the estimated RHF energy of ^3P O. The figure shows as expected that the O_2 near RHF wave function does not dissociate to two RHF atomic wave functions. More important, the figure shows that the first-order potential curve is very similar in shape to the experimental curve. Although the basis set used (342) gives a molecular SCF energy ~0.53 eV above the estimated RHF limit, the same basis set yields SCF atomic energies less than 0.01 eV above the estimated RHF energies. Thus it appears that the first-order potential curve for O_2 obtained with a complete basis would yield a dissociation energy ~0.5 eV greater than the calculated value seen in Table III-23, 4.72 eV. Such a value of D_e would be in quantitative agreement with experiment. Table III-23 also shows that the first-order spectroscopic constants in addition to D_e are in most cases considerably improved over both the minimum basis full valence CI and the near RHF values. An exception is seen for ω_e, for which the minimum basis result is in fortuitously close agreement with experiment.

Use of first-order wave functions has the advantage over the OVC approach that the most important configurations are determined by the calculation, rather than specified beforehand. In Table III-24 the seven most important configurations are seen for short, equilibrium, and longer bond distances. This table

Table III-22. Orbital occupancies included in the first-order wave function for O_2 (342). All $^3\Sigma_g^-$ configurations arising from each orbital occupancy are included in the calculations. $n\sigma_g$ refers to the $4\sigma_g$, $5\sigma_g$, ... orbitals and $m\pi_u$ to the $2\pi_u$, $3\pi_u$, ... orbitals.

Type excitation

$1\sigma_g^2\ 1\sigma_u^2\ 2\sigma_g^2\ 2\sigma_u^2\ 3\sigma_g^2\ 1\pi_u^4\ 1\pi_g^2$

$2\sigma_u \to 3\sigma_u$	$1\pi_u^2 \to 3\sigma_g^2$
$3\sigma_g \to n\sigma_g$	$1\pi_u^2 \to 1\pi_g^2$
$1\pi_u \to m\pi_u$	$3\sigma_g^2 \to 3\sigma_u\ n\sigma_u$
$1\pi_g \to m\pi_g$	$3\sigma_g^2 \to 1\pi_g\ m\pi_g$
$2\sigma_g^2 \to 3\sigma_u^2$	$3\sigma_g\ 1\pi_u \to 3\sigma_u\ m\pi_g$
$2\sigma_u^2 \to 3\sigma_u^2$	$3\sigma_g\ 1\pi_u \to 1\pi_g\ n\sigma_u$
$2\sigma_g\ 3\sigma_g \to 3\sigma_u^2$	$3\sigma_g\ 1\pi_g \to 3\sigma_u\ m\pi_u$
$2\sigma_g\ 1\pi_u \to 3\sigma_u\ 1\pi_g$	$1\pi_u^2 \to 3\sigma_u\ n\sigma_u$
$3\sigma_g^2 \to 3\sigma_u^2$	$1\pi_u^2 \to 1\pi_g\ m\pi_g$
$3\sigma_g\ 1\pi_u \to 3\sigma_u\ 1\pi_g$	$1\pi_u\ 1\pi_g \to 3\sigma_u\ n\sigma_g$
	$3\sigma_g^2\ 1\pi_u^2 \to 3\sigma_u^2\ 1\pi_g^2$

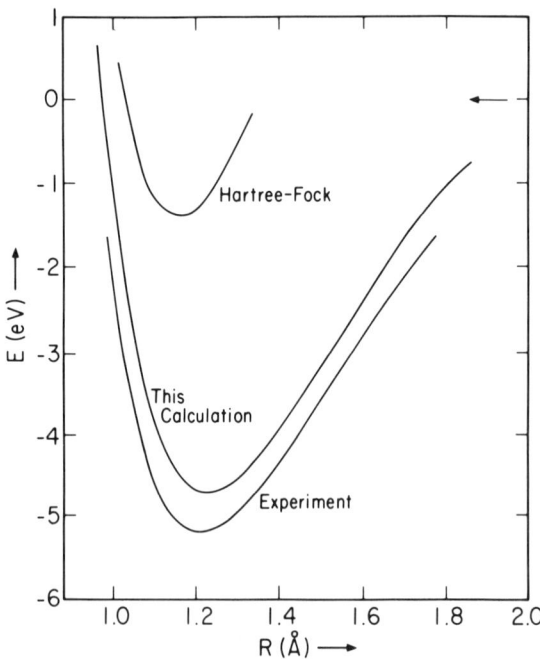

Figure III-6. Comparison between near Hartree-Fock, first-order, and experimental potential energy curves for the ground state of O_2.

---

provides quantitative information concerning the electronic structure of O_2. It can be shown that as $R \to \infty$, the $^3\Sigma_g^-$ O_2 wave function which describes two separate oxygen atoms in the SCF approximation is

$$\psi = \frac{1}{\sqrt{8}} 3\sigma_g^2 \, 1\pi_u^4 \, 1\pi_g^2 + \frac{1}{\sqrt{8}} 3\sigma_u^2 \, 1\pi_u^4 \, 1\pi_g^2 + \frac{1}{\sqrt{8}} 3\sigma_g^2 \, 1\pi_u^2 \, 1\pi_g^4$$

$$+ \frac{1}{\sqrt{8}} 3\sigma_u^2 \, 1\pi_u^2 \, 1\pi_g^4 + \frac{1}{\sqrt{2}} 3\sigma_g \, 3\sigma_u \, 1\pi_u^3 \, 1\pi_g^3 \qquad (III,31)$$

where the doubly occupied $1\sigma_g$, $1\sigma_u$, $2\sigma_g$, and $2\sigma_u$ orbitals have been omitted. And it is indeed seen that these five configurations, numbers 1,2,3,5, and 7

Table III-23. Theoretical and experimental spectroscopic constants for the ground state of O_2. The total energies given are at 2.3 bohrs.

Calculation	E(hartrees)	D_e(eV)	r_e(Å)	ω_e(cm^{-1})	$\omega_e x_e$(cm^{-1})	B_e(cm^{-1})	α_e(cm^{-1})
Full Valence CI, Minimum Basis	-149.2107	3.81	1.30	1582	14	1.25	0.0127
Near Hartree-Fock	-149.6639	1.43	1.152	2000	5.36	1.588	0.0126
First-order	-149.7944	4.72	1.220	1614	15.80	1.417	0.0169
Experiment	--	5.21	1.207	1580	12.07	1.446	0.0158

in Table III-24, all have large coefficients at 3.5 bohrs. This demonstrates rather clearly the breakdown of the single configuration approximation, since 3.5 bohrs is an intermediate (as opposed to long range) separation. It is worth noting that at R_e the fourth most important configuration for the open-shell ground state of O_2 is a single excitation. In addition, configuration six in Table III-24 is the same split-shell excitation found by Das and Wahl via trial and error to be so important for F_2. This $3\sigma_g 1\pi_g \rightarrow 3\sigma_u 2\pi_u$ excitation is included in the first-order wave function as a matter of course, since only a single electron occupies an orbital ($2\pi_u$) beyond the valence shell.

We conclude that the very difficult problem of accurately predicting dissociation energies can be solved for small molecules in a practical manner using either the OVC or first-order approaches.

Interaction potential between two helium atoms. The potential energy curve for the $^1\Sigma_g^+$ ground state of the He_2 molecule has been the subject of a great deal of experimental effort and considerable speculation. In particular, the repulsive or short-range part of the potential has been studied by Amdur and coworkers using high-velocity molecular beams (8). The long-range attraction between two helium atoms is a problem of particular interest since it represents the simplest case of a van der Waals interaction between two noble gas atoms. A number of experimental techniques, including most recently measurement of transport properties of dilute gases (196) and analysis of low-energy molecular beam scattering experiments (35), have been used in attempts to deduce the form of the van der Waals potential. Our discussion revolves around the fact that

Table III-24. Seven important configurations in the first-order wave function for the ground state of O_2. Results are given for three different internuclear separations, 2.3 bohrs being very close to R_e.

Excitations	Coefficient		
	R=1.8	R=2.3	R=3.5
1. $1\sigma_g^2 1\sigma_u^2 2\sigma_g^2 2\sigma_u^2 3\sigma_g^2 1\pi_u^4 1\pi_g^2$	0.9858	0.9719	0.8377
2. $3\sigma_g 1\pi_u \rightarrow 3\sigma_u 1\pi_g$	0.0734	0.1298	0.2977
3. $1\pi_u^2 \rightarrow 1\pi_g^2$	0.0704	0.1166	0.3187
4. $1\pi_u \rightarrow 2\pi_u$	0.0878	0.0910	0.0463
5. $3\sigma_g^2 \rightarrow 3\sigma_u^2$	0.0403	0.0729	0.2205
6. $3\sigma_g 1\pi_g \rightarrow 3\sigma_u 2\pi_u$	0.0442	0.0561	0.0373
7. $3\sigma_g^2 1\pi_u^2 \rightarrow 3\sigma_u^2 1\pi_g^2$	0.0080	0.0296	0.2233

electronic structure calculations have played a significant role in the search for the correct form of the He_2 potential curve.

An early theoretical calculation on the short-range part of the He-He potential was carried out by Phillipson (295). Phillipson used a double zeta plus polarization basis set (2s 1p) on each helium atom and carried out CI calculations including as many as 64 configurations. Phillipson's calculations were quite controversial because they predicted a potential <u>18 electron volts</u> more repulsive than the experimental curve of Amdur and coworkers (8) at 1/2 bohr internuclear separation. Since very few successful comparisons between <u>ab initio</u> calculations and experiment had been made at that time (1962), experimentalists were understandably skeptical of the Phillipson results.

A more accurate calculation on the repulsive He-He potential was carried out in 1966 by Matsumoto, Bender and Davidson (242). An extended basis set was used and the iterative natural orbital method employed to obtain optimized 50 configuration wave functions. The results of Matsumoto, et al. were in essential agreement with those of Phillipson. At 1/2 bohr separation, the total energy obtained was -2.81998 hartrees, which energy was estimated to lie less than 1 eV above the exact He_2 energy for that bond distance. The predicted repulsive potential was fit to a function of the form ae^{-br}, the result being $a = 237$ eV and $b = 4.23/\text{Å}$.

The significance of the Phillipson and Matsumoto calculations lies in the fact that the "experimental" results discussed above were found to be in error. In the most recent experimental work (188), the repulsive part of the He-He potential is found to be in essential agreement with the theoretical calculations. The experimental values of the potential parameters a and b are 196 eV and $4.21/\text{Å}$, in good agreement with those predicted by Matsumoto, et al. (242).

The reliable prediction of van der Waals forces from rigorous quantum mechanical calculations is a difficult task. For He-He, SCF calculations close to the Hartree-Fock limit have demonstrated that the RHF potential curve is strictly repulsive (206). In fact, there is a general feeling among theoreticians that the Hartree-Fock energy of a molecule composed of two closed-shell atoms is never lower than that of two Hartree-Fock atoms infinitely far apart (206). Thus the van der Waals attraction between two He atoms is due to electron correlation. The real difficulty arises from the fact that the van der Waals attractive well depth is only a small fraction of the total correlation energy of the molecule. For He-He, the well depth is ~ 10 °K or 0.00004 hartrees, while the correlation energy of the He atom is three orders of magnitude greater, 0.042 hartrees. So it would be quite possible to calculate 99% of the He$_2$ correlation energy, yet miss the van der Waals attraction completely. In August of 1970, two calculations (42,340) were reported in the literature which surmount the above difficulties to obtain attractive theoretical potential curves for He$_2$ in reasonable agreement with experimentally-based potentials. Although the two calculations are quite similar, the motivations which determined the form of the wave functions chosen were quite different.

The calculations of Schaefer, et al. (340) are based on pair correlation ideas and may be viewed as an outgrowth of earlier work by Nesbet (272) and by Kestner and Sinanoglu (205). The RHF wave function for He$_2$ is of the form $1\sigma_g^2 1\sigma_u^2$. A transformation to localized orbitals may be carried out using symmetry alone

$$1\sigma_A = \frac{1}{\sqrt{2}} [1\sigma_g + 1\sigma_u]$$

$$1\sigma_B = \frac{1}{\sqrt{2}} [1\sigma_g - 1\sigma_u]$$

(III,32)

The resulting wave function $1\sigma_A^2 1\sigma_B^2$ is identical to $1\sigma_g^2 1\sigma_u^2$ and the localized orbitals $1\sigma_A$ and $1\sigma_B$ are similar, but not identical, to $1s_A$ and $1s_B$ Hartree-Fock atomic orbitals. With the RHF wave function in localized form $1\sigma_A^2 1\sigma_B^2$ there are three symmetry-adapted pair correlation energies $e(1\sigma_A,1\sigma_A)$, $e(1\sigma_A,1\sigma_B)$, and $e(1\sigma_B,1\sigma_B)$. The first and last should correspond to the atomic correlation energies of atoms A and B and $e(1\sigma_A,1\sigma_B)$ to the interatomic correlation energy. The essence of the calculation of Schaefer, et al. (340) is to assume that $e(1\sigma_A,1\sigma_A)$ and $e(1\sigma_B,1\sigma_B)$ do not vary with internuclear separation. If this is true, then the He-He potential may be obtained by adding $e(1\sigma_A,1\sigma_B)$ to the Hartree-Fock potential curve.

The calculations of Bertoncini and Wahl (42) are of the optimized valence configurations (OVC) type, discussed in the previous subsection with respect to F_2. According to the OVC prescriptions, they choose configurations besides the RHF which have vanishing contributions at infinite separation but are important near R_e. For He_2, those configurations with vanishing coefficients at infinite separation are most readily determined after the SCF orbitals are localized, Eq. (III,32). In a localized picture, all excitations with respect to $1\sigma_A^2 1\sigma_B^2$ of the type $1\sigma_A 1\sigma_B \to x_i x_j$ have vanishing coefficients for infinite separation of the two atoms. The He-He potential curve of Bertoncini and Wahl is based on a five-configuration SCF wave function including the RHF configuration plus the orbital occupancies $1\sigma_A 1\sigma_B 2\sigma_A 2\sigma_B$ and $1\sigma_A 1\sigma_B 1\pi_A 1\pi_B$, both of which give rise to two linearly independent $^1\Sigma^+$ configurations.

Carried to their respective limits, the two approaches outlined above should in fact give identical results since they both attempt to calculate all the correlation energy present in the molecule but not in the separated atoms. In fact the former calculations predict a well depth of 12.0 °K at 2.96 Å separation and the latter a well depth of 10.5 °K for the predicted equilibrium internuclear separation 2.99 Å. The difference in well depth arises first of all from the fact that the basis set of Schaefer et al. includes a 4f Slater function on each He atom while that of Bertoncini and Wahl does not. A single (nine-configuration) calculation near R_e by Bertoncini and Wahl including f functions gives a depth of 11.4 °K. In addition, the pair correlation formulation implies that <u>all</u> configurations of the type $1\sigma_A 1\sigma_B \rightarrow x_i x_j$ should be included. For this reason, Schaefer et al. included 346 configurations of this type. The fact that the 9-configuration SCF result of Bertoncini and Wahl yields a well depth only 0.6 °K less than the 347 configuration result of Schaefer et al. may be taken as testimony of the efficacy of the multiconfiguration SCF approach. Schaefer and coworkers estimate that calculations of this type with a complete basis would lower the predicted well depth by ~0.1 °K.

Experimentally-based estimates of the He-He well depth vary from 9 to 12.5 °K, and the most reliable "experimental" values of the equilibrium internuclear separation are between 2.9 and 3.0 Å (35,196,240). Thus it appears that both theoretical potential curves, which attempt to isolate and neglect intra-atomic electron correlation, are in qualitative agreement with experiment. To assess the quantitative validity of this approach will require either more accurate experiments or more elaborate calculations. However, the general idea of neglecting atomic correlation effects is certainly appealing, and the methods described above have now been used to investigate van der Waals interactions in HeH and the polyatomic systems He + B $^1\Sigma_u^+$ H$_2$ and Li + H$_2$.

Table III-25 gives a breakdown of the different correlation effects contributing to the van der Waals attraction between two He atoms at 5.6 bohrs (2.963 Å) internuclear separation. The contribution of σ ($\lambda=0$) orbitals of all atomic symmetries (s,p,d, and f) is 13.4 °K, while that from all types of π orbitals is 7.7 °K. As suggested by the form of the Bertoncini-Wahl wave function, δ orbitals are unimportant. The sum of the σ, π, and δ contributions is 21.1 °K, which must be subtracted from 9.2 °K (by which the near Hartree-Fock potential is repulsive) to give the predicted well depth 11.9 °K. Adding the elements of the four rows of Table III-25 shows that s functions alone yield 0.2 °K of the interatomic correlation energy, while the addition of p, d, and f atomic orbitals to the basis yields 15.9 °K, 4.4 °K, and 0.6 °K, respectively.

<u>Positions of the electronic states of BeO</u>. For none of the alkaline earth oxides BeO, MgO, and CaO has the electronic ground state been determined experimentally. Although a number of singlet states of these molecules have been observed, to date no electronic transitions involving triplet states have been recorded. Beryllium oxide is a particularly suitable molecule for theoretical studies for two reasons, the most obvious of which is that it has only 12 electrons. The second reason is that the definitive experimental work done on BeO was completed prior to 1950 by Lagerqvist (220). The extreme toxicity of BeO vapor caused Lagerqvist to become very ill, and the poisonous nature of BeO remains an obstacle in the path of spectroscopic studies. A rough idea

Table III-25. Breakdown of the He-He correlation energy contributions in °K (340). In the first column s + p indicates that a basis of s and p orbitals on each He atom was used to describe correlation effects.

	σ	π	δ
s	0.2	--	--
s + p	10.7	5.4	--
s + p + d	13.1	7.4	0.02
s + p + d + f	13.4	7.7	0.03

of which electronic states should be low-lying may be obtained by consideration of the isoelectronic C_2 molecule. The experimental positions of the electronic states of C_2, as given by Ballik and Ramsay (18), are seen in Table III-26. Table III-26 shows that there are an unusually large number (5) of electronic states of C_2 lying below 2 eV. In the simplest picture, this proliferation of states is due to the 4σ, 5σ, and 1π orbitals being nearly degenerate. Except for the $^1\Pi$ state (which from Hund's first rule is expected to lie above the $^3\Pi$ state), any one of the five states in Table III-26 might be the electronic ground state of BeO. Lagerqvist has experimentally determined that the $^1\Pi$ state lies 1.17 eV above the $^1\Sigma^+$ state, giving evidence that the electronic states of BeO may be as closely spaced as those of C_2.

The experimental interest in the alkaline earth oxides is reflected in reported SCF calculations on BeO (179), MgO (313), and CaO (67). To some degree, all three of these studies imply that the RHF approximation is inadequate for reliable predictions of the positions of the respective electronic states. For BeO, the calculations of Huo, Freed, and Klemperer (179) are particularly persuasive along these lines. From near RHF wave functions they find the $^1\Pi$ state to lie 0.91 eV <u>below</u> the $^1\Sigma^+$ state, whereas in fact the $^1\Sigma^+$ state is the lower by 1.17 eV. This seems to establish rather clearly that reliable non-empirical predictions of the electronic spectra of the alkaline earth oxides require wave functions which explicitly include electron correlation.

Recently the positions of the electronic states of BeO have been predicted using first-order wave functions (343). It may be useful to remind the reader that the first-order wave function is just a particular type of CI which in general includes all configurations in which no more than one electron occupies an orbital beyond the valence shell (6σ and 2π for BeO). In these calculations a double zeta plus polarization basis of Slater functions is used. For the Be atom, this basis set requires some further comment, since the SCF orbital occupancy is $1s^2\ 2s^2$ and one might be inclined to think that a (4s 1p) set might be of double zeta plus polarization quality. This is not the case for Be since the 2p orbital is so nearly degenerate with the 2s that a reasonable basis must describe these two orbitals with equal precision. To do this, a double zeta basis (4s 2p) was used from an optimized calculation on the $1s^2\ 2s\ 2p\ ^3P$ excited state of the Be atom. The polarization function then was a 3d function on Be and the final basis set (4s 2p 1d) centered on both atoms. Following essentially the prescription given in the first-order O_2 calculation, one finds 157 configurations for the $^1\Sigma^+$ state, 291 configurations for $^3\Sigma^+$, and 591 for $^3\Pi$.

Table III-26. Experimental positions T_e of the low-lying electronic states of C_2 (18). For comparison with heteronuclear BeO, the assigned molecular orbitals do not reflect the g or u symmetry.

Electron configuration	State	T_e (eV)
$1\sigma^2\ 2\sigma^2\ 3\sigma^2\ 4\sigma^2\ 1\pi^4$	$^1\Sigma^+$	0.0
$1\sigma^2\ 2\sigma^2\ 3\sigma^2\ 4\sigma^2\ 5\sigma\ 1\pi^3$	$^3\Pi$	0.09
$1\sigma^2\ 2\sigma^2\ 3\sigma^2\ 4\sigma^2\ 5\sigma^2\ 1\pi^2$	$^3\Sigma^-$	0.80
$1\sigma^2\ 2\sigma^2\ 3\sigma^2\ 4\sigma^2\ 5\sigma\ 1\pi^3$	$^1\Pi$	1.04
$1\sigma^2\ 2\sigma^2\ 3\sigma^2\ 4\sigma\ 5\sigma\ 1\pi^4$	$^3\Sigma^+$	1.65

One finds in general that more configurations are required to describe open- than closed-shell molecules to a specific accuracy. In addition Π states require about twice as many configurations as Σ states since the + and - distinction is not possible for states with a non-zero eigenvalue of the axial component of the angular momentum operator.

Table III-27 and Figure III-7 summarize the BeO predictions. Perhaps the strongest evidence that the predictions are reliable comes from the comparison with experiment for the $^1\Sigma^+$ state. In particular the calculated dissociation energy D_e is 6.58 eV, while the experimental value is 6.7 ± 0.4 eV. The experimental uncertainty is due to the necessary extrapolation of the observed $^1\Sigma^+$ and $^1\Pi$ vibrational levels. As an additional comparison, the near Hartree-Fock D_e of Yoshimine is 4.13 eV (401).

Table III-27. Predicted and experimental (in parentheses) spectroscopic constants for $^9Be^{16}O$.

	T_e (eV)	R_e (Å)	ω_e (cm^{-1})	B_e (cm^{-1})
$^1\Sigma^+$	0.00	1.313 (1.331)	1629 (1487)	1.699 (1.651)
$^3\Pi$	0.73	1.463	1270	1.365
$^1\Pi$	(1.17)	(1.463)	(1144)	(1.366)
$^3\Sigma^+$	1.94	1.384	1234	1.527
$^3\Sigma^-$	Repulsive			

A result predicted for BeO qualitatively different from that observed for C_2 concerns the shape of the $^3\Sigma^-$ potential curve. Although this state is bound and lies at 0.80 eV in C_2, it is clearly repulsive for BeO. Actually, the $^3\Sigma^-$ potential curve seen in Figure III-7 is the result of an SCF calculation on the $1\sigma^2\ 2\sigma^2\ 3\sigma^2\ 4\sigma^2\ 5\sigma^2\ 1\pi^2$ orbital occupancy. However the RHF approximation is exceptional for this state in that the molecular RHF wave function dissociates properly to Be and O ground state atomic RHF wave functions. Limited consideration of other configurations, e.g., $1\sigma^2\ 2\sigma^2\ 3\sigma^2\ 4\sigma^2\ 1\pi^3\ 2\pi$, shows that in fact the $^3\Sigma^-$ potential curve is repulsive.

The importance of the results of Table III-27 is that they provide what should be a reliable <u>a priori</u> prediction of the positions of the electronic states of a molecule of importance prior to experiment. The $^1\Sigma^+$ state certainly

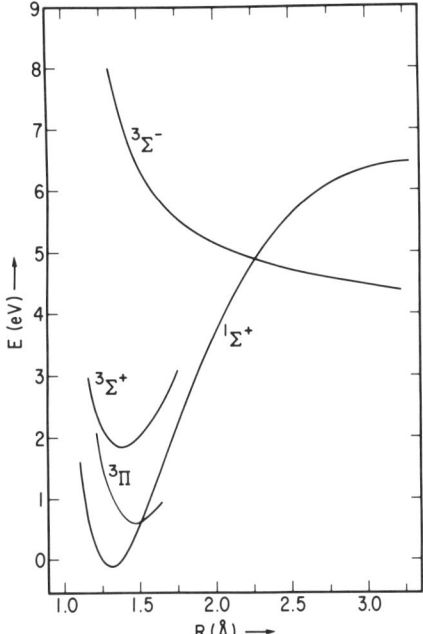

Figure III-7. Theoretical potential curves for the low-lying states of BeO.

appears to be the ground state with the $^1\Sigma^+$ - $^3\Pi$ separation much larger than in C_2. The $^3\Sigma^+$ state lies about the same amount above the ground state as in C_2 and experimental data on the $^1\Pi$ state is included for completeness. These results may also prove helpful in the interpretation of the spectra of the heavier alkaline earth oxides.

Existence and Properties of the NaLi Molecule. As early as 1928, the alkali-metal diatomic molecules Li_2, Na_2, K_2, Rb_2, Cs_2, LiK, LiRb, LiCs, NaK, NaRb, NaCs, KRb, KCs, and RbCs had all been observed experimentally (386). However, as of 1970 and despite several attempts, the NaLi molecule, which completes the above series, had not been observed. The spectroscopic identification of NaLi is complicated by the existence of Na_2 and Li_2 band systems in the regions of the spectrum where NaLi might be observed. The observed alkali diatomics all appear to be weakly bound (dissociation energy of the order of 1 eV) with relatively long bond distances and small force constants. A reliable theoretical study of NaLi would clearly be an important step toward a more complete understanding of the alkali diatomics.

Das, Bertoncini and Wahl have undertaken a systematic theoretical study of the alkali diatomics using the method of optimized valence configurations (95, 41). To date, work has been completed on Li_2, NaLi, and Na_2. The results for the experimentally characterized Li_2 and Na_2 molecules show that the NaLi predictions may be considered quite reliable. In all cases, extended basis sets (slightly larger than double-zeta-plus-polarization) were used.

For NaLi, one may anticipate the existence of five low-lying electronic states, identified with the following orbital occupancies:

$$1\sigma^2\ 2\sigma^2\ 3\sigma^2\ 4\sigma^2\ 5\sigma^2\ 1\pi^4 \qquad X\ {}^1\Sigma^+ \qquad (III,33)$$

$$1\sigma^2\ 2\sigma^2\ 3\sigma^2\ 4\sigma^2\ 5\sigma\ 6\sigma\ 1\pi^4 \qquad {}^3\Sigma^+, {}^1\Sigma^+ \qquad (III,34)$$

$$1\sigma^2\ 2\sigma^2\ 3\sigma^2\ 4\sigma^2\ 5\sigma\ 1\pi^4\ 2\pi \qquad {}^3\Pi, {}^1\Pi \qquad (III,35)$$

In the simplest ground state picture, the 1σ orbital is 1s (Na), 2σ is 2s (Na), 3σ is 1s (Li), 4σ is $2p\sigma$ (Na), 1π is $2p\pi$ (Na), and 5σ is the bonding orbital. Bertoncini, Das, and Wahl (41) have carried out calculations on four of the five electronic states, excluding the ${}^1\Sigma^+$ state arising from (III,34) since it is the second state of ${}^1\Sigma^+$ symmetry and thus presents some special problems.

The calculations on the X $^1\Sigma^+$ are 4-configuration SCF calculations within the OVC framework. The other three states are treated in the single configuration or RHF formulation, since the molecular RHF wave functions dissociate to RHF wave functions describing the correct states of the isolated atoms.

Table III-28 displays the OVC and (except for NaLi) experimental spectroscopic constants for several alkali diatomic ground states. For both Li$_2$ and NaLi, the single configuration SCF values of the bond distance R_e and vibrational frequency ω_e are not far from the OVC values. However, the SCF dissociation energy for Li$_2$ is only 0.17 eV and for NaLi even less, 0.05 eV. This shows once again the necessity of explicit inclusion of electron correlation in D_e predictions.

Figure III-8 gives the theoretical potential curves. Included in the figure is the calculated SCF potential curve for the $^2\Sigma^+$ ground state of NaLi$^+$, which arises from orbital occupancy $1\sigma^2\ 2\sigma^2\ 3\sigma^2\ 4\sigma^2\ 5\sigma\ 1\pi^4$. It is of particular interest to note that the SCF dissociation energy of NaLi$^+$ is 0.92 eV, greater than that obtained by the more extensive OVC calculation on X $^1\Sigma^+$ NaLi. The analogous phenomenon has been observed experimentally for other alkali-metal diatomics, e.g., the dissociation energy of Na$_2$ is 0.75 ± 0.03 eV while that of Na$_2^+$ is 1.02 ± 0.04 eV (132). In the simplest empirical MO picture this result is not expected since NaLi may be said to have an "electron pair ($5\sigma^2$) bond," while NaLi$^+$ has only a "one-electron (5σ) bond." In simple pictures of valence theory electron-pair bonds are expected to be stronger than one-electron bonds. It is clear both from experiment (132) and the theoretical dissociation energies of NaLi and NaLi$^+$ (41) that the above simple argument is not valid.

Table III-28. Spectroscopic constants for the lowest $^1\Sigma^+$ states of Li_2, NaLi, and Na_2 from multiconfiguration SCF wave functions (41,95). Experimental values are in parentheses.

	D_e(eV)	R_e(bohrs)	ω_e(cm^{-1})
Li_2	0.99 (1.03)	5.089 (5.051)	345 (351)
NaLi	0.85	5.548	249
Na_2	0.72 (0.76)	5.931 (5.818)	156 (159)

Table III-29 shows the four configurations included in the OVC wave function for the ground state of NaLi. All three configurations in addition to the SCF involve replacement of the $5\sigma^2$ pair. Put in another way, the calculations show that for the alkali-metal diatomics, nearly quantitative potential curves may be obtained by correlating only the electron-pair bond. For short internuclear separations, the fourth configuration in Table III-29, $5\sigma^2 \to 2\pi^2$, becomes very important. This is understandable, since the united atom limit (zero internuclear separation) is the ^1S state of the silicon atom, for which the SCF configuration $1s^2\ 2s^2\ 2p^6\ 3s^2\ 3p^2$ is closely related to the fourth configuration in Table III-29. For large separation the second configurations, $5\sigma^2 \to 6\sigma^2$, becomes very important. This is because the linear combination $(1/\sqrt{2}\ 5\sigma^2 - 1/\sqrt{2}\ 6\sigma^2)$ is required to describe a 2s orbital on Li and a 3s orbital on Na at infinite separation.

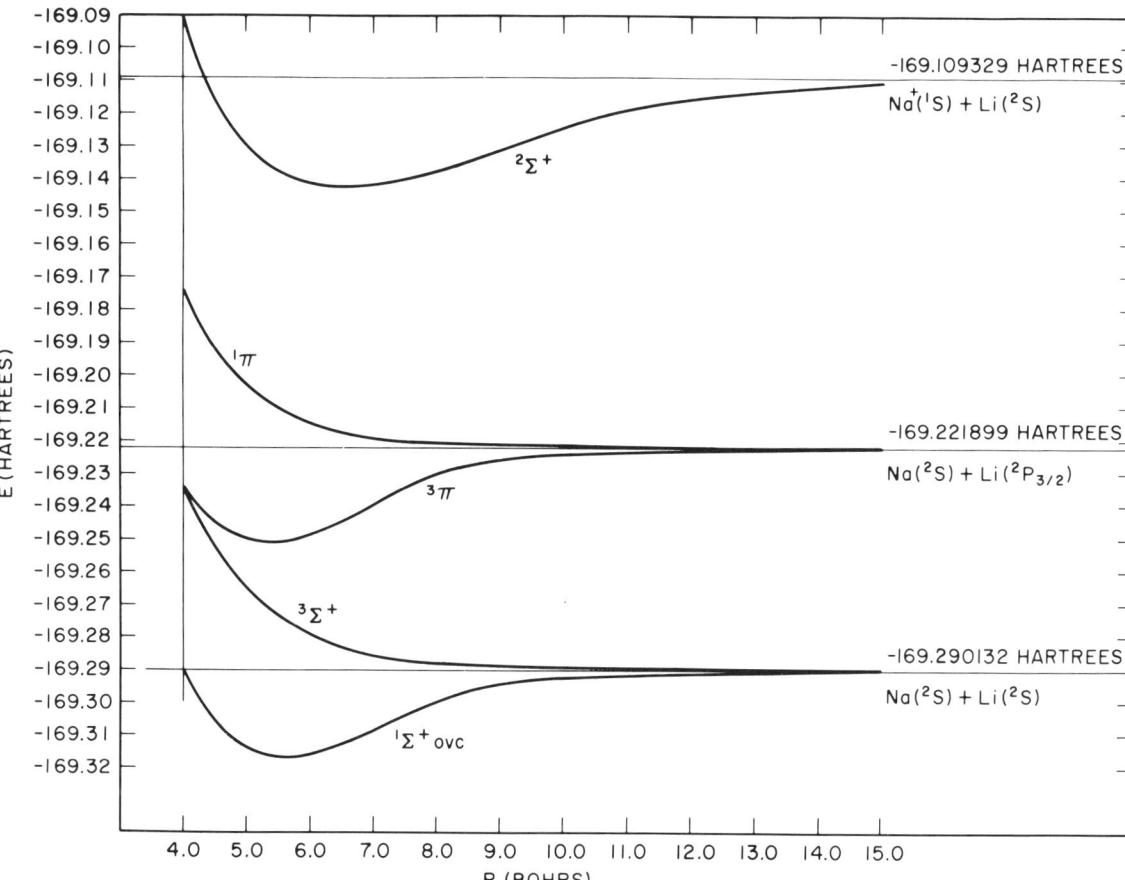

Figure III-8. Predicted potential curves for four states of NaLi and the ground state of NaLi$^+$.

There is an interesting postscript to the NaLi calculations of Bertoncini, Das, and Wahl. The spectrum of NaLi was produced by laser excitation by Hessel (164). Hessel found ω_e to be 248-250 cm^{-1} and B_e to be 0.361 cm^{-1}, compared to the theoretical values 257 and 0.396 cm^{-1}. This type of interplay between theory and experiment seems nearly ideal: a) experiments suggest a problem, b) a theoretical study is carried out, and c) the theoretical predictions are verified by experiment.

Larger diatomics: KrF and KrF$^+$. Among five or six experimentally observed molecules containing krypton are two diatomics, KrF and KrF$^+$. In 1964 Falconer, Morton, and Streng (119) reported what appeared to be the ESR spectrum of KrF in crystals of KrF$_4$ subjected to γ radiation. Since KrF$_4$ does not appear to be stable, a more likely interpretation of their experiment would involve KrF in crystals of KrF$_2$ (22). The evidence for the existence of KrF$^+$ seems indisputable, and Berkowitz and Chupka have measured the dissociation energy D_0 to be ≥ 1.58 eV (36). Liu and Schaefer (229) have reported calculations including electron correlation for both KrF and KrF$^+$ and this work is of interest a) because of its relation to the experiments mentioned above, and b) because a theoretical method capable of treating electron correlation in large molecules is demonstrated.

Two different extended basis sets were used in the calculations on KrF and KrF$^+$. The first of these, Kr (8s 6p 5d 2f) and F(4s 3p 2d 1f), was designed to yield molecular energies differing from complete basis results only by the atomic SCF errors in the basis (0.110 hartrees for Kr and 0.0012 hartrees for F).

Table III-29. Form of the MCSCF wave function for NaLi in the repulsive (3.5 bohrs), near R_e (5.5 bohrs) and long-range (15 bohrs) regions of internuclear separation (41).

Configuration	Coefficients		
	R = 3.5	R = 5.5	R = 15.0
1. $1\sigma^2\ 2\sigma^2\ 3\sigma^2\ 4\sigma^2\ 5\sigma^2\ 1\pi^4$	0.9392	0.9511	0.7277
2. $1\sigma^2\ 2\sigma^2\ 3\sigma^2\ 4\sigma^2\ 6\sigma^2\ 1\pi^4$	-0.1371	-0.1571	-0.6856
3. $1\sigma^2\ 2\sigma^2\ 3\sigma^2\ 4\sigma^2\ 7\sigma^2\ 1\pi^4$	0.0674	0.1145	0.0163
4. $1\sigma^2\ 2\sigma^2\ 3\sigma^2\ 4\sigma^2\ 1\pi^4\ 2\pi^2$	0.3075	0.2402	0.0129

This first basis was used both for SCF and CI calculations. The second basis, Kr(10s 9p 7d 2f) and F(6s 4p 3d 2f), was expected to yield SCF energies and properties very close to the RHF limit. The primary purpose of the second basis set was to determine whether in fact the smaller first basis performed as desired at the SCF level. For KrF, the SCF potential curves obtained with the two basis sets were parallel to within 0.1 eV. For the ground state of KrF$^+$ the two SCF curves were parallel to within 0.17 eV. Other properties also appear to be reliably predicted by the smaller basis. For example the smaller basis yields SCF dipole and quadrupole moments 0.103 and 2.47 atomic units while the larger basis predicts 0.106 and 2.37. These comparisons are important since it was not feasible to perform CI calculations with the larger set, which consists of 43 σ, 27 π, and 14 δ functions.

Electron correlation was included in the work of Liu and Schaefer (229) using first-order wave functions. For KrF, the RHF $^2\Sigma^+$ ground state orbital occupancy is

$$1\sigma^2\ 2\sigma^2\ 3\sigma^2\ 4\sigma^2\ 5\sigma^2\ 6\sigma^2\ 7\sigma^2\ 8\sigma^2\ 9\sigma^2\ 10\sigma^2\ 11\sigma\ 1\pi^4\ 2\pi^4\ 3\pi^4\ 4\pi^4\ 5\pi^4\ 1\delta^4 \quad (III,36)$$

Even using as selective a method for choosing configurations as the first-order approach, it would not be a good idea to try to correlate all 45 electrons of KrF. What was in fact done was to correlate only the valence orbitals of KrF. In an atomic picture, these are the Kr 4p and F 2p orbitals. For the molecule, the related orbitals are 10σ, 11σ, 4π and 5π. In this way correlation involving only the 11 outermost electrons is considered. The remaining 34 electrons always occupy the 1σ -9σ, 1π - 3π, and 1δ orbitals. With these restrictions and the smaller basis set, there are 158 configurations in the first-order wave function. The orbital occupancy for the lowest $^1\Sigma^+$ state of KrF$^+$ differs from (III,35) only in the absense of the 11σ orbital, and the corresponding first-order wave function includes 253 configurations.

The results for KrF are surprising in that completely repulsive potential curves are obtained by both SCF and CI methods. Since the Kr-F bond distance in KrF$_2$ is ~3.6 bohrs, if KrF has a chemical bond, one would expect an attractive potential in this region. However even at 4 bohrs the SCF and CI potential curves are repulsive by 1.1 and 0.7 eV, respectively. Excluding the van der Waals attraction (which is probably (38) of the order of 0.01 eV for KrF), the authors conclude that KrF does not exist in the gas phase. This result appears to be in direct conflict with the ESR measurements of Falconer, et al. (119). However a possibility which the theoretical calculations do not preclude is that the KrF radical is stabilized in the KrF_2 crystal structure.

For KrF^+ the SCF (large basis) and CI (smaller basis) results are seen in Table III-30. The predicted dissociation energy D_0 is 1.90 eV, in qualitative agreement with the experimental value, $\geq$ 1.58 eV, of Berkowitz and Chupka (36). After the calculations were completed, it was learned that McKee and Bartlett had synthesized and measured the Raman spectrum of the compound $KrSbF_{12}$ (243). From the crystal structure of the corresponding xenon compound and the similarity of the two Raman spectra, the structure of $KrSbF_{12}$ is almost certainly $KrF^+SbF_{11}^-$. The Raman spectrum showed the Kr-F stretch to occur at 626 cm^{-1}, in very close agreement with the CI value in Table III-30, 621 cm^{-1}. This agreement must be considered in part fortuitous since a) the first-order ω_e is only reliable to ~10% or 60 cm^{-1}, and b) one would expect the gas phase KrF^+ stretching frequency to be somewhat different from that observed in a crystal of $KrF^+ SbF_{11}^-$. Nevertheless, the agreement with experiment for both D_0 and ω_e would seem to suggest that nonempirical calculations of semi-quantitative reliability are now possible for large diatomic molecules.

Table III-30. Predicted properties of the KrF$^+$ ion (229). Unless indicated, properties are in atomic units.

Property	Near Hartree-Fock	First-Order
Energy	-2850.9391	-2850.9743
D_e (eV)	-0.02	1.94
R_e (Å)	1.680	1.752
ω_e (cm^{-1})	810	621
$\omega_e x_e$ (cm^{-1})	6.3	8.3
B_e (cm^{-1})	0.386	0.355
α_e (cm^{-1})	0.0029	0.0044
Dipole moment at Kr	0.106	0.201
Quadrupole moment at Kr	2.37	2.73
Field gradient at Kr	-17.03	-16.17
Field gradient at F	-6.20	-5.75
$< 1/r_{Kr} >$	185.1648	185.1661
$< 1/r_F >$	37.2692	37.2604
$< r^2_{Kr} >$	136.4207	136.8786
$< r^2_F >$	411.7337	412.8075

IV
LINEAR POLYATOMIC MOLECULES

As may be seen in the recent book by Herzberg (162), spectroscopists tend to place linear polyatomic molecules in a distinct class. This is because the spectra of polyatomic linear molecules are likely to be intermediate in complexity between those of diatomic molecules and polyatomic molecules of general geometry. In an analogous manner, theoreticians have found that linear molecules are somewhat more difficult to study via rigorous electronic structure calculations than are diatomic molecules. However, it is also true that _ab initio_ calculations on a particular polyatomic molecule can be much easier for linear than for nonlinear geometries. And it is often the case that only linear geometries need be considered to solve a particular chemical problem. For example, many atom plus diatom $A + BC \rightarrow AB + C$ reactions are found to have reaction coordinates contained in the linear part of the respective three-dimensional potential surfaces. In order to predict the barrier height (which is related to the activation energy), only linear geometries need be considered.

Electronic structure calculations on linear polyatomic molecules differ from diatomic calculations in only one way. Namely, three- and four-center electron-repulsion or two-electron integrals must be evaluated. Four-center integrals $(ij|k\ell)$, where i,j,k and ℓ are functions centered on four different nuclei, are the most difficult to evaluate. Over the past fifteen years, McLean and coworkers (245,251,253) have developed very sophisticated and efficient methods for the evaluation of that restricted class of three- and four-center integrals (involving Slater functions) which occurs in linear molecule calculations. Using these methods it is usually possible to compute all the integrals required for a linear molecule calculation in a time within a factor of two of that required for a diatomic calculation with the same number of Slater functions.

The evolution of McLean's computer programs is interesting to follow since at each of three stages these programs have typified the state of the art. McLean's first program was written for the Univac 1103 computer and completed in 1959, at which time minimum basis set SCF results were reported for acetylene and carbon dioxide (245). A second program, written for the IBM 7094 computer by McLean and Yoshimine, was completed in 1966 (251). SCF calculations with <u>extended</u> basis sets for molecules as large as cyanogen (C_2N_2 with 26 electrons) were carried out using this second program. A third program was written for the IBM 360/91 computer by Liu, Bagus, Yoshimine and McLean and completed in 1970 (253). With this most recent program, extended basis calculations are currently feasible for linear molecules as large as xenon difluoride (63 electrons).

It should be pointed out that when extended basis sets are used, the solution of the SCF equations sometimes requires more computation time than the evaluation of two-electron integrals. And if large scale CI is undertaken, the computation of the Hamiltonian matrix elements (I, 51) between configurations is usually the most time-consuming step in the entire calculation. Both of the above processes are significantly simpler for linear than for nonlinear polyatomic molecules. This is in large part due to the fact that, for a basis set of given size, there are usually many fewer distinct nonzero two-electron integrals for a linear system.

Suppose a basis set of 8s, 6p, 6d, and 4f functions is employed in a polyatomic calculation. If the molecule is linear, there are 24σ, 16π, 10δ, and 4ϕ orbitals, yielding 434,502 distinct nonzero integrals $(ij|k\ell)$. However, suppose the molecule is nonlinear and has no symmetry (point group C_1). Then there will be 84 orbitals, all of a (the a here is the irreducible representation) symmetry and 6,374,235 distinct nonzero two-electron integrals. Since the duration of both the SCF and CI steps discussed at the beginning of this paragraph is proportional to the number of integrals which must be manipulated, the linear calculation would be much less time-consuming than the comparable nonlinear calculation. In addition the number of configurations to be included for a linear calculation is usually substantially less than that required for the comparable nonlinear calculation. Let us use the same basis set and, for simplicity, consider full CI for a two-electron closed-shell system. For the linear case four types of orbital occupancies ($\sigma_i \sigma_j$, $\pi_i \pi_j$, $\delta_i \delta_j$, and $\phi_i \phi_j$) can give rise to configurations of $^1\Sigma^+$ symmetry, and the full CI includes 501 $^1\Sigma^+$ configurations. For C_1 symmetry the orbital occupancies are of a single type $a_i a_j$ and there are 3570 ^1A configurations in the full CI. Thus the size of the CI problem to be considered in the linear case is less than one-seventh that needed for the nonlinear case.

Due to the simplifications described above, a considerable number of polyatomic linear molecule studies have been carried out which, without the assumed linearity, would have been either very difficult or not possible at all. This chapter discusses the implications of a number of these electronic structure calculations for linear molecules.

A. SINGLE CONFIGURATION WAVE FUNCTIONS

<u>Dipole moments and dissociation energies</u>. McLean and Yoshimine (250) have carried out a systematic study of dipole moments and dissociation energies calculated within the restricted Hartree-Fock formalism for closed-shell ground states of linear molecules. This work is important first because it allows a serious evaluation of the RHF approximation for these two properties. Secondly well-calibrated predictions are made for a number of molecules for which experimental values are not available.

The McLean-Yoshimine calculations are unusual in that large basis sets are used throughout. Even on the largest molecules considered, for example NCCCH, a double-zeta-plus-polarization basis is used. This means that a (4s 2p 1d) basis of Slater functions is centered on N and each of the three C atoms. For hydrogen a (2s 1p 1d) basis is used. For the smaller molecules, for example CO_2, extended basis sets are used. For CO_2 a (5s 4p 1d 1f) set is centered on each atom. By evaluating the effects of additional basis functions, McLean and Yoshimine are able to estimate the energy difference between a particular SCF result and the RHF limit (251). For CO_2 the SCF total energy from the double-zeta-plus-polarization calculation is estimated to lie 0.021

hartree above the RHF energy, while the estimated deviation for the extended basis SCF energy is only 0.002 hartree.

SCF and experimental dipole moments are seen in Table IV-1. Where both double-zeta-plus-polarization and extended basis set results are available it is seen that the two calculated dipole moments differ by at most 0.12 debye. This is encouraging since the use of an extended basis set may be prohibitively expensive for large molecules. For HCN, FCN, and ClCN, SCF dipole moments are in fairly good agreement with experiment. It seems likely that the dipole moments of XCN molecules, where X is a halide atom, are well represented in the RHF approximation. This behavior is reminiscent of that seen in Section IIIA for the alkali halides.

Although dipole moments for the substituted acetylenes FCCH and ClCCH may appear to differ in sign from the experimental values, this is not necessarily so. The sign of the dipole moment is usually not determined experimentally. Since near RHF dipole moments have an expected reliability of about 1 debye, the calculations in some cases may be considered to determine the sign. Such a case is the NCCCH molecule, for which there is no possibility of the SCF and experimental values differing by 7.7 debyes. With this in mind, it appears that the predicted dipole moment of FCCH is qualitatively correct and for ClCCH quantitative agreement is obtained. The fact that the dipole moment of N_2O differs from experiment by a factor of three is <u>not</u> surprising since the absolute error is only 0.46 debye, about as good as can be expected.

Table IV-2 compares SCF and experimental dissociation energies. As was the case for dipole moments, the extended basis set results are in essential

Table IV-1. Electric dipole moments (in debyes) of linear molecules predicted from self-consistent-field calculations with different basis sets (250). A positive entry means + - polarity for the molecule as written. In general the sign of the experimental dipole moment is undetermined.

Molecule	Basis Set		Experiment
	Double Zeta Plus Polarization	Extended	
HCN	3.20	3.29	2.95
NNO	0.52	0.64	0.18
OCN⁻ [a]	7.25	7.22	
FCN	2.24	2.28	2.17
SCO	0.99		0.71
SCN⁻ [a]	6.93		
ClCN	3.05		2.80
FCCH	−0.85	−0.91	0.73
ClCCH	−0.42		0.44
NCCCH	−4.13		3.6 ± 0.2

[a] The dipole moment of an ion depends on the origin with respect to which it is calculated. The calculated dipole moments of OCN⁻ are with respect to oxygen nucleus as origin while that for SCN⁻ is for sulfur nucleus as origin.

agreement with the double-zeta-plus-polarization values. However, for dissociation energies, even the near Hartree-Fock results are in uniformly poor agreement with experiment. For N_2O the calculated dissociation energy is only 34% of experiment. Table IV-2 shows clearly that <u>a priori</u> predictions of dissociation energies are likely to be of little value unless electron correlation is explicitly included.

<u>Direct calculation of molecular polarizabilities</u>. The polarizabilities of a molecule represent the response of the molecule to an applied electric field. The role of polarizabilities in perturbation theories of intermolecular forces is well known (166). The development of lasers and the field of nonlinear optics has made possible the observation of double-quantum light scattering by molecules (40). These experiments may be interpreted in terms of the hyperpolarizability tensor of the molecule in question. The reliable prediction of polarizabilities is a particularly appropriate goal for the theoretician, since a relatively small number of polarizabilities have been experimentally determined with precision.

In the presence of an electric field directed along the molecular axis, the total dipole moment m_z (z is molecular axis) of a linear molecule may be written (55)

$$m_z = \mu_z + \alpha_{zz}F_z + \frac{1}{2}A_{z:zz}F_{zz}' + \frac{1}{2}\beta_{zzz}F_z^2 + \frac{1}{2}B_{zz:zz}F_zF_{zz}'$$
$$+ \frac{1}{2}\gamma_{zzzz}F_z^3 + \ldots \qquad (\text{IV},1)$$

In this equation, μ_z is the permanent dipole moment, F_z the electric field, and F_{zz}' the electric field gradient. α_{zz} is the (dipole) polarizability,

Table IV-2. Dissociation energies (in eV) from SCF calculations on $^1\Sigma^+$ ground states of linear molecules (250). Each calculated dissociation energy is obtained by subtracting the molecular SCF energy from the sum of the atomic SCF energies obtained with the same basis.

Molecule	Basis Set		Experiment
	Double Zeta Plus Polarization	Extended	
HCN	8.82	8.85	13.55 ± 0.05
CO_2	11.12	11.30	16.85
NNO	3.93	3.97	11.72
FCN	7.65	7.66	
SCO	9.28		14.4 ± 0.2
ClCN	7.25		
FCCH	11.88	11.91	
ClCCH	11.43		
NCCCH	16.88		

β_{zzz} the (first) hyperpolarizability, and γ_{zzzz} the second hyperpolarizability. Similarly the quadrupole moment T_{zz} in an electric field is

$$T_{zz} = \theta_{zz} + A_{z:zz}F_z + \frac{3}{2}C_{zz:zz}F_{zz}' + \frac{1}{2}B_{zz:zz}F_z^2 + \ldots \quad (IV,2)$$

in which θ_{zz} is the permanent quadrupole moment of the molecule. A, B, and C are usually referred to as quadrupole polarizabilities.

There are a number of theoretical formulations in which it is possible to calculate polarizabilities. One is the coupled Hartree-Fock method, described briefly in our discussion of paramagnetism in BH. Another method, recently applied with some success by Liebmann and Moskowitz (227), is the uncoupled Hartree-Fock approach of Langhoff, Karplus and Hurst (221). The approach used by McLean and Yoshimine (248) is more straightforward than either of the above two methods. McLean and Yoshimine simply solve the self-consistent-field equations including the electric field in the Hamiltonian. The field is generated by fixing point charges on the molecular axis. This must be done in such a manner that the molecular wave function is negligible at the positions of the charges. Molecular dipole and quadrupole moments are computed for several different external fields. These moments are then combined with Equations (IV,1) and (IV,2) to provide two sets of linear equations, which are solved to give the predicted polarizabilities. A check on the reliability of this method is given by the fact that $A_{z:zz}$ and $B_{zz:zz}$ can be obtained from either the set of equations describing the total dipole moment m_z or the set describing T_{zz}.

Table IV-3 shows a comparison of polarizabilities obtained using different basis sets for CO and N_2O. It should be mentioned that all the basis sets in Table IV-3 are relatively large and one would normally expect calculated properties to differ little from one basis to the next. For example, the SCF quadrupole moment θ_{zz} of CO is calculated to be -1.665, -1.593, and -1.634 using the three basis sets in Table IV-3. The lead terms α_{zz} and $A_{z:zz}$ in the expansion of the dipole (IV,1) and quadrupole (IV,2) moment expansions are

Table IV-3. Calculated polarizabilities as a function of basis set for CO and N_2O. All values are in atomic units.

Basis Set	CO			N_2O	
	(4s 2p 1d 1f)	(4s 3p 1d 1f)	(5s 4p 1d 1f)	(4s 2p 1d 1f)	(4s 3p 1d 1f)
SCF Energy	-112.7790	-112.7860	-112.7891	-183.7444	-183.7567
α_{zz}	11.28	11.88	14.24	24.60	27.46
β_{zzz}	13	10	22	32	42
γ_{zzzz}	-3	-9	200	-100	-500
$A_{z:zz}$	-6.6	-8.9	-10.8	-9.6	-10.7
$B_{zz:zz}$	-31	-28	-134	-41	-100
$\frac{3}{2} C_{zz:zz}$	33	31	49	77	91

the least sensitive to basis set. The hyperpolarizability β_{zzz} and the term $C_{zz:zz}$ are stable to a factor of 2 in going from the double-zeta-plus-polarization to the extended basis sets. However, the method and/or basis sets used do not appear to yield values of γ_{zzzz} and $B_{zz:zz}$ which are reliable. Nevertheless, this should be considered a relatively minor failing of the approach. As pointed out by McLean and Yoshimine (248), γ_{zzzz} and $B_{zz:zz}$ give rise to induced effects which, for fields and field gradients typically present in polar liquids, are at least an order of magnitude less than those induced by α_{zz}, β_{zzz}, $A_{z:zz}$, and $C_{zz:zz}$.

Any evaluation of theoretical methods for the prediction of polarizabilities is made difficult by the paucity of reliable experimental data. For the simplest molecule, H2, a comparison is possible with the nearly exact theoretical treatment of Kolos and Wolniewicz (210). The McLean-Yoshimine H2 polarizability α_{zz} is 6.55 atomic units, in close agreement with the accurate theoretical value, 6.38. As mentioned in the preface, the methods of Kolos and Wolniewicz are not readily extended to molecules with more than two electrons. Hirschfelder, Curtiss, and Bird recommend values of 17.5 and 32.8 atomic units for the polarizabilities α_{zz} of CO and N_2O (166). The McLean and Yoshimine values, 14.24 and 27.46, are 81% and 84% of "experiment" for these two molecules. If the values recommended by Hirschfelder, Curtiss, and Bird are correct, the above comparisons imply that the method of McLean and Yoshimine is capable of qualitatively correct α_{zz} predictions. It is not possible to evaluate the reliability of the higher polarizabilities, since no experimental values are available.

Table IV-4 shows additional polarizabilities calculated by McLean and Yoshimine. Values of γ_{zzzz} and $B_{zz:zz}$ are not reported since, as noted, these higher polarizabilities do not appear stable with respect to basis set. For each molecule the polarizabilities given in Table IV-4 were obtained with the largest basis set used by McLean and Yoshimine (251). For HF, HCl, and HCN, the predicted values agree nicely with the experimental values given by Hirschfelder, Curtiss, and Bird (166). Taken with the earlier comparison for H_2, N_2O, and CO, there is considerable reason for optimism concerning the qualitative reliability of the predictions in Table IV-4.

There are a number of chemically reasonable predictions seen in Table

Table IV-4. Predicted polarizabilities of linear molecules. These results of McLean and Yoshimine are unpublished but have been reported in the compendium of Krauss (216). Values in parentheses are the recommended "experimental" values (166).

Molecule	Polarizability $\alpha_{zz}(10^{-25} cm^3)$	Hyperpolarizability $\beta_{zzz}(10^{-33} cm^5/esu)$	$A_{z:zz}(10^{-34} cm^4)$	$\frac{3}{2} C_{zz:zz}(10^{-42} cm^5)$
HF	8.1 (9.6)	-74	32	34
HCl	23.4 (31.3)	-288	137	167
BF	23.6	57	-330	513
AlF	45.0	-800	-660	1060
SiO	39.6	350	-49	580
PN	49.7	150	215	506
OCN⁻	41.1	-100	170	590
SCN⁻	73.5		280	1140
HCN	33.2 (39.2)	200	-94	360
FCN	32.5	-100	267	515
ClCN	53.5	-140	378	978
FCCH	46.4	-800	610	1070
ClCCH	76.1	-1700	1020	2070

IV-4. The polarizability α_{zz} of AlF is about twice that of BF, and that of SiO is about twice the calculated value for CO. In the same way SCN^- is almost twice as polarizable as OCN^-, ClCN is more polarizable than FCN, and chlorinated acetylene is predicted to be more polarizable than fluorinated acetylene. In general, a molecule containing a second-row atom will be more polarizable than the molecule in which the second-row atom has been replaced by the first-row atom in the same column of the periodic table. The isoelectronic molecules AlF, SiO, and PN are seen to have about the same polarizability α_{zz}. The calculated higher polarizabilities are more difficult to classify according to chemically reasonable trends. However, some correlations are again possible; for example, the values of β_{zzz}, $A_{z:zz}$, and $C_{zz:zz}$ for ClCCH are all just about twice the FCCH values.

Since we will not return to the subject of polarizabilities in chapter V, a very brief discussion of the results of Liebmann and Moskowitz (227) for H_2O, CH_4, CO, and H_2CO is given here.
These calculations employ extended (but significantly less complete than those of McLean and Yoshimine) gaussian basis sets and the uncoupled Hartree-Fock formulation of Langhoff, et al. (221). The results are seen in Table IV-5. To comprehend Table IV-5 it must be understood that the polarizability $\underline{\alpha}$ and hyperpolarizability $\underline{\beta}$ are second and third rank tensors. This was not mentioned in the discussion of the linear molecule results since McLean and Yoshimine report only those polarizability components parallel to the molecular axis.

For CO it is possible to compare the results of Liebmann and Moskowitz with those of McLean and Yoshimine for α_{zz} and β_{zzz}. In the units of Table

IV-5, Liebmann and Moskowitz find 32.6 and -1213 for α_{zz} and β_{zzz}, while McLean and Yoshimine report 21.1 and -193. Both values of α_{zz} agree qualitatively with the recommended value 26.0, but the hyperpolarizabilities differ by a factor of 6. For H_2O, there is an experimental value 14.5 of the rotationally averaged quantity $<\alpha>$

$$<\alpha> = \frac{1}{3}(\alpha_{xx} + \alpha_{yy} + \alpha_{zz}) \qquad (IV,3)$$

This value agrees closely with the value 14.43 predicted by Liebmann and Moskowitz. In addition Table IV-5 shows that the calculated methane polarizability appears to be qualitatively correct. To conclude, these calculations (227) provide further evidence that it is possible to predict qualitatively reliable polarizabilities α, but raise questions concerning the reliability of predicted hyperpolarizabilities β.

Predicted properties of LiOH and Li_2O. Two interesting molecules about which little is known experimentally are lithium hydroxide and dilithium monoxide. Lithium hydroxide is of particular interest since it is the simplest ionic polyatomic molecule. However, not even the geometry of the gaseous LiOH molecule has been determined experimentally. Nonempirical SCF calculations by Buenker and Peyerimhoff predict LiOH to be linear (56). Li_2O has been observed in the gas phase and found to have a linear LiOLi structure with Li-O bond distance 1.54 Å (54) or 1.58 Å (393). LiOH and Li_2O thus form a reasonable starting point for theoretical studies of the properties of molecules containing lithium.

Table IV-5. Polarizabilities α and hyperpolarizabilities β for H_2O, CH_4, CO, and H_2CO (227). "Experimental" values are in parentheses. Values of α are in 10^{-25} cm^3, and those of β in 10^{-33} cm^5/esu.

	α_{xx}	α_{yy}	α_{zz}	β_{xxz}	β_{yyz}	β_{zzz}	β_{xyz}
H_2O	12.3	16.5	14.5	27	123	152	--
CH_4	31.8	31.8	31.8	--	--	--	189
	(26.0)	(26.0)	(26.0)				
CO	20.1	20.1	32.6	125	--	-1213	--
	(16.25)	(16.25)	(26.0)				
H_2CO	23.1	27.1	45.5	60	138	-1676	--

The calculations discussed here were undertaken by McLean and Seung (252) as an illustrative example for a seminar series on the computation of atomic and molecular structures. This course was given in autumn of 1968 by Clementi, McLean and Nesbet in the chemistry department of the University of California at Berkeley. One consequence of the instructive purpose of the calculations was that the LiOH calculations represent the most exhaustive comparison to date of the effect of different basis sets on predicted properties. Both molecules are small enough that the SCF calculations using the largest basis sets should be very close to the Hartree-Fock limit. Since the results of these calculations (252) have never been published, it seems particularly appropriate to discuss the work herein.

In the present discussion of lithium hydroxide, we refer to SCF results obtained with the following 9 basis sets:

a) Li(2s), O(2s 1p), H(1s). Minimum basis set of Slater functions with exponents chosen by Slater's rules (361). These exponents are $1s(Li) = 2.7$, $2s(Li) = 0.65$, $1s(O) = 7.7$, $2s(O) = 2p(O) = 2.275$, $1s(H) = 1.0$.

b) Li(2s), O(2s 1p), H(1s). Minimum basis optimized for LiOH. The optimum exponents ζ were $1s(Li) = 2.696$, $2s(Li) = 0.888$, $1s(O) = 7.652$, $2s(O) = 2.150$, $2p(O) = 2.271$, $1s(H) = 1.105$. The largest difference between this optimized basis and that obtained from Slater's rules is for the 2s(Li) function, much more contracted in the optimized set.

c) Li(4s), O(4s 2p), H(2s). Double zeta Slater basis set optimized for the isolated atoms. For hydrogen the two 1s exponents on hydrogen were optimized for LiOH and found to be 1.100 and 1.366.

d) Li(4s 1p 1d), O(4s 2p 1d 1f), H(2s 1p 1d). Double-zeta-plus-polarization Slater basis. The exponents of the polarization functions were optimized by repeated SCF calculations on LiOH.

e) Li(5s), O(5s 4p), H(2s). Extended basis set of Slater functions excluding polarization functions. The two hydrogen 1s functions are as in c).

f) Li(5s 1p 1d), O(5s 4p 1d 1f), H(2s 1p 1d). Extended Slater basis set including the polarization functions determined in d).

g) Li(5s 2p 1d 1f), O(5s 4p 2d 1f), H(3s 1p 1d). This is an <u>ionic</u> Slater basis set obtained from optimized near Hartree-Fock calculations on Li^+ and OH^-. Polarization functions on Li were obtained from an optimized LiF calculation.

h) Li(10s 2p / 5s 2p), O(10s 6p 1d / 5s 3p 1d), H(6s 1p / 3s 1p). This is the smaller of two contracted gaussian sets.

i) Li(10s 3p / 6s 3p), O(11s 7p 2d / 7s 4p 2d), H(6s 2p / 4s 2p). This is the larger of two contracted gaussian sets and is an extended basis including polarization.

A unique feature of this work is the use of rival atomic (f) and ionic (g) basis sets, both of which should be sufficiently complete to yield SCF properites approaching the RHF limit. Since the two basis sets are in detail very different, they provide a check on the reliability of predicted properties.

Table IV-6 gives the energies obtained with the 9 basis sets. The geometry assumed in all these calculations was that predicted by Buenker and Peyerimhoff (56). Dissociation energies are obtained by subtracting the LiOH energies from the sum of the Li, O, and H energies obtained with comparable basis sets. The extended basis set (f) including polarization is seen to yield the lowest SCF energy. This is not surprising, since the polarization functions used were specifically optimized for LiOH. However, the ionic calculation (g) is a close second (0.0022 hartree higher), even though no exponent optimization was carried out for LiOH with this basis. The larger of the two gaussian sets yields an energy 0.0089 above the lowest Slater function energy, and a more careful contraction (see section ID) of the gaussian basis would almost certainly decrease this difference.

Table IV-6. Self-consistent field energy quantities (in hartrees) for LiOH obtained with different basis sets. Basis sets are described in detail in the text. All calculations are for R(Li-O) = 3.0236 bohrs, R(O-H) = 1.833 bohrs.

Basis Set	a Minimum	b Optimized Minimum	c Double Zeta	d Double Zeta Plus Polarization	e Extended	f Extended Plus Polarization	g Ionic	h Smaller Gaussian	i Larger Gaussian
Total Energy	-82.3547	-82.4633	-82.8624	-82.9448	-82.9206	-82.9555	-82.9533	-82.9313	-82.9466
Potential Energy	-167.2565	-164.2989	-166.0215	-165.7998	-165.8694	-165.8131	-165.8001		
Kinetic Energy	84.9018	81.8356	83.1591	82.8550	82.9488	82.8576	82.8478		
V/T	-1.97000	-2.00767	-1.99643	-2.00108	-1.99966	-2.00118	-2.00127		
Dissociation Energy (eV)		0.14	3.45	5.71	4.86	5.81	5.75		
Orbital Energies									
1σ	-20.0588	-20.4413	-20.2715	-20.4127	-20.3862	-20.4273	-20.4387	-20.4189	-20.4254
2σ	-2.4358	-2.4186	-2.4313	-2.4178	-2.4357	-2.4188	-2.4273	-2.4338	-2.4253
3σ	-0.9733	-1.0505	-1.0606	-1.1540	-1.1379	-1.1655	-1.1747	-1.1574	-1.1616
4σ	-0.3174	-0.4041	-0.4400	-0.5350	-0.5165	-0.5463	-0.5540	-0.5389	-0.5436
1π	-0.1510	-0.2176	-0.2862	-0.3668	-0.3502	-0.3754	-0.3829	-0.3690	-0.3732

The virial ratio is seen to approach 2.0 rather closely as the size of the basis is increased. The final dissociation energy 5.81 eV is probably within 0.1 eV of the Hartree-Fock value, and both the double-zeta and especially the minimum basis values of D_e are significantly in error. The orbital energies are not as well stabilized as the total energy, but calculation (f) appears to yield ε's within 0.01 hartree or ~0.25 eV of the Hartree-Fock values. The minimum basis orbital energies are very far from the accurate results and the double-zeta values of ε are still poor though improved. The primary deficiency of the minimum and double-zeta basis sets is that they do not include a 2p function centered on lithium. If minimum basis sets are to be used for the Li and Be atoms, the 2s-2p near degeneracy should be recognized and a 2p function added to the basis. Similarly, a 3p function should be included in a minimum basis set for Na and Mg.

Many of the predicted one-electron properties seen in Table IV-7 are familiar. The values of $<1/r>$, proportional to the diamagnetic shielding of each nucleus, are very insensitive to basis set, with even the minimum sets yielding reasonable values. The dipole moment of LiOH is large and surprisingly insensitive to basis set. The double-zeta-plus-polarization dipole moment is quite close to the larger basis set results. The predicted quadrupole moments show clearly the inadequacy of the minimum basis sets, and even the double-zeta-plus-polarization value of θ differs by ~10% from the larger basis results. The electric field gradients are very sensitive to basis set. In particular the field gradient at Li is difficult to assess since the values obtained from the two largest sets differ by almost a factor of 2. Comparison with

Table IV-7. Properties of lithium hydroxide from SCF calculations using a variety of basis sets. Unless indicated, all properties are in atomic units.

	a	b	c	d	e	f	g
Basis Set	Minimum	Optimized Minimum	Double Zeta	Double Zeta Plus Polarization	Extended	Extended Plus Polarization	Ionic
$<1/r_{Li}>$	8.6012	8.5959	8.6119	8.6206	8.6049	8.6200	8.6116
$<1/r_O>$	24.4131	23.8947	24.1452	24.0136	24.0453	24.0049	23.9941
$<1/r_H>$	6.1681	6.1567	6.1743	6.1064	6.0986	6.0915	6.0855
Dipole Moment μ (debyes)	-5.168	-5.692	-5.684	-4.868	-5.268	-4.764	-4.856
Quadrupole Moment θ (10^{-26} esu·cm^2)	4.87	7.96	9.18	9.34	10.74	10.27	10.63
Magnetic Susceptibility (ergs/gauss^2·mole)							
ξ_s	-3.3×10^{-5}	-3.0×10^{-5}	-3.0×10^{-5}	-3.2×10^{-5}	-3.1×10^{-5}	-3.2×10^{-5}	-3.3×10^{-5}
ξ_r	-4.8×10^{-6}	-4.5×10^{-6}	-4.7×10^{-6}	-5.4×10^{-6}	-5.3×10^{-6}	-5.6×10^{-6}	-5.9×10^{-6}
Rotational g Factors							
g_s	0.5225	0.5225	0.5225	0.5225	0.5225	0.5225	0.5225
g_r	-0.1651	-0.1037	-0.0969	-0.1324	-0.1069	-0.1256	-0.1343
Forces on nuclei							
F(Li)	0.1941	0.2503	0.2541	0.0723	0.2225	0.0562	0.0282
F(O)	-1.2331	-1.3415	-0.8323	-0.2217	-0.3336	0.1127	0.0710
F(H)	0.0987	0.0520	0.0544	-0.0494	0.0523	-0.0481	-0.0446
Sum	-0.9403	-1.0392	-0.5238	-0.1987	-0.0587	0.1208	0.0546
Field Gradients							
q(Li)	-0.0259	-0.0342	-0.0348	-0.0234	-0.0264	-0.0236	-0.0420
q(O)	4.215	1.542	2.435	1.401	1.771	1.397	1.423
q(H)	0.530	0.542	0.541	0.471	0.544	0.435	0.465
Polarizabilities							
α_{zz}	31.7	11.8	8.3	17.0	6.6	15.0	14.1
β_{zzz}	530	160	38	95	-20	60	71
$A_{z:zz}$	6.8	11.6	20.0	21.5	19.3	22.7	22.6

the double-zeta-plus-polarization result seems to imply that q(Li) = -0.0236 atomic units is the more reasonable value and that the ionic basis is somehow biased. Polarizabilities calculated with the minimum basis sets are quite unreliable. However the two largest bases (f) and (g) give qualitatively consistent results. From the latter two calculations it appears that α_{zz} and $A_{z:zz}$ should be within 10% of the RHF values and β_{zzz} within 25%.

ξ_s and ξ_r are computed lower and upper reference values for the magnetic susceptibility (249). ξ_s is the value in the limit that the electronic cloud does not rotate with the nuclear frame, while ξ_r is the value in the limit that the electronic cloud rotates rigidly with the nuclear frame. Table IV-7 shows that ξ_s is very insensitive to basis and ξ_r is rather insensitive. The most accurate calculation implies that the magnetic susceptibility of LiOH lies in the region of -3.2×10^{-5} and -5.6×10^{-6} ergs/gauss^2·mole. Using the same two approximations g_s and g_r give limits on the true molecular rotational g factor. The expressions from which g_s and g_r are computed are:

$$g_s = (M/I) \sum_A (Z_A e) z_A^2 \qquad (IV,4)$$

$$g_r = g_s - (M/I) < z^2 - \tfrac{1}{2}(x^2+y^2) > \qquad (IV,5)$$

where M is the proton mass, I the moment of inertia of the molecule, Z_A the nuclear charge on the Ath nucleus, and z_A the distance of the Ath nucleus from the center of mass. g_s is the same for all basis sets in Table IV-7 since it depends only on the positions of the nuclei. But g_r is quite sensitive to basis set, since it is obtained by subtracting nearly equal nuclear and electronic contributions.

For a molecule at its equilibrium geometry, the forces on all nuclei are zero. For a Hartree-Fock wave function, Kern and Karplus (203) have shown that at any geometry the sum of the forces on all the nuclei is zero. Thus, inspection of forces predicted from SCF wave functions provides insight into closeness both to the RHF and to the exact wave function. The force F_A on a nucleus is given by the product of the nuclear charge and the electric field at the nucleus. For linear molecules, the force is the expectation value of the operator

$$Z_A \left[- \sum_{B \neq A} Z_B/R_{AB}^2 + e \sum_i z_{iA}/r_{iA}^3 \right] \qquad (IV,6)$$

where R_{AB} is the distance between nuclei A and B, z_{iA} is the (positive or negative) position of the ith electron relative to nucleus A, and r_{iA} is the absolute value of z_{iA}. Table IV-7 shows that the SCF forces become progressively smaller (in absolute value) with increasing size of basis set. In particular, the addition of polarization functions significantly lowers the force at each nucleus. The smallest forces are obtained with the ionic basis, which indicates that in at least one sense the ionic basis is more appropriate than the energetically superior basis set (f).

Table IV-8 shows the predicted properties of Li_2O. SCF results obtained with three different basis sets are shown to give an idea of the reliability of the computed properties. The basis sets used are (c), (d), and (f) described above, except that a diffuse 2p function with exponent 0.6 has been added to the oxygen basis sets (d) and (f). The double-zeta energy results, as was the case for LiOH, are unreliable. The $2\sigma_u$ orbital energy is particularly poor. The quadrupole moment from the double zeta calculation also is in poor agreement with the larger basis set results. The rotational g_r factor is very small and as a result very sensitive to the basis set. The double zeta electric field gradient at oxygen appears to be quite far from the correct value.

A general conclusion which may be made on the basis of the LiOH and Li_2O results is that polarization functions (at least 2p on Li and H; at least 3d on oxygen) must be included in a basis set if reliable molecular properties are to be obtained. In general, the properties obtained at the double-zeta-plus-polarization level are rather close to those predicted from the largest basis set used. To conclude, it would appear that LiOH and Li_2O are two molecules concerning which a great deal more is reliably known from theory than from experiment.

Relative stabilities of lithium cyanide and isocyanide. The gas-phase geometry of neither lithium cyanide (LiCN) nor lithium isocyanide (LiNC) has been determined experimentally. However, there are some interesting differences between the crystalline properties of LiCN (geometry not specified) and those of NaCN and KCN, thought to be classic ionic compounds. For example, the melting point of the lithium compound is 160°, whereas NaCN and KCN have much higher melting points, 564° and 634°. In addition, the crystal structure implies that lithium is essentially bonded to nitrogen and therefore the proper name of the solid is lithium isocyanide. In concert with efforts by Bak to obtain experimentally the gas-phase spectrum of LiCN or LiNC, Bak, Clementi, and Kortzeborn (16) undertook a theoretical study to predict the relative stabilities of these two molecules.

Table IV-8. Ground state properties of dilithium monoxide (Li_2O) in the self-consistent-field approximation. The assumed Li-O bond distance is 3.13 bohrs.

	c Double Zeta	d Double Zeta Plus Polarization	f Extended Plus Polarization
Total Energy	-89.6366	-89.8052	-89.8127
Potential Energy	-179.7935	-179.4518	-179.5059
Kinetic Energy	90.1569	89.6466	89.6932
V/T	-1.99423	-2.00177	-2.00133
Dissociation Energy (eV)	-0.69	3.69	3.75
Orbital Energies			
$1\sigma_g$	-20.2054	-20.3485	-20.3407
$2\sigma_g$	-2.3704	-2.3578	-2.3527
$3\sigma_g$	-0.9333	-1.0164	-1.0106
$1\sigma_u$	-2.3702	-2.3577	-2.3526
$2\sigma_u$	-0.1809	-0.3374	-0.3322
$1\pi_u$	-0.2237	-0.2717	-0.2681
Quadrupole Moment (10^{-26} esu·cm^2)	16.62	25.61	25.31
$<1/r_{Li}>$	8.8729	8.8656	8.8713
$<1/r_O>$	24.5905	24.4609	24.4734
Magnetic Susceptibility (ergs/gauss^2·mole)			
ξ_s	-5.7×10^{-5}	-5.5×10^{-5}	-5.4×10^{-5}
ξ_r	-6.9×10^{-6}	-7.8×10^{-6}	-7.4×10^{-6}
Rotational g Factors			
g_s	0.4309	0.4309	0.4309
g_r	-0.0365	-0.0056	-0.0014
Force on Li Nucleus	0.2326	0.0914	0.0851
Field Gradients			
q(Li)	-0.0293	-0.0430	-0.0414
q(O)	2.3250	-0.7000	-0.7044

The single-configuration SCF calculations of Bak, et al. (16) employ a basis of gaussian functions, designated (see section ID) Li(11s 3p / 5s 3p), C(11s 7p / 5s 4p), N(11s 7p / 5s 4p). It should be noted that while this basis is extended, no polarization functions are included. If assumed linear the SCF wave function orbital occupancy for LiCN or LiNC (these are just two different geometries corresponding to the same electronic state) is $1\sigma^2\, 2\sigma^2\, 3\sigma^2\, 4\sigma^2\, 5\sigma^2\, 6\sigma^2\, 1\pi^4$. After some exploratory work using small gaussian basis sets, the basis set described above was used in SCF calculations for more than 20 different LiCN and LiNC geometries.

These calculations predict the LiCN isomer to have bond distances R(Li-C) = 3.632 bohrs and R(C-N) = 2.192 bohrs and corresponding total energy -99.75559 hartrees. Similarly the predicted LiNC structure is R(Li-N) = 3.335 bohrs, R(N-C) = 2.186 bohrs, with total energy -99.77077 hartrees. An interesting feature of the predictions is that the CN bond distance in the two molecules differs by only 0.006 bohrs or 0.003 Å. These predicted CN bond distances are in harmony with the fact, pointed out by Bak et al., that for a variety of organic and inorganic compounds, experimental C≡N triple bond distances are quite close to 2.19 bohrs. The isolated CN radical has equilibrium internuclear separation 2.21 bohrs. It should also be noted that the predicted Li-C distance in CN is significantly longer than the Li-N distance in LiNC. These points are emphasized here because geometry predictions are likely to be much more reliable than energy difference predictions from SCF calculations.

The calculations predict LiNC to be the more stable isomer by 0.0141 hartree = 0.38 eV = 8.8 kcal/mole. The authors are quick to point out that this energy difference is only a small fraction of the correlation energy, so the prediction of LiNC being the more stable is open to question. However, a simple estimate is made which suggests that LiNC has a larger correlation energy than LiCN. If true this would lower the energy of LiNC further with respect to LiCN.

An interesting aspect of the calculations is the breakdown of the total energy into contributions from different molecular orbitals. A seldom-mentioned identity relating the total electronic energy E, the orbital energies ϵ_i, and one-electron integrals $I(i|i)$ is

$$E(SCF) = \sum_i [\epsilon_i + I(i\ i)] = \sum_i \eta_i \quad (IV,7)$$

Eq (IV,7) holds for closed-shell SCF wave functions and the sum over i only includes n/2 terms, one for each $\alpha\beta$ pair of spin orbitals. Bak, Clementi, and Kortzeborn (16) have denoted the quantity η the <u>total orbital energy</u>, and Table IV-9 shows these quantities for LiCN and LiNC. It is interesting to note that the ordering of the total orbital energies η_i is not the same as that for the usual orbital energies ϵ_i. For LiCN, the 3σ and 4σ orbital values of ϵ are -2.4991 and -1.1693 hartrees, while the η values are -9.7921 and -10.4654 hartrees. Table IV-9 makes possible a division of the total energy into contributions from a) σ orbitals, b) π orbitals, and c) nuclear repulsion. Although the nuclear repulsion is greater for LiNC, both the σ and π energies are also larger in absolute value. The latter electronic stabilization of

LiNC is sufficient to overcome the increased electron repulsion and predict LiNC to lie 8.8 kcal/mole lower in energy than LiCN.

A tool frequently used for qualitative electronic structure discussions is the population analysis. Following Mulliken (269), overlap integrals between basis functions are used to analyze an SCF wave function so that a certain number of electrons is assigned to each basis function. Table IV-10 shows population analyses for LiCN and LiNC. For comparison with the total s and p populations in Table IV-10, the separated Li, C, and N atom SCF wave functions include 11 "s electrons" and 5 "p electrons." Thus one could say in some simple picture that 1.508 electrons have shifted from s to p orbitals during formation of LiCN. Of particular interest are the total atomic populations, which may be expressed for LiCN $Li^{+0.475} C^{-0.294} N^{-0.180}$ and for LiNC $Li^{+0.817} N^{-0.812} C^{-0.004}$. These total atomic populations suggest that LiCN and LiNC have quite different electronic structures. LiNC is characterized by a large transfer of electron density from Li almost entirely to the adjacent N atom. The amount of electron transfer in LiCN is much less and is shared by both the carbon and nitrogen atoms. Crystallographers also try to use their data (experimental, of course) to assign charges to atoms in a molecule. The crystal structure of LiNC has been interpreted to yield a point-charge formula $Li^{0.8} N^{-0.9} C^{0.1}$ (224). Although the theoretical and "experimental" populations agree nicely, it should be understood that in reality there is no nonarbitrary way of assigning charges to the atoms in a molecule.

Returning to observables, vibration frequencies were predicted for both LiCN and LiNC by Bak, et al. For $^6Li\ ^{12}C\ ^{14}N$ the calculated linear stretching

Table IV-9. Total orbital energies $\eta_i = \epsilon_i + I(i|i)$ for LiCN and LiNC near their predicted geometries. The orbitals are ordered by their usual orbital energies ϵ_i.

	LiCN	LiNC
1σ	-43.2157	-43.5808
2σ	-33.1519	-32.8950
3σ	-9.7921	-10.1199
4σ	-10.4654	-10.5693
5σ	-7.7182	-8.5048
6σ	-7.6484	-6.9529
$1\pi_x$	-7.7941	-7.9958
$1\pi_y$	-7.7941	-7.9958
$\sum_\sigma \eta$	-111.9918	-112.6246
$\sum_\pi \eta$	-15.5881	-15.9916
Nuclear Repulsion	27.8243	28.8454
Total Energy	-99.7556	-99.7708

Table IV-10. Comparison of population analyses for lithium cyanide and isocyanide (16). The molecular axis is the z axis.

Molecule	Nucleus	Types				Total Population
		s	p_z	p_y	p_x	
LiCN	Li	2.271	0.240	0.007	0.007	2.525
	C	3.542	0.966	0.893	0.893	6.294
	N	3.679	1.301	1.100	1.100	7.180
Total s or p		9.492		6.508		
LiNC	Li	2.041	0.096	0.022	0.022	2.183
	N	3.528	1.524	1.380	1.380	7.812
	C	3.810	0.999	0.597	0.597	6.004
Total s or p		9.379		6.621		

mode frequencies were $\nu_1 = 2375$ and $\nu_2 = 670$ cm^{-1}. For ^{6}Li ^{14}N ^{12}C the predictions are $\nu_1 = 2367$ and $\nu_2 = 785$ cm^{-1}. In the CN radical the CN stretching frequency is 2069 cm^{-1} and in other cyano compounds 2250 cm^{-1} is typical. The Li-C and Li-N stretching frequencies are both small. The bending frequency for the predicted ground state LiNC isomer was predicted to be ~250 cm^{-1}. The predicted dipole moments are nearly the same, 3.8 debyes for LiCN and 3.5 for LiNC. The availability of these theoretical predictions certainly will provide additional incentive to those currently attempting to experimentally determine the gas-phase properties of LiNC (or is it LiCN?).

B. INCLUSION OF ELECTRON CORRELATION

The linear HeH_2^+ potential energy surface. The field of ion-molecule reactions is a rapidly growing part of chemistry. It is now becoming possible to obtain particularly detailed information concerning the dynamics of ion-molecule reactions by molecular beam studies (239). With increasingly reliable theoretical potential energy surfaces becoming available, there is reason for optimism concerning the possibility of fruitful collaboration between theory and experiment. For the very simplest polyatomic ion-molecule reaction

$$D^+ + H_2 \rightarrow DH + H^+ \qquad \Delta H = 0 \qquad (IV,8)$$

an accurate ground state potential surface has been calculated (93). However the use of this ground state surface has been found not to allow an adequate description of the dynamics of reaction (IV,8). This is because a second potential surface, which includes the products of the reaction

$$D^+ + H_2 \rightarrow DH^+ + H \qquad \Delta H = 1.835 \text{ eV} \qquad (IV,9)$$

must also be considered. An "avoided crossing," or near degeneracy of the two surfaces for certain geometries, is responsible for the inadequacy of the ground state surface by itself (307). The simplest polyatomic ion-molecule reaction which can be studied with a single potential surface is

$$He + H_2^+ \rightarrow HeH^+ + H \qquad (IV,10)$$

A reliable study including electron correlation of the ground state HeH_2^+ potential surface has been reported by Edmiston and coworkers (113).

The method of calculation for these HeH_2^+ calculations was analogous to that used in the He_2^+ calculations of Edmiston and Krauss discussed in section IC as the first example of the pseudonatural orbital method. An uncontracted gaussian basis set, He(6s 3p) and H(4s 3p), was used. The first step in the calculations was the determination of the SCF wave function, which for linear HeH_2^+ is of the form $1\sigma\alpha\ 1\sigma\beta\ 2\sigma\alpha$. Then pseudonatural orbitals were calculated for the $1\sigma^2$ pair, and these PSNO used in final CI calculations including 65 configurations. For the separated two-electron systems He and HeH^+, about 90% of the correlation energy is accounted for.

The most likely reaction coordinate for Eq.(IV,10) involves a linear He H H^+ arrangement, and Edmiston, et al. (113) report calculations for 22 different linear He H H^+ geometries. Figure IV-1a shows the predicted reaction coordinate for this linear arrangement. The right hand side of the figure represents reactants He plus H_2^+ and it can be seen that the H_2^+ bond distance is ~2.0 bohrs. Similarly the left hand side shows the products HeH^+ + H and the HeH^+ equilibrium separation is seen to be ~1.5 bohrs.(it is actually 1.4632). As He approaches H_2^+ along the reaction coordinate, the H_2^+ bond distance increases only slightly until the He-H distance reaches ~2.0 bohrs. Then the He atom becomes "bonded" to the proton and the H-H distance increases steadily until products HeH^+ and H remain.

Figure IV-1b shows the SCF and CI energies along the linear He H H^+ reaction coordinate. The two most obvious features of this figure are 1) the reaction is endothermic, and 2) there is no barrier to the reverse reaction HeH^+ + H $\rightarrow$ He + H_2^+. Confidence in the reliability of the predicted surface

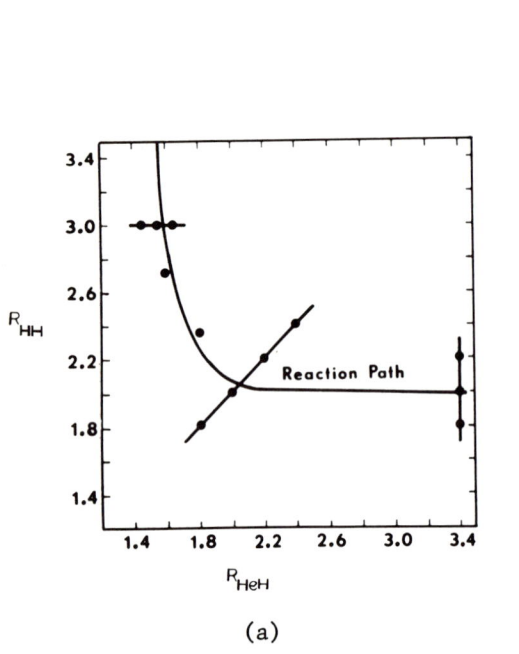 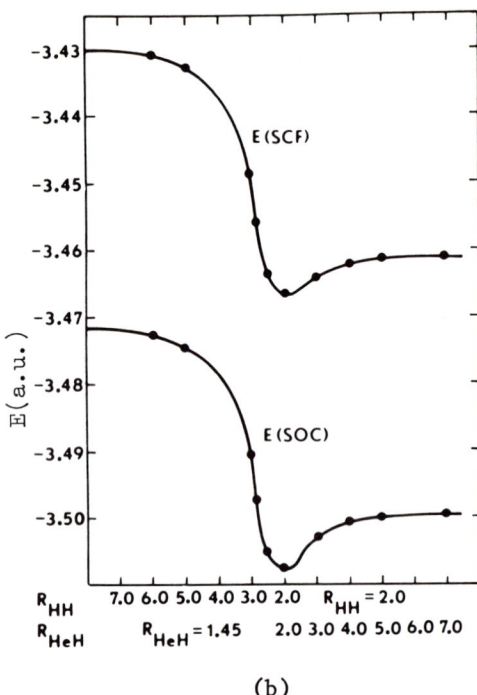

Figure IV-1. (a) Minimum energy path for $He + H_2^+ \rightarrow HeH^+ + H$. Distances are in bohrs. (b) SCF and CI energies in hartrees along the reaction coordinate for linear $He + H_2^+$. SOC refers to superposition of configurations, which is another way of saying CI.

is given by comparison of the theoretical endothermicity 0.82 eV and the experimental value 0.80 eV. Of particular interest is the prediction of a "hydrogen bonded" He H H$^+$ complex in which both He-H and H-H bond distances are ~2.0 bohrs. This complex lies about 5 kcal lower in total energy than the separated reactants He + H$_2^+$. Although this He H H$^+$ complex has not been identified experimentally, the result is not surprising. The primary feature expected to characterize potential surfaces for ion-molecule reactions is the strong, long-range attraction which arises in a perturbation theory picture from the dipole moment induced in the polarizable molecule by the approaching ion. Neutral A + BC → AB + C reactions, on the contrary, are usually expected to proceed via a reaction coordinate containing a finite barrier.

There is a second possible linear arrangement of one helium and two hydrogen atoms, namely H He H. If the reaction coordinate for He + H$_2^+$ were to begin by a symmetrical approach of the helium atom, then a linear H He H$^+$ might precede breakup to HeH$^+$ + H. Figure IV-2 summarizes the results of Edmiston and coworkers (113) for linear H He H$^+$. In this figure, the left-hand side represents HeH$^+$ + H and the right-hand side the lowest possible energy for linear symmetric H He H$^+$. The latter arrangement occurs when both H-He bond distances are about 2.1 bohrs and the CI energy is -3.4317 hartrees. However, this most favorable symmetric geometry may be seen to lie about 25 kcal above the products. The calculations would therefore seem to rule out a C$_{2v}$ or perpendicular approach of He to H$_2^+$.

From a purely theoretical viewpoint, the most interesting aspect of this HeH$_2^+$ work is seen in Figure IV-1b. Namely, the SCF and CI potential surfaces

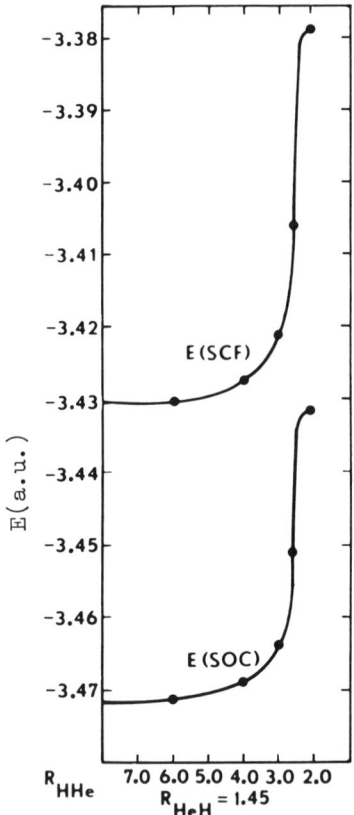

Figure IV-2. CI and SCF energies along the reaction coordinate for linear H He H$^+$ → H + He H$^+$. This path differs from that of Figure IV-1 in that the He is constrained to lie between the two H atoms. The dot farthest to the left corresponds to R(H-He) = 6.0 and R(He-H) = 1.45 bohrs, while that farthest to the right corresponds to both distances being 2.1 bohrs.

for the linear approach (which contains the reaction coordinate) of He to H_2^+ are essentially parallel. Put in another way, the correlation energy is nearly constant as a function of geometry. This is at first a surprising result since it is now becoming abundantly evident that for A + BC neutral surfaces the RHF approximation is rather poor. However, it should be remembered that the dominant feature of ion-molecule potential surfaces is often the ion-induced dipole attraction. And several theoretical studies, including that of Lester (225) for Li^+ + H_2, have shown that this effect is well accounted for in the Hartree-Fock approximation.

It can be tentatively concluded that SCF potential surfaces may be adequate for the study of certain ion-molecule reactions. However, there are a large number of surfaces for which the RHF approximation is of little value due to asymptotic difficulties (217). Reaction (IV,10) presents no asymptotic difficulties for the RHF approximation since both reactants and products include one closed-shell species. For large He-H distances the molecular SCF wave function for HeH_2^+ becomes the antisymmetrized product of SCF wave functions for He and H_2^+. And for large H-H distances, the HeH_2^+ SCF wave function becomes the product of HeH^+ and H SCF wave functions. However, consider the reaction

$$He(^1S) + O_2^+(^2\Pi_g) \rightarrow HeO^+(^2\Pi) + O(^3P) \qquad (IV,11)$$

On the left-hand side the RHF approximation is adequate since for large He-O separations the HeO_2^+ wave function goes over to SCF wave functions for the ground states of He and O_2^+. However, for large O-O distances the HeO_2^+ SCF wave function does not become the product of SCF wave functions for the open-shell ground states of HeO$^+$ and O. Since the RHF wave function for HeO$_2^+$ cannot describe the products of reaction (IV,11) it would not make sense to attempt a self-consistent-field study of this reaction. The problem discussed here is quite analogous to the fact, discussed in chapter III, that the RHF wave functions for N$_2$, O$_2$, and F$_2$ do not dissociate to RHF wave functions for the separated atoms.

The F + H$_2$ → FH + H chemical reaction. Among neutral A + BC → AB + C chemical reactions, the F + H$_2$ reaction has received much attention from experimentalists. This reaction has been studied experimentally by laser spectroscopy (287), infrared chemiluminescence (302) and crossed molecular beams (344). From these experiments emerges about as detailed a picture of reaction dynamics as is available for any chemical reaction. It is natural enough, then, that after looking at relatively simple potential surfaces such as H$_3^+$, HeH$_2^+$, H$_3$ and H$_4$, the first larger system to which theoreticians turned was FH$_2$. The first a priori calculations on F + H$_2$ were the SCF calculations of Newton (284). He established first that the reaction coordinate is linear and second that the RHF approximation predicts a barrier height much greater than the experimental activation energy 1.7 kcal/mole. In this subsection we discuss the results of Bender and Schaefer (31), who have considered the effect of electron correlation on the FH$_2$ potential surface.

The first calculations by Bender and Schaefer were carried out using a contracted gaussian, F(9s 5p / 4s 2p) and H(4s /2s), double zeta basis set. To be able to study nonlinear as well as linear geometries, only a plane of symmetry (C_s point group) was assumed. The basis set above then consists of twelve a' (in the plane) and two a'' (out of the plane) basis functions and the RHF wave function has orbital occupancy $1a'^2 2a'^2 3a'^2 4a'^2 5a' 1a''^2$. First-order wave functions (see sections IIC and IIIB) including 214 configurations were used to assess the effects of electron correlation on the potential surface. For FH_2, the first-order wave function recovers that part of the correlation energy due to the unpaired 5a' orbital and the 6a' valence orbital which is not occupied in the SCF approximation. Those configurations involving replacement of, for example, the $4a'^2$ pair are of the types $4a'^2 \rightarrow 5a'6a'$, $5a'7a',\ldots5a'12a'$ and $4a'^2 \rightarrow 6a'^2, 6a'7a',\ldots6a'12a'$. Following an SCF calculation, natural orbital iterations (section IC) were carried out on the 214 configuration wave function until the total energy stabilized. The purpose of the iterative natural orbital method of course is to guarantee that the 214 configuration wave functions are nearly optimum with respect to the form of the molecular orbitals.

Calculations were carried out for about 150 linear geometries. These calculated energies were used to predict the <u>minimum energy path</u> (or reaction coordinate) for $F + H_2 \rightarrow FH + H$. One point on the minimum energy path is found by fixing the FH distance and varying the H-H distance to find a minimum in the energy. The minimum energy path is rather well described by the twelve points seen in Table IV-11. The reaction exothermicity is predicted to be 0.6 kcal/mole from the SCF calculation and -20.4 kcal from the CI, compared to experiment ~ -31.2 kcal. Thus the SCF approximation does not predict the reaction to be exothermic at all, while the CI gives a qualitatively reasonable

Table IV-11. Minimum energy paths for the chemical reaction $F + H_2 \rightarrow FH + H$ obtained with a double zeta basis set. Internuclear separations are given in bohrs and energies in kcal/mole relative to the reactants.

	Self-Consistent-Field			214 Configurations	
R(FH)	R(HH)	Energy	R(FH)	R(HH)	Energy
6.0	1.380	0.03	6.0	1.425	0.00
4.0	1.381	1.48	4.0	1.429	0.82
3.5	1.401	3.90	3.5	1.439	2.09
3.0	1.397	10.07	3.0	1.459	4.22
2.8	1.403	14.33	2.9	1.470	4.73
2.6	1.428	20.04	2.8	1.484	5.20
2.4	1.466	27.41	2.7	1.490	5.48
2.2	1.502	32.40			
			2.58	1.54	5.75*
2.01	1.54	34.33*			
			2.412	1.6	5.41
1.930	1.7	29.11	2.229	1.7	3.36
1.890	1.8	25.99	2.095	1.8	0.22
1.845	1.9	21.91	2.005	1.9	-2.87
1.811	2.0	18.40	1.931	2.0	-5.67
1.792	2.1	15.41	1.874	2.1	-8.14
1.782	2.2	12.93	1.840	2.2	-10.17
1.776	2.3	10.88	1.826	2.3	-11.83
1.771	2.4	9.16	1.817	2.4	-13.24
1.768	2.5	7.72	1.811	2.5	-14.46
1.750	3.0	3.12	1.809	3.0	-18.12
1.743	4.0	0.53	1.799	4.0	-20.43
1.742	5.0	0.37	1.798	5.0	-20.61
1.742	10.0	0.58	1.798	10.0	-20.40

result. The saddle point may be associated with the geometry of the "activated complex" in transition state theory (187) and is rigorously defined in a quantum mechanical calculation as the point of highest energy along the minimum energy path. The barrier height then is the difference in energy between the saddle point geometry and that of the isolated reactants. The FH_2 SCF calculation predicts the barrier height to be 34.3 kcal, while the CI barrier is much smaller, 5.8 kcal/mole. These two numbers might be compared with the experimental activation energy E_a, 1.7 kcal, obtained by measuring the rate constant as a function of temperature and fitting the expression $k = A\, e^{-(E_a/RT)}$. Unfortunately, however, when the experimental activation energy is as small as 1.7 kcal, serious questions arise as to whether E_a is nearly equal to the barrier height, even if the latter were known from an exact calculation (348). At any rate, the measured activation energy implies that the true barrier height is probably 5 kcal or less and the CI calculations are seen to represent a vast improvement over SCF in this regard.

Two-dimensional potential surfaces for linear A + BC reactions are traditionally displayed via contour maps, such as that seen in Figure IV-3. The same $F + H_2$ surface is seen in three-dimensional plots in Figure IV-4. Using the tabulated minimum energy path of Table IV-11 and the three graphical representations, one obtains a fairly clear picture of the potential surface predicted by the 214 configuration calculations. The SCF saddle point occurs near an F-H distance of 2.01 bohrs (1.06 Å) and H-H distance of 1.54 bohrs (0.81 Å) while the CI predicts 2.58, and 1.54 bohrs (1.37 and 0.81 Å). For comparison, the isolated diatomic bond distances are 0.92 and 0.74 Å. Thus both calculations predict the saddle point to occur rather early (significantly lengthened F-H distance) for a slightly expanded H-H distance. More generally, the SCF and CI reaction coordinates are quite similar, the primary difference in the calculations being in the predicted relative energies along the path.

The CI surface of Figures IV-3 and IV-4 is of the type often referred to as "repulsive". Using this type of potential surface, Polanyi and coworkers (303) have made classical trajectory calculations which predict a "mixed-energy release", or sharing of exothermicity between product translation and vibration. This analysis appears to be consistent with the findings of experimental studies of the $F + H_2 \rightarrow FH + H$ reaction (287, 302, 344).

The minimum energy path (Table IV-11) divides the surface into two regions. In the first, a very gentle uphill grade is traversed as the F atom approaches the nearly unperturbed H_2 molecule. At $R(F-H) = 2.7$ bohrs the H-H separation is only 1.49 bohrs. The second region of the surface is the downhill part in which simultaneously a) the $R(F-H)$ is decreased until it reaches r_e, b) the H_2 molecule separates, and c) the energy goes monotonically downward to that of the products, HF + H.

Note the prediction by both calculations of a small long range attraction (~0.2 kcal/mole) between HF and H.

It has been shown that first-order wave functions calculated for diatomic molecules using basis sets without polarization functions do not yield reliable dissociation energies. Since the predicted $F + H_2 \rightarrow FH + H$ exothermicity from the double zeta basis set was 10 kcal/mole too small, Bender

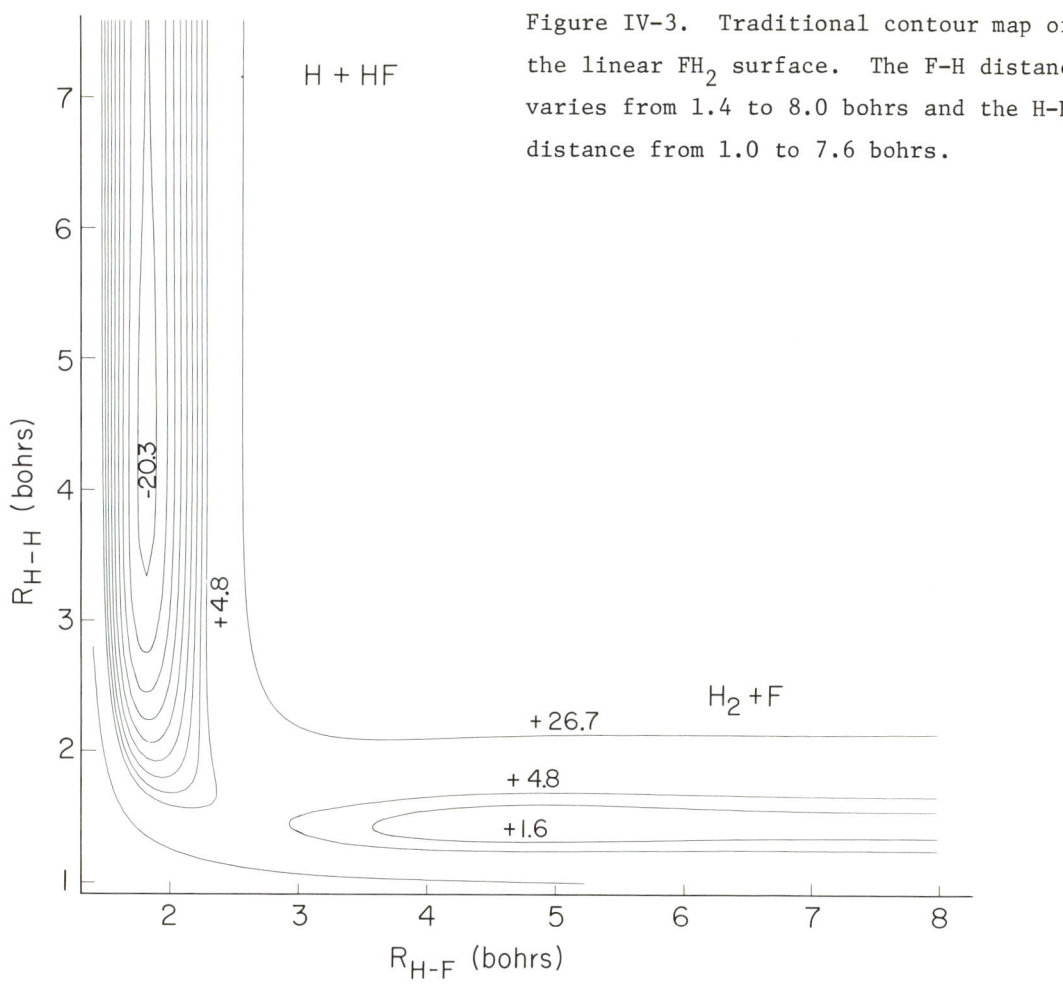

Figure IV-3. Traditional contour map of the linear FH_2 surface. The F-H distance varies from 1.4 to 8.0 bohrs and the H-H distance from 1.0 to 7.6 bohrs.

and Schaefer (31) decided to carry out a second set of calculations using a larger F(9s 5p 2d / 4s 2p 1d), H(4s 1p / 2s 1p) basis. For reasons of economy these calculations were designed to take advantage of linear symmetry. The linear first-order wave function from this basis includes 338 configurations, and the minimum energy path is seen in Table IV-12. There it is seen that

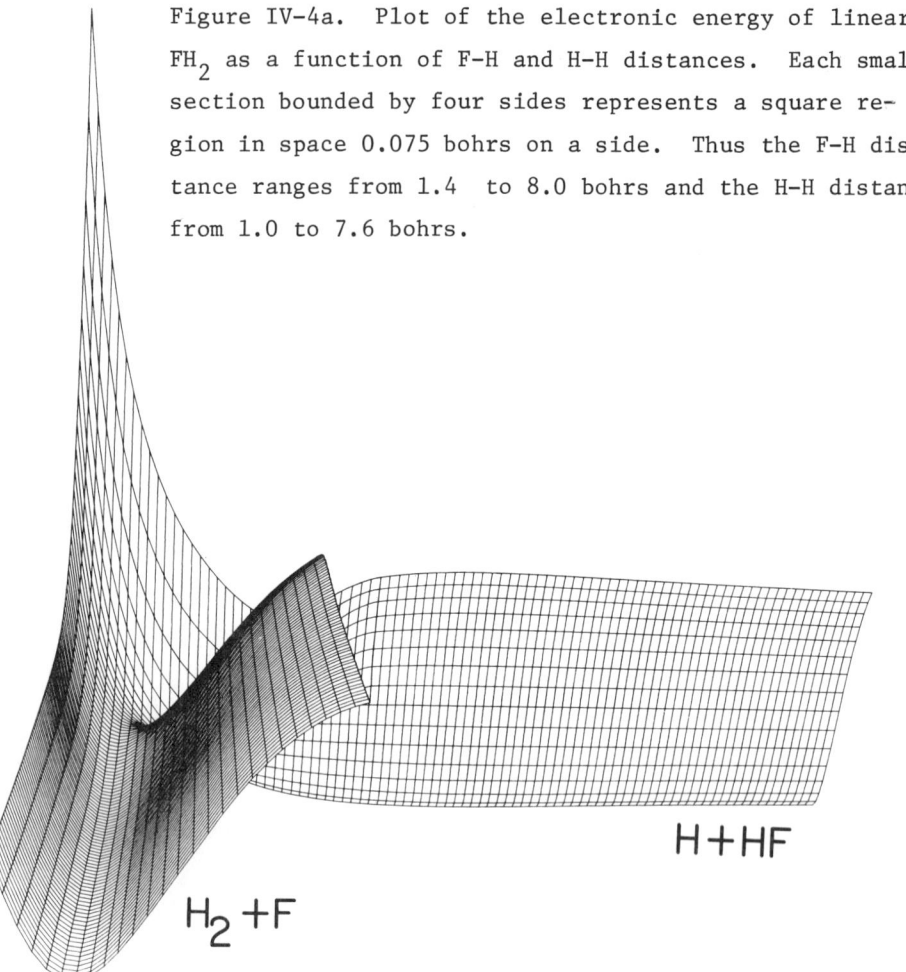

Figure IV-4a. Plot of the electronic energy of linear FH_2 as a function of F-H and H-H distances. Each small section bounded by four sides represents a square region in space 0.075 bohrs on a side. Thus the F-H distance ranges from 1.4 to 8.0 bohrs and the H-H distance from 1.0 to 7.6 bohrs.

the addition of polarization functions to the basis greatly improves the agreement with experiment for both the barrier height and exothermicity. In fact the predicted barrier height, 1.66 kcal/mole, is virtually indistinguishable from experiment. The exothermicity, -34.4 kcal/mole, is about 10% larger than experiment. It seems reasonable to conclude that the 338 con-

Figure IV-4b. Same as 4a but viewed from the exit channel FH + H.

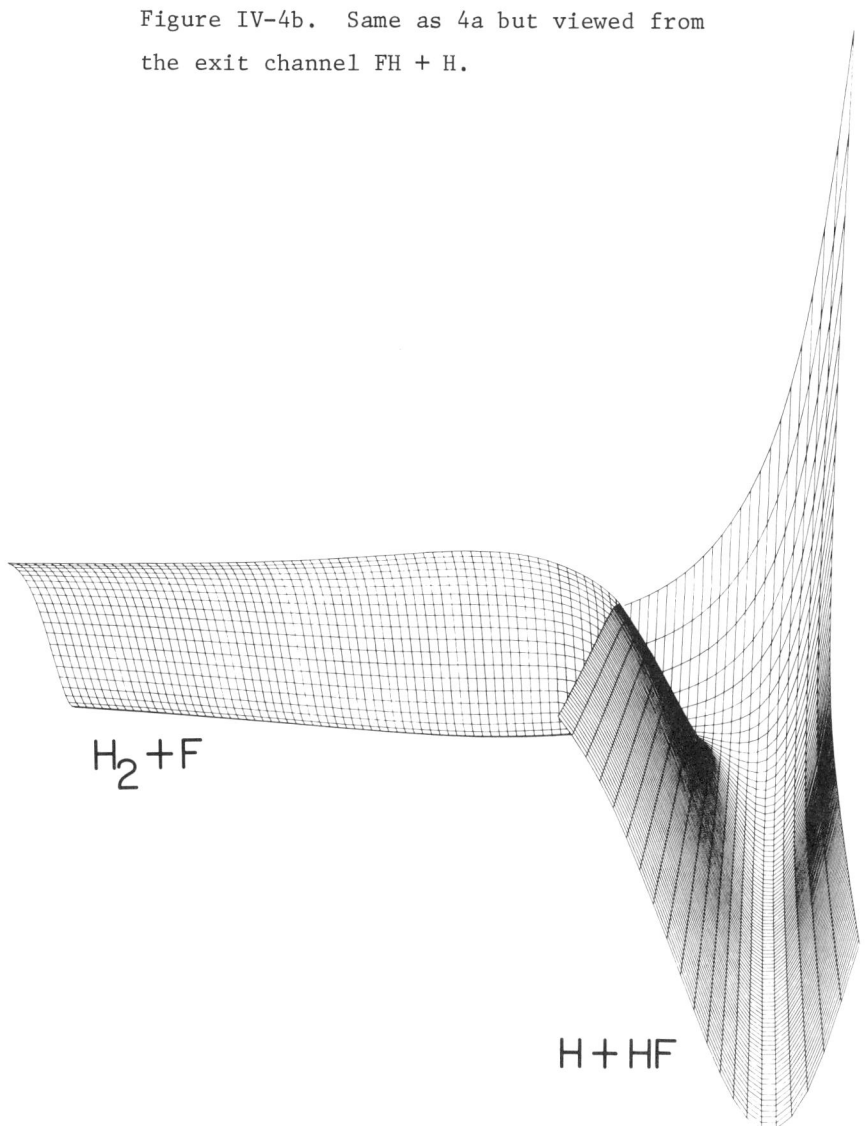

Table IV-12. Double-zeta-plus-polarization minimum energy path for $F + H_2 \rightarrow FH + H$. Distances are in bohr radii. 338 configuration first-order wave functions were used.

R(F-H)	R(H-H)	E(kcal/mole)	
∞	1.42	0.00	Reactants
6.0	1.42	0.00	
4.0	1.42	0.25	
3.6	1.43	0.44	
3.4	1.43	0.75	
3.2	1.44	1.27	
3.0	1.44	1.61	
2.90	1.45	1.66	Saddle Point
2.8	1.48	1.59	
2.63	1.5	1.18	
2.36	1.6	-2.15	
2.03	1.8	-12.30	
1.85	2.0	-20.70	
1.80	2.2	-25.30	
1.78	2.5	-29.15	
1.77	3.0	-32.91	
1.76	∞	-34.44	Products

figuration potential surface may be quite realistic. Encouragingly, the qualitative features of the surface are similar to those predicted by the double-zeta calculation. One of the noticeable differences is that the saddle point in the 338 configuration calculation occurs for a longer F-H distance and shorter H-H distance.

Krypton difluoride. The chemistry of krypton centers about KrF_2 (22). Unlike the analogous xenon compounds, KrF_4 and KrF_6 have not been made. KrF_2 stands on the fringe of stability, being bound by about 23 kcal or 1 eV (141) with respect to Kr + F + F. KrF_2 is thus not stable with respect to Kr + F_2. Furthermore, although XeF_2 is relatively stable, ArF_2 is thought not to exist. A satisfactory treatment of the electronic structure of KrF_2 would clearly go a long way towards explaining why noble gas atoms form chemical bonds. However, from our experience on diatomic molecules, it would seen unlikely that the single-configuration Hartree-Fock approximation would predict binding for KrF_2. Liu, Bagus, and Schaefer (231) have carried out ab initio calculations on KrF_2 with explicit inclusion of electron correlation.

A question of great interest with respect to noble gas compounds concerns the importance of "outer" or "higher" orbitals (22). For Kr, these "outer" orbitals would be 4d, 5s, and 5p. Since these orbitals are not occupied in the ground state Kr SCF wave function, $1s^2\ 2s^2\ 2p^6\ 3s^2\ 3p^6\ 4s^2\ 3d^6\ 4p^6$, it is consistent with the discussion of Chapter I to refer to the Kr 4d, 5s, and 5p orbitals as "polarization functions." The possibility of these outer orbitals being particularly important was raised due to the failure of simple empirical models to rationalize the bonding in rare gas molecules. The first nonempirical investigation of this question was that of Collins, Cruickshank, and Breeze (82). They used a minimum basis of Slater functions, each expanded as a linear combination of three gaussians. Then 4d, 5s, and 5p functions were added to the basis to test their energetic importance. Addition of the 4d functions lowered the SCF energy by 0.306 hartree = 8.3 eV. The 5s and 5p functions proved to be less important. Collins et al. concluded that "4d orbitals must be included to give an adequate description of the bonding."

This same question was pursued by Liu et al. using a much more adequate basis set, that used for the CI calculations on KrF and KrF^+ discussed in Chapter III. Their results are summarized in Table IV-13. For comparison, the lowest total energy obtained by Collins et al. (82) was -2921.922 hartrees, or 29.812 hartrees above the lowest energy in Table IV-13. The table shows that polarization functions in fact have far less energetic importance than indicated by the calculations of Collins. This illustrates a point of general importance-- namely, when a minimal basis set is used, almost any function added to the basis set will appear quite important. In order to assess the importance of "outer orbitals" one must use an adequate basis of inner orbitals. In the more realistic calculations, the addition of two 4d functions centered on krypton lowers the SCF energy by 0.0574 hartree = 1.56 eV. Thus 4d functions are less than one-fifth as important as indicated by Collins et al. Table IV-13 shows that most of the energy lowering due to polarization functions come from Kr 4d functions and F 3d functions. 4f functions added to either atom are relatively unimportant.

The question of whether 4d orbitals are "necessary" for an adequate description of bonding in KrF_2 is somewhat elusive. This is partly due to the

Table IV 13. Effect of polarization functions ("outer orbitals") on the self-consistent-field energy of KrF$_2$ at 3.5 bohrs bond distance. All energies are in hartrees.

Basis Set	Description	Energy	Lowering
Kr (8s 6p 3d) F (4s 3p)	Atomic functions	-2950.6347	--
two 3d, one 4f on F	Fluorine polarization	-2950.6905	0.0558
two 4d on Kr		-2950.6921	0.0574
two 4d, two 4f on Kr	Krypton polarization	-2950.7104	0.0757
two 4d on Kr, two 3d on F		-2950.7219	0.0872
two 4f on Kr, one 4f on F		-2950.6561	0.0214
Kr(8s 6p 5d 2f) F(4s 3p 2d 1f)	Full polarization	-2950.7341	0.0994

fact that the Hartree-Fock approximation is not capable of predicting a bound (with respect to Kr + F + F) KrF$_2$ molecule. Further, we know from calculations on molecules as simple as O$_2$ and F$_2$ that 3d functions must be included to obtain dissociation energies within 10% of experiment. And no one has ever suggested that 3d orbitals must be invoked to "explain" the bonding in O$_2$ and F$_2$. To conclude it would appear that Kr 4d functions are no more or less vital to a description of the bonding in KrF$_2$ than are 3d functions centered on C, N, O, and F in molecules containing these atoms.

The second question of greatest interest concerning KrF$_2$ is "what is the simplest reliable theoretical approach which can predict binding?" The SCF wave function for KrF$_2$ is of the form (excluding the inner 38 electrons)

$$5\sigma_u^2 \; 3\pi_u^4 \; 2\pi_g^4 \; 8\sigma_g^2 \; 4\pi_u^4 \tag{IV,12}$$

The orbitals included in (IV,12) can all be constructed from Kr 4p and F 2p orbitals. In order for the Kr F$_2$ wave function to dissociate to SCF wave functions for Kr + F + F, a second configuration must be included, $8\sigma_g^2 \to 6\sigma_u^2$, or

$$5\sigma_u^2 \; 3\pi_u^4 \; 2\pi_g^4 \; 6\sigma_u^2 \; 4\pi_u^4 \tag{IV,13}$$

In addition to single configuration SCF calculations on (IV,12), Liu et al. carried out two configuration SCF calculations including both (IV,12) and (IV,13). In light of the asymptotic difficulties inherent in the single configuration approach, this two configuration wave function is the simplest from which one could expect to obtain a qualitatively reasonable potential surface.

Six other orbital occupancies (yielding 6 $^1\Sigma_g^+$ symmetry eigenfunctions) may also be constructed from the KrF$_2$ valence orbitals.

$$5\sigma_u \; 3\pi_u^4 \; 2\pi_g^4 \; 8\sigma_g^2 \; 4\pi_u^4 \; 6\sigma_u \tag{IV,14}$$

$$3\pi_u^4 \; 2\pi_g^4 \; 8\sigma_g^2 \; 4\pi_u^4 \; 6\sigma_u^2 \tag{IV,15}$$

$$5\sigma_u^2 \; 3\pi_u^2 \; 2\pi_g^4 \; 8\sigma_g^2 \; 4\pi_u^4 \; 6\sigma_u^2 \tag{IV,16}$$

$$5\sigma_u^2 \; 3\pi_u^4 \; 2\pi_g^2 \; 8\sigma_g^2 \; 4\pi_u^4 \; 6\sigma_u^2 \tag{IV,17}$$

$$5\sigma_u^2 \; 3\pi_u^4 \; 2\pi_g^4 \; 8\sigma_g^2 \; 4\pi_u^2 \; 6\sigma_u^2 \tag{IV,18}$$

$$5\sigma_u^2 \; 3\pi_u^3 \; 2\pi_g^4 \; 8\sigma_g^2 \; 4\pi_u^3 \; 6\sigma_u^2 \tag{IV,19}$$

The largest CI calculations carried out by Liu, Bagus, and Schaefer were of the first-order type (see F + H$_2$ discussion above). Single and double excitations with respect to (IV,12) through (IV,19) were included, with the usual first-order restriction that no more than one electron occupy an orbital beyond the valence shell, i.e., beyond $6\sigma_u$. In addition, correlation effects involving δ and ϕ orbitals were neglected. In all, 993 configurations were included in the CI calculations.

Figure IV-5 shows potential curves for the KrF$_2$ symmetric stretch using the three approximations described above. The SCF, two configuration SCF, and 993 configuration calculations yield remarkably dissimilar results. The SCF potential curve has its minimum at 1.813 Å, in reasonable agreement with the experimental values (22) 1.875 and 1.889 Å. However the minimum of the SCF potential curve lies 0.1095 hartree above the comparable SCF energies for Kr + F + F. That is, the SCF calculation predicts KrF$_2$ to be unbound by 2.98 eV. The two configuration potential curve dissociates properly but is repulsive, although it contains an interesting inflection point. Thus the correlation effects which lead to binding in KrF$_2$ are much more subtle than those found by Das and Wahl (94) for F$_2$, where the analogous two-configuration treatment yields an attractive potential curve.

Only the 993 configuration calculation yields a physically reasonable potential curve. The predicted bond distance is 1.906 Å, 0.017 Å greater than one of the two experimental values. The calculated dissociation energy, the difference between the energy at R_e and that at infinite separation, is 0.39 eV, only about one-third the experimental value. However, calculations on KrF$^+$ with

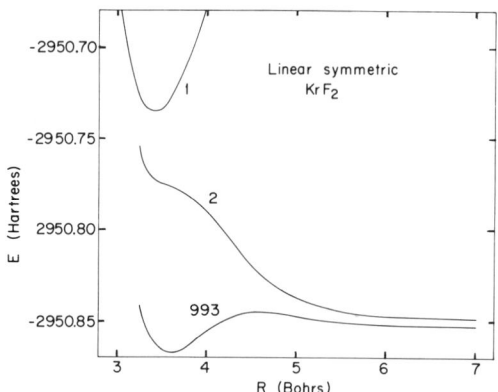

Fig. IV-5. Potential energy curves for the symmetric dissociation of KrF_2. The numbers 1, 2, and 993 refer to the number of configurations included in the three calculations.

a much larger basis set (229) indicate that the KrF_2 basis may be about 0.35 eV better for the separated atoms than for the molecule near R_e. The additional fact that δ and ϕ orbitals were not included in the CI makes it difficult to determine the degree to which the small absolute error in the dissociation energy is due to the model adopted, the first-order wave function. The most important point in this regard is that the molecule *is* predicted to be bound. A very interesting feature of the first-order potential curve is the existence of a potential maximum at 2.39 Å, lying 0.23 eV above the dissociation limit. Although the height of this barrier will be lowered by a more complete calculation, the fact that the curve is repulsive at large R implies that the maximum does correspond to reality. Since a potential maximum has never been established or predicted for the ground state of any neutral molecule, this prediction for KrF_2 is unique. It is interesting to point out that this maximum can be rationalized in terms of the ionic-covalent model of Coulson (91), in which an avoided crossing occurs between the $Kr + F + F$ and $Kr^+ + F + F^-$ curves.

V
NONLINEAR POLYATOMIC MOLECULES

Early efforts to calculate <u>ab initio</u> wave functions for nonlinear polyatomic molecules were plagued by errors in two-electron integrals, particularly the three- and four-center integrals. It now appears that the first such <u>correct</u> wave function reported in the literature for a molecule containing four atoms was that of Foster and Boys (123) for formaldehyde (H_2CO). The SCF wave function of Foster and Boys was obtained from a minimum basis set of Slater functions, each expanded as a linear combination of gaussians.

Despite the success of Foster and Boys, most of the effort towards polyatomic wave functions during the late 50's and early 60's involved the development of direct methods for the evaluation of multicenter integrals over Slater functions. A center for this type of work was M.I.T., where the method employed was that of Barnett and Coulson (19) in which each Slater function is expanded in a series of spherical harmonics about a common center. The culmination of this work was the series of minimum basis set SCF wave functions of Pitzer and Lipscomb (298) for ethane.

The systematic study of the electronic structure of polyatomic molecules only began after the acceptance of gaussian functions. The first such systematic study was that of Krauss (215) on the series of hydrides H_3, CH_2, NH_2, H_2O, H_2O^+, BH_4^-, BH_4^+ and CH_4. Krauss's work was characterized by relatively large uncontracted basis sets, e.g., C(9s 5p) and H(3s) for the CH_2 molecule.

SCF calculations may now (1971) be straightforwardly carried out by even a novice for molecules with as many as 40 electrons. For just this purpose there are available a number of standardized and widely distributed computer programs with interesting names: POLYATOM (92), IBMOL (77), MOSES (330), and MOLE (323). As will be seen in this chapter, SCF calculations have recently been performed on molecules with as many as 136 electrons. However, such a calculation presses both machine and computational methods to their present (1971) limits and may be likened to the most intricate molecular beam experiment, made possible by only very recent technological advances.

The calculation of polyatomic wave functions including electron correlation has seriously lagged behind SCF developments. In fact, prior to 1970, not a single wave function of beyond Hartree-Fock accuracy for a polyatomic molecule with more than 4 electrons had been reported in the literature. However, due to CI methods primarily developed for atomic and diatomic calculations (chapters II and III), this situation is now changing. Several groups are actively pursuing the correlation problem in polyatomic molecules, and some of their recent results are discussed in this chapter.

A. TRIATOMIC MOLECULES

<u>Magnetic hyperfine structure of NO_2</u>. During the period 1950-1966 microwave spectroscopists studied the NO_2 radical in great detail (223). This experimental

work was coupled with sophisticated theoretical analysis (228) to yield accurate values of the NO_2 magnetic hyperfine structure (hfs) parameters. In the restricted Hartree-Fock approximation, the orbital occupancy for the $^2A_1$ ground state of NO_2 is

$$1a_1^2\ 1b_2^2\ 2a_1^2\ 3a_1^2\ 2b_2^2\ 4a_1^2\ 5a_1^2\ 3b_2^2\ 1b_1^2\ 4b_2^2\ 1a_2^2\ 6a_1 \qquad (V,1)$$

Since only the $6a_1$ orbital is unpaired, in this approximation the hfs parameters depend only on the form of the $6a_1$ orbital. For this reason, the experimental hfs parameters have been used to predict the character of this unpaired molecular orbital. From our discussion of atomic hyperfine structure in chapter II, we tend to be wary of the usefulness of the RHF approximation for hfs predictions. Nevertheless, the intense experimental interest in this problem led Schaefer and Rothenberg (341) to examine the hfs of NO_2 using SCF wave functions.

The contracted gaussian double-zeta N and O basis sets of Dunning (105) were used and a set of 3d functions with exponent $\alpha = 0.8$ added to each atom. From the basis set studies of chapters I (CO and H_2O) and IV (LiOH and Li_2O), it is likely that most SCF one-electron properties obtained with this basis will be within 10% of the true Hartree-Fock values. The calculated total energy without d functions is -204.0679 hartrees, which is probably about 0.05 hartree above the RHF energy.

Table V-1 compares the predicted SCF hfs parameters to the experimental values of Curl and coworkers (223). For a brief introduction to hfs, the reader is referred to the discussion of chapter II. The magnetic hfs parameters at the nitrogen nucleus were measured using the isotopic molecule $^{14}N^{16}O_2$ (^{14}N nuclear spin I = 1) and those at the two equivalent oxygen nuclei using

$^{14}N^{16}O^{17}O$ (^{16}O has I = 0, but ^{17}O has I = 5/2). As in chapter II, $|\psi(0)|^2$ is the spin density or Fermi contact interaction at one of the nuclei. The parameters $(3\hat{r}_a^2 - 1)/r^3$ and $(\hat{r}_b - i\hat{r}_c)/r^3$ characterize the so-called dipole-dipole interaction between the electronic and nuclear magnetic moments (69). The off-diagonal parameter $3\hat{r}_a\hat{r}_b/r^3$ at the oxygen nucleus has not been determined experimentally. a, b, and c are the molecular axes, where b is perpendicular to the plane of the molecule and c bisects the ONO angle.

The predicted hfs parameters in Table V-1 are in surprisingly good agreement with experiment, especially for the parameters at the nitrogen nucleus. The addition of d functions to the basis set has relatively little effect on the predicted properties. The near-perfect agreement for the nitrogen spin density $|\psi(0)|^2$ must be considered at least somewhat fortuitous in light of the great importance of correlation effects in atomic spin densities. However, the excellent agreement can be partially rationalized by the fact that the NO_2 nitrogen spin density is almost entirely a molecular effect, being five times larger than that of the free nitrogen atom. The O^{17} parameters are in rough agreement with experiment but are all of insufficient magnitude.

A population analysis of the $6a_1$ orbital indicates 17% nitrogen s character, 38% nitrogen p character, less than 1% oxygen s character, and 44% oxygen p character. Thus the $6a_1$ MO has a much larger percentage of nitrogen basis function population than does the molecule as a whole. One effect of configuration interaction would be to allow other orbitals (with more oxygen character) to appear in the wave function in an unpaired manner. This is perhaps the simplest way in which one can imagine the theoretical values of the oxygen nucleus hfs parameters increasing.

Table V-1. Predicted (SCF) and experimental magnetic hyperfine structure parameters for NO_2 (341). All entries are in units of $10^{24} cm^{-2}$.

	SCF no d functions	SCF with d functions	Experiment
Properties at the nitrogen nucleus			
$\|\psi(0)\|^2$	2.597	3.101	3.08
$(3\hat{r}_a^2 - 1)/r^3$	-4.125	-4.122	-3.88
$(\hat{r}_b - i\,\hat{r}_c)/r^3$	4.360	4.211	3.36
Properties at the oxygen nucleus			
$\|\psi(0)\|^2$	0.230	0.300	0.71
$(3\hat{r}_a^2 - 1)/r^3$	-3.462	-3.178	-4.71
$(\hat{r}_b - i\,\hat{r}_c)/r^3$	3.843	3.585	5.04
$3\,\hat{r}_a\,\hat{r}_b/r^3$	1.524	1.579	

The NO_2 results imply that at least for certain polyatomic free radicals, the RHF approximation can yield qualitatively correct hfs parameters. However, in order to assure better than factor of 2 reliability in hfs parameters, it appears that CI must be carried out. The inadequacy of the RHF approximation for quantitative purposes also weighs heavily against the use of experimental hfs parameters to "determine" the nature of an unpaired RHF orbital. In such exercises, the model being used, the single configuration wave function, is inherently incapable of reproducing the experimental results.

Molecular quadrupole moments and other properties of sulfur dioxide. The molecular quadrupole moment tensor θ provides an important measure of the charge distribution in a molecule. In an xyz coordinate system there are in general six distinct elements of the quadrupole moment tensor: θ_{xx}, θ_{xy}, θ_{xz}, θ_{yy}, θ_{yz}, and θ_{zz}. θ_{xx}, for example, is given as the expectation value of the operator

$$\tfrac{1}{2}\left\{\sum_A Z_A (3x_A^2 - r_A^2) - \sum_i (3x_i^2 - r_i^2)\right\} \qquad (V,2)$$

The A sum goes over nuclei and the i sum over electrons. The value of the quadrupole moment depends on the origin chosen, and for continuity the center of mass is usually taken as $(x,y,z,) = (0,0,0)$. Thus x_A is the distance in the x-direction of the Ath nucleus from the center of mass. As of 1966, very few quadrupole moments had been reliably determined from experiment (218). Among available values, those obtained from microwave linewidth data were thought to be most reliable. Fortunately, this situation has changed radically during the past five years. Primarily due to the development by Flygare and coworkers of the rotational Zeeman effect, reliable experimental molecular quadrupole moments are now available for a large number of molecules (121). SO_2 is a molecule for which particularly accurate values of the quadrupole moments have been obtained by Flygare and coworkers (301). Since SO_2 is perhaps the largest molecule (32 electrons) for which an SCF study (with adequate basis set) of molecular properties has been reported, this subsection reviews the calculations of Rothenberg and Schaefer (322) on SO_2.

The basis set used in this work was double-zeta-plus-polarization, using contracted functions on sulfur (12s 9p 1d / 6s 4p 1d) and oxygen (10s 5p 1d / 4s 2p 1d). SCF calculations were carried out at the experimental geometry both with and without the polarization or d functions. Table V-2 shows the predicted properties for SO_2.

The coordinate system is chosen such that x bisects the OSO angle and z is perpendicular to the plane containing the molecule. In this coordinate system only the θ_{xx}, θ_{yy}, and θ_{zz} components of the quadrupole moment tensor are non-zero. Two essential points should be made concerning the calculated quadrupole moments: a) the SCF values without d functions are quite different from the values obtained when the basis set was expanded to include polarization functions, and b) the values with d functions are in good agreement with experiment. Point a) reiterates the findings of earlier work, that polarization functions must be included in a basis set to obtain molecular properties near the RHF limit. Point b) gives real hope that molecular quadrupole moments may be reliably predicted within the RHF approximation.

Some comments concerning the other properties predicted in Table V-2 are appropriate. The dissociation energy 3.85 eV agrees poorly with experiment and the Hartree-Fock value of D_e is not likely to be more than 6 eV. This provides additional evidence of the inadequacy of RHF wave functions in predicting dissociation energies. The dipole moment prediction is improved by the addition of d functions, and additional polarization functions (e.g., 4f functions on S

Table V-2. Self-consistent-field and experimental molecular properties for SO_2 (322).

	Calculated		
Property	Without d functions	With d functions	Experiment
Total energy	-546.9512	-547.2089	---
Dissociation energy (eV)	-3.17	3.85	10.71, 10.97
Ionization potential (eV)	13.41	13.38	12.34
Dipole moment (debyes)	2.83	2.28	1.63
Second moments ($10^{-16} cm^2$) of the electronic charge distribution			
Q_{xx}	9.09	8.89	8.7 ± 0.2
Q_{yy}	31.10	30.42	30.0 ± 0.2
Q_{zz}	4.30	4.25	4.2 ± 0.2
Quadrupole moment tensor (10^{-26} esu·cm^2)			
θ_{xx}	2.52	1.71	1.3 ± 0.3
θ_{yy}	-9.56	-6.88	-5.3 ± 0.4
θ_{zz}	7.04	5.17	4.0 ± 0.6
Octupole moment tensor (10^{-34} esu·cm^2)			
Ω_{xxx}	7.02	4.38	---
Ω_{xyy}	-8.71	-4.89	---
Ω_{xzz}	1.69	0.51	---
Diamagnetic susceptibility tensor (10^{-6} erg/G^2·mole)			
χ_{xx}^d	-150.2	-147.1	-145.1
χ_{yy}^d	-56.8	-55.8	-54.9
χ_{zz}^d	-170.5	-166.8	-164.3
χ_{av}^d	-125.8	-123.2	-121.4
Force at nucleus (a.u.)			
$F_x(S)$	-2.214	-0.952	0.0
$F_x(O)$	1.287	0.397	0.0
$F_y(O)$	2.384	0.761	0.0

cont.

| | Calculated | | |
Property	Without d functions	With d functions	Experiment
Diamagnetic shielding tensor (ppm)			
$\sigma_{xx}^d(S)$	-1134.1	-1134.6	---
$\sigma_{yy}^d(S)$	-1261.6	-1264.1	---
$\sigma_{zz}^d(S)$	-1057.7	-1056.8	---
$\sigma_{av}^d(S)$	-1151.1	-1151.8	---
$\sigma_{xx}(O)$	-489.2	-486.9	---
$\sigma_{yy}(O)$	-687.3	-690.4	---
$\sigma_{zz}(O)$	-416.9	-415.7	---
$\sigma_{xy}(O)$	-102.1	-104.8	---
$\sigma_{av}^d(O)$	-531.2	-531.0	---
Electric field gradient at nucleus (a.u.)			
$q_{zz}(S)$	2.258	2.1587	---
$q_{yy}(S)$	-0.1334	-0.0464	---
$q_{zz}(S)$	-2.1239	2.1119	---
$q_{\alpha\alpha}(O)$	2.0845	1.5280	---
$q_{\beta\beta}(O)$	-1.2585	-0.7023	---
$q_{\gamma\gamma}(O)$	-0.8260	-0.8256	---

and O) might very well lower the SCF dipole moment to 2.0 debyes. The remaining error, 0.37 debye, would be fairly typical of RHF results. The predicted elements of the diamagnetic susceptibility tensor are very close to experiment. This is particularly true after the inclusion of d functions. More generally, the diamagnetic susceptibility is proportional to expectation values like $<r>^2$ and these are very reliably predicted in the RHF approximation. The forces at the S and O nuclei are greatly improved by the addition of polarization functions.

As mentioned earlier, calculated electric field gradients and experimental quadrupole coupling constants may be combined to predict nuclear electric-quadrupole moments. From the SO_2 calculations discussed here and from similar H_2S and OCS calculations, Schaefer and Rothenberg predict $Q(^{33}S) = -0.062$ and $Q(^{35}S) = 0.043$ barns (1 barn = 10^{-24} cm^2). A reliability of 10% is suggested for these nuclear moments.

A point of some controversy concerns the importance of d orbitals in molecules containing Si, P, and S atoms. Ever since Pauling's 1931 paper (289) chemists have been using d orbitals, in one hybridization scheme or another, to "explain" the properties of unusual molecules. From a mathematical point of view, the addition of atomic d functions to a basis set is just another step toward the attainment of a complete set of functions. For SO_2 there does seem to be some substance to the feeling that d functions are of unusual importance. The basis for this statement is a comparative study of SO_2 and the corresponding molecule not containing a second row atom, O_3. For example, Table IV-2 shows that addition of d functions lowers the SO_2 SCF energy by 0.258 hartree. However, the comparable calculation on ozone yields an energy lowering of only 0.102 hartree. From another point of view, a population analysis of the SO_2 wave function shows the S atom d function population to be 0.425 electrons, while the central oxygen atom in O_3 has a d function population of only 0.150 electrons. By both of the above criteria, d functions on sulfur are two to three times more important than d functions on oxygen.

Zero-point vibrational effects in the water molecule. Molecular properties (such as those for NO_2 and SO_2 above) are generally predicted from a single theoretical calculation at the experimentally known geometry. Occasionally, a geometry variation is carried out and properties calculated for the predicted geometry. However, even this latter procedure is not completely satisfactory, since the observed molecular properties also depend on the vibrational wave function. For a molecule in its ground vibrational state (all vibrational quantum numbers zero) there will be zero-point vibrational corrections to the properties predicted from an electronic wave function. Kern, Matcha, and Ermler (204, 114) have made an important study of nuclear corrections to the predicted one-electron properties of H_2O.

In order to evaluate vibrational effects for water, one first needs a potential energy surface, preferably expressed in a suitable analytic form. In the work of Ermler and Kern (114) the potential energy surface is obtained from SCF calculations using a contracted gaussian basis O(9s 5p 2d / 4s 3p 2d) and H(4s 1p / 2s 1p). This basis set yields SCF energies about 0.015 hartree above the estimated RHF limit, and the SCF potential surface thus obtained should be nearly parallel to the true Hartree-Fock surface. Calculations were carried out for 45 carefully chosen geometries near the predicted SCF energy minimum. Harmonic and cubic force constants were obtained by fitting the calculated energies to the expression

$$E = E_e + \tfrac{1}{2} f_r (\Delta r_1^2 + \Delta r_2^2) + \tfrac{1}{2} f_\alpha \Delta\alpha^2 + f_{r'} \Delta r_1 \Delta r_2$$
$$+ f_{r\alpha} (\Delta r_1 + \Delta r_2)\Delta\alpha + \frac{1}{r_e} \left[f_{rrr}(\Delta r_1^3 + \Delta r_2^3) + f_{\alpha\alpha\alpha} \Delta\alpha^3 \right.$$
$$+ f_{rrr'} (\Delta r_1 + \Delta r_2)\Delta r_1 \Delta r_2 + f_{rr\alpha} (\Delta r_1^2 + \Delta r_2^2) \Delta\alpha$$
$$\left. + f_{rr'\alpha} \Delta r_1 \Delta r_2 \Delta\alpha + f_{r\alpha\alpha} (\Delta r_1 + \Delta r_2) \Delta\alpha^2 \right] \qquad (V,3)$$

In (V,3),

$$\Delta r_1 = r_1 - r_e \qquad \Delta r_2 = r_2 - r_e \qquad \Delta\alpha = r_e(\alpha - \alpha_e) \qquad (V,4)$$

where the factor of r_e in the expression for $\Delta\alpha$ allows all force constants to be in units of force/distance.

Table V-3 compares the SCF geometry and force constants to experiment. Perhaps the most obvious point in this table is the good agreement between SCF and experimental geometries. However the truly unique feature of Table V-3 is the prediction of force constants through cubic and comparison with experiment. All four harmonic force constants are in reasonable agreement with experiment, the deviation being 13.0 to 21.4 percent. In addition, the two diagonal cubic force constants f_{rrr} and $f_{\alpha\alpha\alpha}$ agree very nicely with experiment. It certainly comes as no surprise that three of the four cubic force constants representing the various coupled motions are in poor agreement with experiment. In light of our awareness of the importance of electron correlation, the sort of agreement seen for the first six force constants in Table V-3 is a source of optimism concerning the qualitative reliability of RHF force constants. Ermler and Kern also report harmonic vibration frequencies for the asymmetric stretch, bend, and symmetric stretch in H_2O, D_2O, and HDO. All theoretical values are within 8.3% of experiment. For H_2O, for example, the three SCF vibration frequencies are 4258, 1778, and 4148 cm^{-1}, compared to experiment 3943, 1648, and 3832 cm^{-1}.

Like the energy (V,3), each electronic property may be expanded as a function of bond angle and bond distance displacements. In the three reduced normal coordinates q_1, q_2, and q_3 for H_2O, the electronic property P is expanded

Table V-3. Self-consistent-field and experimental geometry and force constants for water (114). All force constants are in millidynes/Å.

Parameter	SCF	Experiment	% Error
r_e	0.9413 Å	0.9572 Å	-1.7
α_e	106.11°	104.52°	1.5
harmonic force constants			
f_r	9.876	8.454	16.8
f_α	0.881	0.761	15.8
f_r'	-0.079	-0.101	21.4
$f_{r\alpha}$	0.258	0.228	13.0
cubic force constants			
f_{rrr}	-10.51	-9.44 ± 0.03	-10.0
$f_{\alpha\alpha\alpha}$	-0.145	-0.14 ± 0.01	-3.6
$f_{rrr'}$	-0.0004	-0.32 ± 0.08	99.9
$f_{rr\alpha}$	-0.033	0.16 ± 0.02	-120.6
$f_{rr'\alpha}$	-0.526	-0.66 ± 0.01	20.3
$f_{r\alpha\alpha}$	-0.149	0.15 ± 0.10	-199.2

$$P = P_e + \sum_i \alpha_i q_i + \sum_{ij} \beta_{ij} q_i q_j + \sum_{ijk} \gamma_{ijk} q_i q_j q_k + \ldots \quad (V,5)$$

The constants α_i, β_{ij}, and γ_{ijk} relate the property P at an arbitrary geometry to P_e and the normal coordinates. Kern and Matcha have used a perturbation approach to derive an expression for the vibrationally averaged property $<P>$. For the ground vibrational state ($v_1 = v_2 = v_3 = 0$), $<P>$ is given in terms of the constants α, β, and γ in Eq. (V,5), plus the harmonic frequencies (ω_1, ω_2, and ω_3) and cubic potential constants k_{ijk} (204). From a practical point of view this means that in addition to the energy expression with coefficients in Table V-3, the property P must be calculated at a sufficient number of geometries to determine the coefficients α, β, and γ in (V,5). For water, properties P were computed for 45 different geometries.

Table V-4 summarizes the results of Ermler and Kern (114). For property P the zero-point correction is just the difference between the vibrationally averaged property $<P>$ and P_e, the calculated value at the theoretically predicted geometry. The vibrational corrections for H_2O, D_2O, and HDO are different because the harmonic frequencies and cubic potential constants differ for the three isotopic molecules. Typically, the zero-point corrections amount to about 1% of the predicted one-electron properties. This means that the errors in the RHF approximation itself are usually much larger than those due to nuclear corrections. A striking exception to this 1% rule occurs for the zz component of the quadrupole coupling tensor at the oxygen nucleus. There the corrections to the values for H_2O, D_2O, and HDO are 23%, 17% and 20% of the calculated electronic values. For HDO, P_e is -0.788, $<P>$ is -0.943, and the experimental value is -1.28 ± 0.01 MHz.

Table V-4. Zero-point vibrational corrections to certain one-electron properties of water.

Property	Units	P_e	Vibrational Corrections		
			H_2O	D_2O	HDO
Dipole moment	debyes	1.9975	0.0061	0.0046	0.0051
Quadrupole moments					
θ_{xx}	10^{-26} esu·cm^2	-2.5101	-0.0612	-0.0445	-0.0529
θ_{yy}		2.4291	0.0526	0.0374	0.0454
θ_{zz}		0.0811	0.0085	0.0071	0.0075
Octupole moments					
Ω_{zzz}	10^{-34} esu·cm^2	-1.3361	-0.0238	-0.0168	-0.0200
Ω_{xxz}		-1.1255	-0.0624	-0.0458	-0.0540
Ω_{yyz}		2.4616	0.0863	0.0626	0.0740
Diamagnetic susceptibility					
χ^d_{av}	10^{-6} erg/G^2·mole	-15.139	0.0218	0.0158	0.0189
Diamagnetic shielding					
$\sigma^d(O)$	ppm	416.515	0.019	0.014	0.016
$\sigma^d(H_1)$		103.983	0.1153	0.0857	0.1123
$\sigma^d(H_2)$		103.983	0.1153	0.0857	0.0895
Quadrupole coupling constants					
$eq_{xx}Q(^{17}O)$	MHz	11.348	0.225	0.164	0.193
$eq_{yy}Q(^{17}O)$		10.560	-0.045	-0.132	-0.039
$eq_{zz}Q(^{17}O)$		-0.788	-0.179	-0.132	-0.155

The vibrational correction is seen to bring the theoretical value into significantly improved agreement with experiment.

From the theoretical data in Table V-4, it is possible to predict isotope shifts. Since the predicted isotope shift is obtained as the difference of two quantities computed from the same SCF wave functions, it can be hoped that errors due to neglect of electron correlation will cancel out. The isotope shift between H_2O and D_2O for the quadrupole moment tensor has been measured experimentally and for θ_{xx}, θ_{yy}, and θ_{zz} the shifts (in 10^{-26} esu·cm^2) are -0.098, -0.094, and 0.191. The predicted values are -0.118, -0.086, and 0.204. The rather good agreement seems to imply that the calculations of Ermler and Kern (114) provide an essentially correct description of zero-point vibrational effects in water.

Rydberg states of H_2O. For many closed-shell molecules, most of the observed electronic states are Rydberg states (162). The electronic structures of the Rydberg states of a particular molecule are qualitatively rather similar. The outermost electron occupies a diffuse (or Rydberg) orbital which is bound to a positive ion core of (n-1) electrons and n protons. For a first-row atom, the diffuse or Rydberg orbitals are the 3s, 3p, 3d, 4s, 4p, 4d, 4f, ... orbitals. Thus the lowest Rydberg states of the carbon atom correspond to orbital occupancies $1s^2$ $2s^2$ 2p 3s, $1s^2$ $2s^2$ 2p 3p, etc., in which one of the 2p orbitals is replaced by a Rydberg orbital. In general one expects the relative positions of Rydberg states to be well-described by the Hartree-Fock approximation. There should be very little correlation energy associated with a Rydberg orbital since it lies in a region of space distinct from the other SCF orbitals. In addition the positive ion core (C^+ in the carbon atom example cited above) is expected to remain virtually unchanged throughout the entire Rydberg series. Thus the energy differences between the various Rydberg states in a particular series will be primarily determined by the form of the outermost or diffuse orbitals. This amounts to a one-electron problem, for which the RHF approximation is admirably suited.

Recall that the solution of the SCF equations for closed-shell systems yields a set of n/2 spatial orbitals, which doubly-occupied produce the single determinant wave function of lowest energy for that system. However, as conventionally carried out, the solution of the SCF equations yield an additional set of orbitals, the <u>virtual orbitals</u>. The virtual SCF orbitals are orthogonal to the occupied SCF orbitals and, like the occupied SCF orbitals, are eigenfunctions of the so-called Hartree-Fock Hamiltonian. The virtual orbitals are ordered by their orbital energies, and the virtual orbital ε's are greater than the corresponding occupied SCF orbital energies. Therefore one might hope that the virtual orbitals obtained from an SCF calculation would correspond to the excited state orbitals for the molecule in question. Unfortunately this is not true. The virtual SCF orbitals are on the contrary known to be appropriate for the excited states of the (n+1) electron system, the negative ion (197). In addition the virtual orbital energies ε_i are not related to ionization potentials by a Koopmans' type of theorem.

Hunt and Goddard (174) have reformulated the SCF equations in such a way that the virtual orbitals are variationally correct approximations to the SCF orbitals for the excited states. This is done by removing the so-called self coulomb and exchange operators (197) from the Hartree-Fock Hamiltonian. Let us consider the water molecule to discuss the implications of this method. The $^1A_1$ ground state orbital occupancy for H_2O is $1a_1^2\ 2a_1^2\ 1b_2^2\ 3a_1^2\ 1b_1^2$, and the first excited singlet state is of $^1B_1$ symmetry and orbital occupancy $1a_1^2\ 2a_1^2\ 1b_2^2\ 3a_1^2\ 1b_1\ 4a_1$. Unfortunately, the ordinary $4a_1$ virtual orbital, or regular virtual orbital (RVO), is optimum for the $^2A_1$ state of H_2O^- arising from $1a_1^2\ 2a_1^2\ 1b_2^2\ 3a_1^2\ 1b_1^2\ 4a_1$. By "optimum" we mean that, given the $1a_1$, $2a_1$, $1b_2$, $3a_1$, and $1b_1$ orbitals from the neutral molecule ground state SCF calculation, the $4a_1$ RVO yields the lowest energy single determinant wave function for the $^2A_1$ state of H_2O^-. However, the $4a_1$ improved virtual orbital (IVO) of Hunt and Goddard is such that, given the occupied SCF orbitals, the energy of the $^1B_1$ wave function arising from the $1a_1^2\ 2a_1^2\ 1b_2^2\ 3a_1^2\ 1b_1\ 4a_1$ orbital occupancy is a minimum. For the $^1B_1$ state in question, the use of the $4a_1$ regular virtual orbital gives a single configuration energy 11.71 eV above the $^1A_1$ ground state, while the $4a_1$ improved virtual orbital gives an energy 8.53 eV above the ground state. Since the lowest $^1B_1$ state of H_2O is known to lie 7.50 eV above the ground state, the IVO approach is significantly better in practice, as the above formulation would strongly imply.

Table V-5 summarizes the results of Hunt and Goddard (174). These results were obtained with an O(9s 5p / 5s 3p) and H(4s / 3s) contracted gaussian basis augmented by functions to describe the Rydberg orbitals. For oxygen, for example, the most diffuse of the above nine s functions has exponent $\alpha = 0.28461$ (182). Three diffuse s functions on oxygen, with exponents 0.07, 0.0175, and 0.0044, were added to the basis to describe the 3s, 4s, and 5s Rydberg orbitals. Including the diffuse functions, then, the basis set used was O(12s 8p / 8s 6p) and H(7s 1p /

6s 1p). It should be noted that the IVO results for the four $^1B_1$ states in Table V-5 come from a single calculation, which yields the four improved virtual orbitals $4a_1$, $5a_1$, $6a_1$, $7a_1$. However the $^3B_1$ IVO results arise from a different calculation and these four triplet IVO's $4a_1$, $5a_1$, $6a_1$, and $7a_1$ differ from the singlet IVO's due to the use of a different Hartree-Fock Hamiltonian. The $4a_1$, $5a_1$, $6a_1$, and $7a_1$ RVO's of course are the same for both the singlet and triplet states.

Of the excited states shown in Table V-5, only those of $^1B_1$ symmetry have been observed experimentally. The IVO excitation energies are seen to be at least in qualitative agreement with experiment. However these excitation energies do not give a fair representation of the strength of this theoretical approach. The first line in Table V-5 shows that the Koopmans' theorem ionization potential of H_2O is 1.32 eV greater than experiment, and this error carries over to the predicted excitation energies. Comparison of the IVO and experimental ionization potentials shows excellent agreement with experiment. That is, the differences between the excited state energies and the energy of H_2O^+ are quite accurately predicted. Subtraction of 1.32 eV from each of the IVO excitation energies in Table V-5 should provide very reliable predictions of the positions of the unobserved electronic states of H_2O. The above line of reasoning also reinforces our confidence in the description of a Rydberg state as a loosely bound electron attached to a positive ion.

It is interesting to ask how the descriptions 3s, 3p, ... of the Rydberg states seen in Table V-5 were obtained. One approach, of course, is to look at the coefficients of the different gaussian functions in the Rydberg orbital.

Table V-5. Energies (in eV) of the excited states of H_2O predicted from single configuration wave functions. RVO refers to a calculation using regular virtual orbitals and IVO to the improved virtual orbitals.

Orbital Occupancy	Symmetry	Description	Excitation Energy			Ionization Potential	
			RVO	IVO	Expt.	IVO	Expt.
$1a_1^2\ 2a_1^2\ 1b_2^2\ 3a_1^2\ 1b_1^2$	$^1A_1$		0.00	0.00	0.00	13.88	12.56
$1b_1 \rightarrow 4a_1$	$^3B_1$	3s	11.70	7.88	---	5.99	---
$\rightarrow 5a_1$	$^3B_1$	3p	12.42	11.24	---	2.63	---
$\rightarrow 6a_1$	$^3B_1$	4s	11.27	11.81	---	2.06	---
$\rightarrow 7a_1$	$^3B_1$	4p	11.57	12.56	---	1.21	---
$1b_1 \rightarrow 4a_1$	$^1B_1$	3s	11.71	8.53	7.50	5.34	5.06
$\rightarrow 5a_1$	$^1B_1$	3p	12.43	11.32	10.00	2.54	2.56
$\rightarrow 6a_1$	$^1B_1$	4s	11.31	11.93	11.00	1.93	1.56
$\rightarrow 7a_1$	$^1B_1$	4p	11.62	12.68	11.37	1.18	1.19
$1b_1 \rightarrow 2b_2$	$^3A_2$	3p	12.48	9.98	---	3.79	---
$\rightarrow 3b_2$	$^3A_2$	4p	11.85	12.09	---	1.69	---
$1b_1 \rightarrow 2b_2$	$^1A_2$	3p	12.48	10.35	---	3.42	---
$\rightarrow 3b_2$	$^1A_2$	4p	11.85	12.22	---	1.55	---

Table V-6 shows a second approach, the computation of a variety of properties of the Rydberg orbitals. For each of the lowest four $^1B_1$ states of H_2O, a property for the entire wave function is obtained by adding the Rydberg orbital contribution to the H_2O^+ value given on the bottom line. Several interesting points may be seen in Table V-6: a) $\delta(\vec{r})$, the charge density at the nucleus, is much larger for the s than the p Rydberg orbitals, b) for s Rydberg orbitals, the dipole moment is shifted towards the hydrogen atoms, while the p orbitals reinforce the H_2O^+ dipole moment in the oxygen atom direction, and c) the 3s and 4s Rydberg states have $<x^2> \approx <y^2> \approx <z^2>$, i.e., they are almost spherically symmetric. However, the 3p and 4p orbitals extend distinctly farther in the z direction. We have added a question mark to Hunt and Goddard's description of the lowest $^1B_1$ state as 3s, since the spatial extent of this orbital is not much greater than that of H_2O^+.

Geometry of $^3B_1$ methylene and the singlet-triplet separation. The CH_2 or methylene radical is one of the basic building blocks of organic chemistry and has been postulated as a short-lived intermediate in various chemical reactions. As such it is one of the most important of triatomic molecules. However, prior to 1959, CH_2 had not been observed spectroscopically. Herzberg's work (159) established in 1961 that CH_2 has two low-lying electronic states, one triplet and one singlet. If the molecule is assumed bent, these two states arise from orbital occupancies

$$1a_1^2 \; 2a_1^2 \; 1b_2^2 \; 3a_1 \; 1b_1 \qquad {}^3B_1 \qquad (V,6)$$

$$1a_1^2 \; 2a_1^2 \; 1b_2^2 \; 3a_1^2 \qquad {}^1A_1 \qquad (V,7)$$

Table V-6. Properties (in atomic units) of the improved virtual orbitals for the $^1B_1$ excited states of H_2O.

Orbital	$\delta(\vec{r})$	dipole moment	$<y^2>$	$<x^2>$	$<z^2>$	Type of orbital
$4a_1$	0.6895	-1.813	7.01	5.82	7.94	3s (?)
$5a_1$	0.0271	3.373	13.33	13.34	30.96	3p
$6a_1$	0.1637	-5.282	47.92	47.80	60.30	4s
$7a_1$	0.0037	7.172	75.94	75.95	190.56	4p
Properties of the H_2O^+ wave function	294.6	1.056	6.78	4.19	6.31	

Two very important questions which theoretical chemists have played an important role in answering are a) the geometry of the CH_2 ground state and b) the energy difference between electronic states (V, 6) and (V, 7).

By far the earliest nonempirical quantum mechanical study of methylene was that of Foster and Boys (124). A C(3s 1p), H(1s) set of Slater functions was expanded in terms of gaussians for ease of integral evaluation. CI calculations including as many as 128 determinants were carried out for the $^3B_1$, $^1B_1$ (also arising from the $3a_1$ $1b_1$ orbital occupancy), and $^1A_1$ states as a function of geometry. Foster and Boys predicted the $^3B_1$ state to be the ground state, with a bond angle of 129°. They predicted the $^1A_1$ state to have a bond angle of 90° and to lie 1.06 eV higher.

Herzberg's 1961 paper showed almost definitely that the triplet state of CH_2 was the ground state, but stated that the symmetry of this state was $^3\Sigma_g^-$, corresponding to the linear orbital occupancy

$$1\sigma_g^2\ 2\sigma_g^2\ 1\sigma_u^2\ 1\pi_u^2 \qquad (V,8)$$

Comparison with (V,6) shows that as the CH_2 molecule is bent, the $1\sigma_g$, $2\sigma_g$, and $1\sigma_u$ orbitals become $1a_1$, $2a_1$, and $1b_2$, and the doubly-degenerate $1\pi_u$ orbital splits into $3a_1$ and $1b_1$ orbitals. One year later (1961) a highly regarded semi-empirical calculation by Jordan and Longuet-Higgens (189) supported Herzberg's determination that the ground state of CH_2 was linear. In 1966 Herzberg and Johns (160) reported a more detailed study of the singlet transitions in CH_2 and found the $^1A_1$ bond angle to be 102°, in fair agreement with the 90° prediction of Foster and Boys.

In 1969, Harrison and Allen (148) reported CI calculations on the three lowest states of CH_2 using a fully contracted C(10s 5p / 2s 1p) and H(5s / 1s) gaussian basis. Like Foster and Boys, they predicted the triplet ground state to be bent, but with bond angle 138°. Harrison and Allen also predicted the $^1A_1$ state to lie 1.39 eV above the $^3B_1$ state. More complete calculations on the ground state were reported by Bender and Schaefer (30) in 1970. They used a double-zeta gaussian basis, C(9s 5p / 4s 2p) and H(4s /2s), and 408 configurations including SCF plus all single and double excitations. The predicted $^3B_1$ bond angle was 135°. At this stage it became evident that there was a seemingly inexplicable conflict between theory and experiment. Three calculations of increasing reliability predicted the triplet ground state to be bent, but Herzberg's CH_2 spectrum implied a transition between two linear states of Σ symmetry.

At this point an experimental breakthrough was made by Bernheim and coworkers (37), who observed the electron paramagnetic resonance (EPR) spectrum of CH_2 trapped in xenon at 4.2°K. The analysis of this spectrum showed that the triplet is the ground state under these circumstances and that the molecule is "slightly bent." It was not possible to determine the $^3B_1$ bond angle. A second EPR study, by Wasserman, et al. (387) appeared within a few months and indicated a definite bond angle 135 ± 8°, in agreement with all three earlier theoretical predictions. The problem with Herzberg's prediction of a linear CH_2 ground state has now been recognized as due to broadening beyond recognition of certain subbands (which occur for nonlinear molecules) by an unusually strong predissociation (163).

Table V-7 summarizes the theoretical predictions for the CH_2 ground state geometry. In light of the uncertainty (8°) in the experimental determination of the CH_2 bond angle, even more complete calculations than those discussed above were undertaken by McLaughlin, et al. (244). A C(10s 6p 2d / 5s 3p 1d) and H(5s 1p / 3s 1p) extended gaussian basis was chosen and first-order wave functions including 617 $^3B_1$ configurations computed. The iterative natural orbital method was used to obtain an optimum set of orbitals. These calculations on CH_2 are among a very small number of polyatomic calculations in which orbitals have been optimized during the CI process. A comparable calculation on H_2O gives a bond angle 1° smaller than experiment, so the final prediction of 134° for the CH_2 bond angle should be very reliable.

The second question of interest concerning methylene involves the magnitude of the singlet-triplet splitting. On the basis of his spectroscopic investigations, Herzberg (159) concluded that this energy difference is <u>less than</u> 0.99 eV. In 1967, Halberstadt and McNesby (144) analyzed the photolysis of ketene (which yields CH$_2$) to obtain a $^3B_1$-$^1A_1$ splitting of 2.5 kcal/mole. From a similar experiment Carr, et al. (68) obtained an energy difference of 1-2 kcal. The essence of both of these analyses is the contention that a very small singlet-triplet splitting is required to account for the production of singlet ($^1A_1$) methylene under the experimental conditions. However, the calculations of Foster and Boys (124) and Harrison and Allen (148) both predict a $^1A_1$-$^3B_1$ splitting of more than 24 kcal. Once again there is an apparent conflict between theory and experiment.

From calculations analogous to those of Bender and Schaefer (30), O'Neil, et al. (285) predicted the singlet-triplet splitting to be 0.96 eV or 22 kcal. Furthermore, they estimated that the actual $^1A_1$-$^3B_1$ splitting is in fact not likely to be less than 12 kcal. Figure V-1 shows O'Neil's theoretical potential energy surfaces (C_{2v} symmetry imposed) for the $^3B_1$, $^1A_1$, and $^1B_1$ states of CH$_2$. These three-dimensional representations show clearly that the predicted ground state bond angle is ~135°. The $^1A_1$ bond angle is smaller (calculated 104°, experiment 102°) and the $^1B_1$ angle is larger (calculated 144°, experimental ~140°). The interpretation of Figure V-1 is aided by noting that the $^3B_1$ state barrier to linearity is 6.7 kcal. The figures shows the $^1A_1$ barrier to be much larger and the $^1B_1$ barrier quite small.

The most extensive calculations reported to date on the $^1A_1$ state are those

Table V-7. Increasingly reliable theoretical predictions of the ground state bond angle and singlet-triplet energy difference (in eV) for methylene.

Authors	$^3B_1$ Energy (hartrees)	$^3B_1$ Angle	$\Delta E(^3B_1 - {}^1A_1)$
Foster and Boys (124)	-38.904	129°	1.06
Harrison and Allen (148)	-38.9151	138°	1.39
Bender, Schaefer, and O'Neil (30,285)	-38.9822	135°	0.96
McLaughlin, Bender, and Schaefer (32,244)	-39.0121	134°	0.60

of Bender and Schaefer (32). The extended gaussian basis set and method of calculation are the same as those of McLaughlin, et al. (244) for the ground state. For the $^1A_1$ state, however, there are two very important configurations, $1a_1^2 2a_1^2 1b_2^2 3a_1^2$ and $1a_1^2 2a_1^2 1b_2^2 1b_1^2$. Therefore a two-configuration SCF calculation was carried out first, followed by first-order calculations including 338 configurations. The results of this calculation, as well as the earlier work, are seen in Table V-7. The predicted singlet-triplet splitting is smaller, but still far above the suggested experimental values of 2.5 and 1-2 kcal. However, Allen (7) has made a detailed analysis of experimental data for a number of carbene compounds CXY (X and Y replace the hydrogens in CH_2) and concluded that

Triatomic Molecules 315

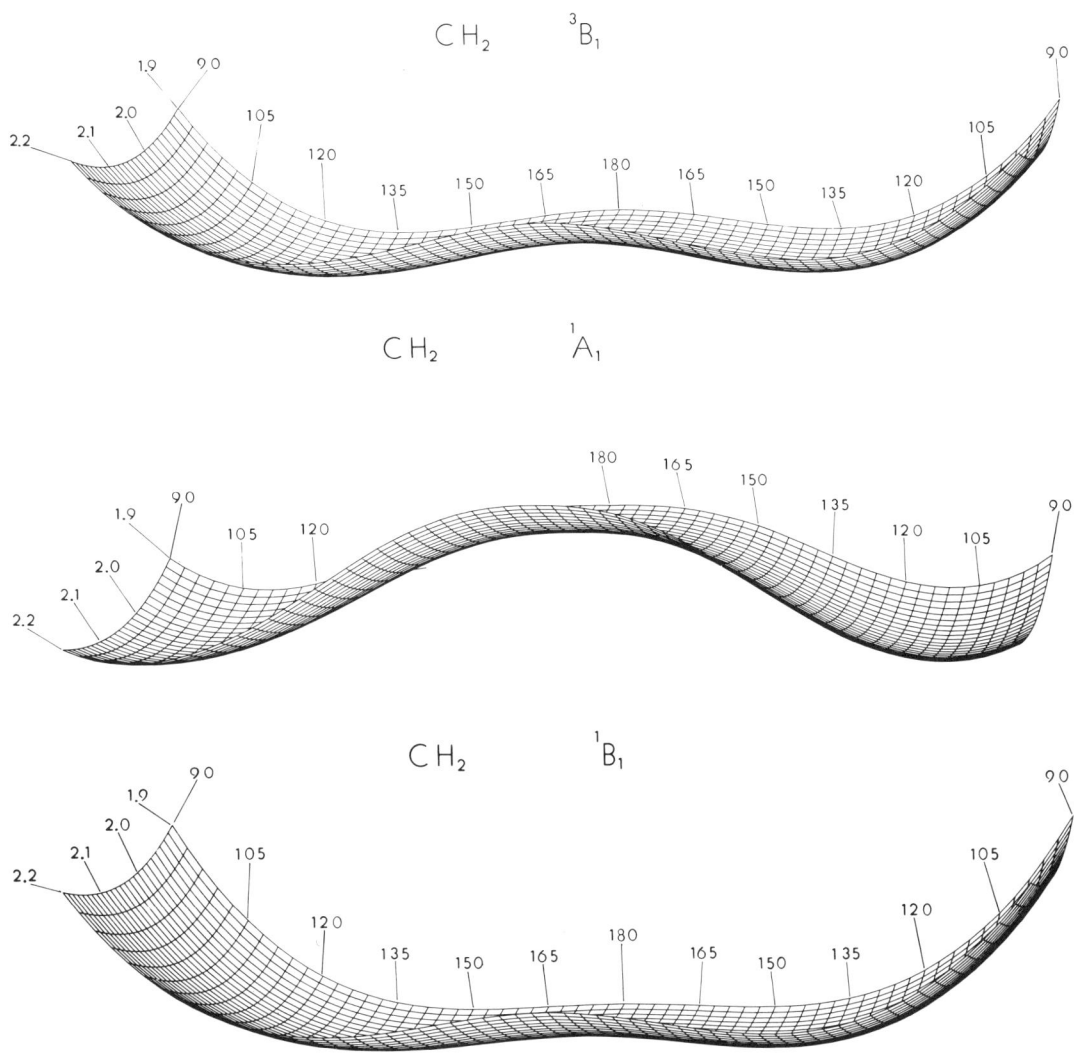

Figure V-1. Theoretical potential energy surfaces for the three lowest electronic states of CH_2. The total energy is plotted as a function of bond distance (in bohrs) and bond angle in degrees.

the singlet-triplet splitting in CH_2 may very well be ~0.5 eV or 11-12 kcal. As can be imagined, the CH_2 singlet-triplet energy difference remains a topic of great interest to theoreticians and experimentalists alike.

The H_3 potential surface. From 1931 to the present, the H_3 potential surface has played an intimate role in the development of chemical reaction theories. The first H_3 surface, of course, was that of Eyring and Polanyi (117), a semi-empirical surface based on London's valence bond expression (232) for the energy of H_3. Over the years a large number of potential surfaces have been proposed for the $H + H_2 \rightarrow H_2 + H$ reaction. The primary purpose of many of these surfaces was to test the transition state theory of Eyring and Polanyi (187). With the development of more sophisticated theories of reaction dynamics (260), the need for accurate potential surfaces has become even more acutely felt. An ultimate goal of theory would be to choose a chemical reaction and a) employ electronic structure theory to calculate the appropriate potential energy surface, b) given the above potential energy surface, use scattering theory to predict detailed dynamical properties, e.g. differential cross sections, and c) perform suitable averaging over the information obtained in b) to predict macroscopic properties, e.g., rate constants as a function of temperature. For the $H + H_2 \rightarrow H_2 + H$ reaction the above procedure is well underway, in large part due to the ab initio potential energy surface of Shavitt, Stevens, Minn, and Karplus (347).

The calculations of Shavitt, et al. used a double-zeta-plus-polarization basis set of Slater functions on each hydrogen atom, H(2s 1p). With this basis, a full CI consists of 200 configurations for $^2\Sigma_u^+$ symmetry (linear symmetric).

402 configurations for $^2\Sigma^+$ symmetry (linear asymmetric), and 680 configurations for $^2A'$ symmetry (C_s point group, only a plane of symmetry). The above comparison illustrates nicely the increase in number of configurations accompanying a move to a point group of lower symmetry. Recall (section IC) that a configuration is a fixed linear combination of Slater determinants, with coefficients chosen to reflect the symmetry of the molecular state being described. In the work of Shavitt, et al., full CI was performed for a large number of both linear and nonlinear H_2 geometries. In addition, vigorous optimization of orbital exponents was carried out. For linear symmetric H_3 near the saddle point the optimized exponents for the terminal hydrogens were (1s) = 0.884, (1s') = 1.233, (2p) = 1.625. The central hydrogen exponents were (1s) = 1.045, (1s') = 1.214, (2p) = 1.671.

One measure of the reliability of any H_3 potential surface is how well it describes the asymptotic limits H and H_2. The H atom is of course no problem, but the quantitative prediction of the H_2 dissociation energy can be difficult. For example, the RHF dissociation energy of H_2 is 84 kcal while the correct value is 109 kcal. The Shavitt calculation, including correlation by CI, gives a value of 106 kcal. In addition the H_2 bond distance R_e predicted is 1.4018 bohrs, compared to experiment, 1.4008. These two H_2 comparisons make it reasonable to assume that the other features, both energetic and geometrical, of the H_3 surface are at least qualitatively correct.

Most of the qualitative questions to be answered concerning the H_3 surface center about the reaction coordinate or minimum energy path. Calculations were carried out for three different angles of approach: θ = 120°, 150°, and 180° (linear) where θ is the H-H-H angle. For each angle of approach a minimum energy path was obtained by varying the H_A-H_B and H_B-H_C distances. For all three minimum energy paths the predicted point of highest energy occurs for a symmetric H_3 arrangement. For the 120° approach, the energy maximum occurs for geometry $R(H_A$-$H_B) = R(H_B$-$H_C) = 1.809$ bohrs, and lies 18.2 kcal/mole above separated $H + H_2$. For 150°, $R(H$-$H) = 1.777$ bohrs was found to lie 12.6 kcal above $H + H_2$. As had been assumed in most earlier work, the lowest minimum energy path was predicted to be linear. The (linear) saddle point occurs at $R(H$-$H) = 1.765$ bohrs and the barrier height is 11.0 kcal.

The H_3 barrier height has long been the subject of controversy, and for many years was thought to be 7-8 kcal. This value was supported by the theoretical work of Conroy and Bruner (84) who predicted a 7.7 kcal barrier based on calculations purported to be nearly exact. Clearly the value 11.0 kcal predicted by Shavitt, Stevens, Minn, and Karplus (SSMK) does not agree with these lower values. The experimental activation energy E_a is 7.44 kcal. However, as mentioned in section IVB, there is not a simple relation between activation energy and barrier height. Shavitt (348) has shown that after scaling down the SSMK surface by 11%, the rigorous application of transition state theory yields rate constants in agreement with the best experimental data. Shavitt concludes that the true barrier height is 9.8 ± 0.2 kcal. Recently, a very reliable series of calculations on linear symmetric H_3 has been reported by Liu (230). A large basis set of Slater functions was used and 321 configurations chosen by the pseudonatural orbital method. The predicted barrier height is 10.1 kcal and Liu estimates the exact barrier height to be 9.6 kcal. Taking all of the above into consideration, the author of this book tends to conclude that the H_3 barrier height is between 9 and 10 kcal.

Figure V-2 shows the linear H_3 potential surface of Shavitt, Stevens, Minn, and Karplus. After a scaling down by ~11%, this surface is expected to be essentially correct. A feature not present in Figure V-2 is a relative minimum in the saddle point region. Since this feature was present in the early semi-empirical surfaces of Eyring and coworkers, it was affectionately referred to as "Eyring's Lake." However, as Professor Eyring pointed out during the August, 1970 Conference on Potential Energy Surfaces at Santa Cruz, it now appears that "Eyring's Lake" was a mirage. A feature which <u>should</u> be present in an H_3 potential surface is the long range (about 6 bohrs separation) van der Waals attraction between H and H_2. However, as was seen in the chapter III discussion of the He-He van der Waals interaction, special methods are required to describe this attraction. In particular a large basis, including more diffuse functions than that of Shavitt, et al. (347) is required.

<u>Electron correlation in the water molecule</u>. For a given number of electrons, the most advanced electronic structure calculations on atoms are likely to be more complete than those on diatomic molecules. For example, the most exhausitve variational calculation reported to date (1971) on the neon atom accounts for ~89% of the correlation energy (63), while for hydrogen fluoride only ~74% of the correlation energy has been recovered (29). Continuing along this iso-electronic sequence, one would expect less than 74% of the correlation energy to have been calculated variationally for the nonlinear polyatomic molecule H_2O. However, due to a remarkable calculation by Meyer (258), this is not the

Figure V-2. Contour maps of the predicted linear potential surface for H₃. (a) shows that part of the surface near the saddle point in detail, while (b) shows the entire surface. Energies are in kcal relative to separated H + H₂. The x in figure (a) shows the position of the saddle point.

(a)
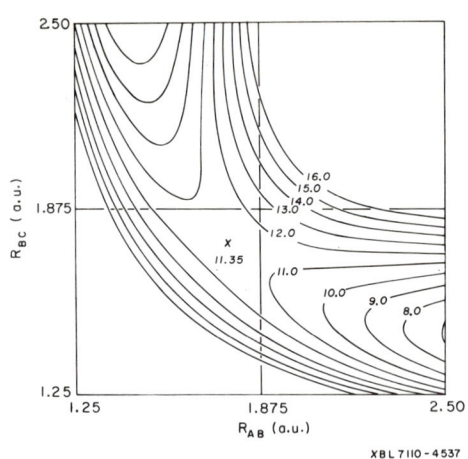

(b)
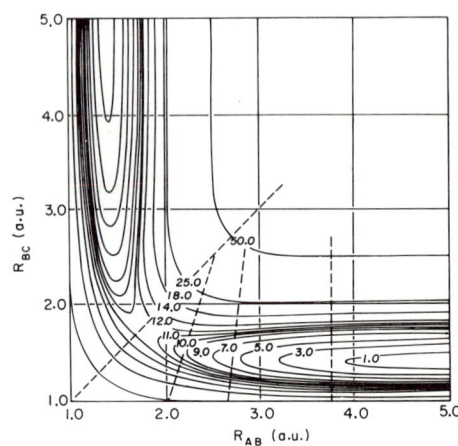

case. In fact Meyer's water calculation accounts for 84% of the correlation energy and thus is nearly as complete as the most exhaustive neon calculation. This calculation provides a great deal of important information concerning the nature of electron correlation in polyatomic molecules.

Meyer uses a large basis set of uncontracted gaussian functions, O(11s 7p 4d 1f) and H(5s 1p), centered on each atom. In addition, a 1s gaussian function is centered in both O-H bonding regions. The SCF energy obtained with this basis is -76.06275 hartrees, compared to the estimated RHF energy -76.066 hartrees. Basis sets this large are rarely used for polyatomic molecules, but experience on atoms and diatomics makes it clear that one must go to such extremes to obtain more than 80% of the correlation energy of a ten-electron system.

The second key to the success of this H_2O calculation is the use of a <u>nonorthogonal</u> set of molecular orbitals. In general, of course, theoretical chemists tend to immediately rule out the use of nonorthogonal orbitals. As discussed in section IC, this is due to the extreme difficulty in evaluating Hamiltonian matrix elements H_{ij}, Eq. (I,51), between configurations i and j. Meyer has surmounted this problem by requiring his wave function to be of a restricted form and deriving specialized but tractable formulas for the nonorthogonal matrix elements H_{ij}. The calculation may be divided into three steps:

    a) Conventional SCF calculation.

    b) Independent calculation of pseudonatural orbitals (section IC) for each pair of SCF orbitals. For the $^1A_1$ ground state of H_2O (SCF orbital occupancy $1a_1^2\ 2a_1^2\ 1b_2^2\ 3a_1^2\ 1b_1^2$), there are 25 different sets of pseudonatural orbitals. There are five diagonal pairs ($1a_1^2$, $2a_1^2$, $1b_2^2$, $3a_1^2$, and $1b_1^2$) and ten off-diagonal pairs of orbitals. However, for each off-diagonal pair there are two different singlet spin configurations. For example the $1a_1\ 2a_1$ PSNO's are calculated from orbital occupancies

$$1a_1\ 2a_1\ 1b_2^2\ 3a_1^2\ 1b_1^2\ x\ y \qquad\qquad (V,9)$$

in which x and y are all other (besides the SCF) orbitals in the basis set. One set of PSNO's is obtained by singlet-coupling $1a_1$ and $2a_1$ and then singlet-coupling all pairs of orbitals x and y. The second set of PSNO's for the $1a_1\ 2a_1$ pair is obtained triplet-coupling $1a_1$ and $2a_1$ and then triplet-coupling x and y to obtain a ten-electron singlet (S=0) function. As discussed in chapter I, the PSNO's obtained in this way provide the fastest convergence for each particular type of electron correlation. However, the overall convergence for the n-electron system is usually lost due to the intractability accompanying the fact that each set of PSNO's is nonorthogonal to the other 24 sets of PSNO's.

c) 10-electron CI using the first few PSNO's from each of the 25 calculations described in b). Several points should be made concerning the nature of this CI: i) no single excitations are included since these have zero H matrix elements with the SCF configuration and will be energetically unimportant, ii) only doubly-excited configurations (besides the SCF) are included, iii) due to the unusual properties of natural orbitals for two-electron systems, each PSNO occurs in only a single configuration.

The nonorthogonality which must be handled is thus only partial nonorthogonality. That is, each PSNO is orthogonal to the SCF orbitals and the other PSNO's arising from the same pair, e.g., the singlet-coupled $1a_1$ $2a_1$ pair. By skillfully exploiting this partial nonorthogonality, Meyer has made possible a very complete H_2O calculation.

The final variational H_2O calculation includes 237 $^1A_1$ configurations, built from 356 PSNO's. The total energy obtained is -76.36834 hartrees, corresponding to a calculated correlation energy 0.30559 hartree. The present author has estimated the correlation energy of H_2O to be 0.364 ± 0.01 hartree. Thus Meyer's calculation is seen to recover about 84% of the correlation energy. The same configurations included in this variational calculation were used to evaluate the symmetry-adapted pair correlation energies, seen in Table V-8. The first point to be made is that the sum of the symmetry-adapted pair correlation energies is 98% of the estimated correlation energy. Thus it would appear that an exact (complete basis set) symmetry-adapted pair correlation treatment would yield 112-115% of the H_2O correlation energy. Use of spinorbital correlation energies (section IID) of course would result in an even more severe overestimation of the total correlation energy. A second point concerns the fact that the valence-electron interpair correlation energies (e.g., $2a_1$ -$3a_1$) are all relatively large. It would be extremely interesting to see whether a pair correlation study carried out in terms of localized orbitals would reduce the interpair correlation energies. In addition, one can reasonably hope (see Table III-14) that use of localized orbitals would

Table V-8. Symmetry-adapted pair correlation energies (in hartrees) for the water molecule (258).

Symmetry-Adapted Pair	Correlation Energy
$1a_1 - 1a_1$	0.0390
$1a_1 - 2a_1$	0.0039
$2a_1 - 2a_1$	0.0122
$1a_1 - 3a_1$	0.0045
$2a_1 - 3a_1$	0.0245
$3a_1 - 3a_1$	0.0267
$1a_1 - 1b_1$	0.0051
$2a_1 - 1b_1$	0.0268
$3a_1 - 1b_1$	0.0452
$1b_1 - 1b_1$	0.0267
$1a_1 - 1b_2$	0.0035
$2a_1 - 1b_2$	0.0275
$3a_1 - 1b_2$	0.0434
$1b_1 - 1b_2$	0.0415
$1b_2 - 1b_2$	0.0264
a) Sum	0.3580
b) SCF Energy	-76.0628
a) + b)	-76.4208
Variational CI Result	-76.3683

bring the sum of the pair energies into closer agreement with the variationally computed correlation energy.

Meyer has carried out related calculations on the five hole states of H_2O obtained by removal of a $1a_1$, $2a_1$, $1b_2$, $3a_1$, or $1b_1$ orbital from the H_2O ground state orbital occupancy. These results are summarized in Table V-9. The direct hole state calculations are of the type first carried out by Bagus on argon (section IIA). In all five cases the agreement between direct hole state CI and experiment must be considered quite good. The calculations allow some insight into the correlation problem in positive ions. For example, the $1a_1$ (essentially 1s oxygen) hole state calculation shows that there is about 0.5 eV more correlation energy in H_2O than in this state of H_2O^+. This result is not entirely expected since the $1a_1^2$ correlation energy of H_2O is (see Table V-8) more than 1 eV and there is no correlation of this type in the $1a_1$ hole state. Table V-9 shows that the $^2A_1$ state of H_2O^+ resulting from ejection of the $2a_1$ electron has ~2 eV <u>more</u> correlation energy than neutral H_2O. This would be a very surprising result indeed had not Bagus predicted the 2s hole state of neon to have a greater correlation energy than neutral neon (10). A detailed analysis of the $2a_1$ hole state calculation shows that this state of H_2O^+ has more correlation than H_2O because of the increased importance of the types of polarization and degeneracy effects included in first-order wave functions (section IIB).

B. MOLECULES WITH FOUR THROUGH EIGHT ATOMS

<u>Geometry predictions from simple basis sets.</u> An interesting conclusion arrived at during the past few years by theoretical chemists is that SCF

calculations, even using quite small basis sets, give rather reliable geometry predictions. An illustration of this point is given by the calculations of Sachs, Geller, and Kaufman (331) on the $^1A_1$ state of CF_2. They optimized the geometry of CF_2 by SCF calculations using (3s 1p) and (5s 2p) uncontracted gaussian sets on each atom. The predicted geometries were r(C-F) = 2.594 bohrs, θ = 106.20° for the small basis and r(C-F) = 2.516 bohrs, θ = 105.84° for the larger set. The experimental bond distance is 2.456 bohrs and bond angle 104.9°. The qualitatively correct geometry prediction from the (3s 1p) basis is certainly an initial surprise when one realizes that this basis yields an F atom SCF energy 5.2675 hartrees = 143 eV = 3305 kcal above the RHF energy. It is seen that very poor total energies can lead to rather reasonable geometry predictions. At this point we discuss two well-tested theoretical approaches which use simple basis sets.

One of the simplest <u>ab initio</u> methods imaginable is the floating spherical gaussian orbital (FSGO) method introduced by Frost (129). For closed-shell molecules n/2 1s gaussian functions are used to construct a single determinant wave function. However, these n/2 spherical gaussians are allowed complete flexibility. That is, the energy of the n-electron wave function is minimized with respect to both the positions and orbital **exponents** α_i of the n/2 functions. Some difficulty is introduced in the calculations due to the nonorthogonality of the different spherical gaussians. However, the formula for the energy of a single determinant constructed from nonorthogonal orbitals is still relatively simple (129).

One of the most appealing aspects of the FSGO method is that the wave function

Table V-9. Theoretical and experimental vertical (calculated for the ground state H_2O geometry) ionization potentials, in eV.

Orbital Ejected	State of H_2O^+	Koopmans' Theorem	Direct Hole State SCF	Direct Hole State CI	Experiment
$1a_1$	$^2A_1$	559.5	539.1	539.6	539.7
$2a_1$	$^2A_1$	36.77	34.22	32.25	32.19
$1b_2$	$^2B_2$	19.50	17.59	18.73	18.55
$3a_1$	$^2A_1$	15.87	13.32	14.54	14.73
$1b_1$	$^2B_1$	13.86	11.10	12.34	12.61

is an antisymmetrized product of localized electron-pair functions. Each electron pair can be immediately identified as inner shell, bonding, lone pair, etc., very much in keeping with the classical chemical concepts of Lewis and Langmuir. For methane, for example, there are five electron pairs. One spherical gaussian describes the carbon 1s orbital and the other four describe the CH bonds. The CH bond orbital is centered 1.256 bohrs from the carbon atom, while the CH bond distance is 2.107 bohrs. Each spherical gaussian may be characterized by an orbital radius ρ_i related to the orbital exponent α by

$$\alpha_i = 1/\rho_i^2 \qquad (V,10)$$

For methane the 1s carbon orbital has radius 0.328 bohrs while the CH bond orbital has a much greater spatial extent, $\rho = 1.694$ bohrs.

The greatest success of the FSGO method has been in the treatment of hydrocarbons. Table V-10 shows the FSGO geometry predictions of Frost and Rouse (130) for methane through cyclopropane. The nine predicted bond distances are in very reasonable agreement with experiment, the largest discrepancy being 0.033 Å for the C-C distance in ethane. The H-C-H angles in ethane and ethylene are very accurately predicted. Even more important are the two chemical trends correctly predicted in Table V-10. In going from ethane to ethylene to acetylene, both the C-H and C-C bond distances decrease, as is observed experimentally. Such a simple model of course will not always give reliable predictions. For the H_2O molecule, for example, the predicted bond angle is 88.4° and bond distance 0.881 Å, as opposed to experiment, 104.5° and 0.957 Å.

A second approach to geometry predictions involves SCF calculations using minimum basis sets of Slater functions, each of which expanded as a linear combination of gaussians. This approach was discussed in section ID and some typical results summarized in Table I-11. Like the FSGO method, the use of a minimum basis is well suited to the investigation of chemical problems, due to the rapid speed of computation. Most of the work reported to date along these lines is that of Pople and coworkers (154). Table V-11 shows the geometry predictions of Newton, et al. (283) for a variety of molecules containing four and five atoms. For each molecule, SCF calculations were repeated as a function of geometry until a minimum energy is reached. Generally there is quite good agreement between calculated and experimental geometries. The average bond distance error is 0.03 Å and bond angles are usually within a few degrees of

Table V-10. Geometries of hydrocarbons predicted by the floating spherical gaussian orbital (FSGO) method. Bond distances are in Å and bond angles in degrees.

Molecule	Parameter	FSGO	Experimental
Methane	R(C-H)	1.115	1.093
Ethane	R(C-H)	1.120	1.093
	R(C-C)	1.501	1.534
	H-C-H Angle	108.2	109.1
Ethylene	R(C-H)	1.101	1.086
	R(C-C)	1.351	1.337
	H-C-H Angle	118.7	117.3
Acetylene	R(C-H)	1.079	1.059
	R(C-C)	1.214	1.205
Cyclopropane	R(C-C)	1.533	1.510

experiment. A notable exception is F_2O_2, where the O-F theoretical bond distance is 0.217 Å too short and the O-O bond distance is 0.175 Å too long. It is worth noting in this regard that SCF calculations based on a larger double zeta basis set yield a much improved FOOF geometry.

Newton, et al. (283) have also calculated quadratic force constants for

Table V-11. Geometry predictions from minimum basis set (each Slater function expanded as a linear combination of three gaussians) SCF calculations. Bond distances are in Å and bond angles in degrees.

Molecule	Parameter	Three-Gaussian	Experiment
CH_4	R(C-H)	1.083	1.085
CH_3F	R(C-H)	1.097	1.105
	R(C-F)	1.384	1.385
	θ(HCH)	108.3	109.9
CH_2F_2	R(C-H)	1.109	1.091
	R(C-F)	1.378	1.358
	θ(HCH)	108.8	112.1
	θ(FCF)	108.7	108.2
CHF_3	R(C-H)	1.119	1.098
	R(C-F)	1.371	1.332
	θ(FCF)	108.6	108.8
CF_4	R(C-F)	1.366	1.317
H_2CO	R(C-H)	1.101	1.101
	R(C-O)	1.217	1.203
	θ(HCH)	114.5	116.5
NH_3	R(N-H)	1.033	1.012
	θ(HNH)	104.2	106.7
NF_3	R(N-F)	1.386	1.365
	θ(FNF)	102.1	102.3
F_2N_2	R(N-F)	1.277	1.384
	R(N-N)	1.373	1.214
	θ(FNN)	111.5	114.5
H_2O_2	R(O-H)	1.001	0.950
	R(O-O)	1.396	1.475
	θ(HOO)	101.1	94.8
	dihedral angle	125 ± 2	111.5
F_2O_2	R(O-F)	1.358	1.575
	R(O-O)	1.392	1.217
	θ(FOO)	104.2	109.5
	dihedral angle	88.1	87.5

a wide variety of molecules using the minimum basis SCF approach. Both stretching and bending force constants are in surprisingly good agreement with experiment, being typically 25% too large. Their results for water may be compared both to experiment and to the extended basis SCF calculations of Ermler and Kern (114) discussed in section VA. For the OH stretch, Newton predicts 9.7 md/Å, compared to Ermler and Kern, 9.876, and experiment 8.454 md/Å. The use of a limited basis set only becomes obvious for the bending force constant, predicted to be 1.32 md/Å from the minimum basis calculation. This compares unfavorably with Ermler and Kern's value 0.881 and experiment 0.761 md/Å.

ls binding energies of BH_3, CH_4, NH_3, H_2O, and HF. The emergence of ESCA or x-ray photoelectron spectroscopy as an important area of chemistry has generated a great deal of interest in the core-electron ionization potentials of molecules (350). Of particular importance is the fact that inner shell ionization potentials (e.g., the carbon ls binding energy) vary with chemical environment, resulting in chemical shifts. The series of molecules CH_4, CH_3F, CH_2F_2, CHF_3 and CF_4, for example have carbon ls binding energies 290.8, 293.6, 296.3, 299.1, and 301.8 eV. It is clear that the substitution of a fluorine atom in the place of hydrogen increases the carbon ls ionization potential. Relative to methane, the carbon ls chemical shifts for the four fluorinated molecules are 2.8, 5.5, 8.3, and 11.0 eV. The same trend is predicted by the use of Koopmans' theorem, the predicted chemical shifts being 2.9, 6.1, 9.4, and 12.7 eV when double zeta basis sets are used (53). It is clear that theoretical chemists are in a position to contribute both to the prediction and understanding of core electron binding energies.

We know from the calculations of Bagus on the hole states of neon and argon that Koopmans' theorem yields inner shell ionization potentials that are too large. However, a direct SCF hole state calculation on the appropriate positive ion allows a comparable treatment of positive ion and neutral molecule and gives quite accurate ionization potentials. The first direct calculations of this type on molecules were carried out by Schwartz (346) for the first-row hydrides BH_3, CH_4, NH_3, H_2O and HF. The ground state SCF wave functions for these five closed-shell molecules are all single determinants. Contracted gaussian basis sets were used, (10s 5p / 6s 3p) on each central atom and (5s / 2s) on hydrogen. Although the absolute energies obtained with this basis are not close to the RHF energies (primarily due to the lack of polarization functions), energy <u>differences</u> are of importance here. And for the neon 1s ionization potential, SCF calculations with this basis yield 868.8 eV, compared to the near RHF result of Bagus, 868.6 eV.

Table V-12 summarizes Schwartz's results (346). Except for BH_3 and hydrogen fluoride the experimental core-electron ionization potentials have been measured by Siegbahn and coworkers (350). For methane Koopmans' theorem places the positive ion 14 eV too high, since the molecular orbitals are not allowed to "relax" after ejection of the 1s electron. Allowing for this relaxation by the direct SCF calculation on the 1s hole state (orbital occupancy $1a_1\ 2a_1^2\ 1t_1^6$, symmetry $^2A_1$), the ionization potential is the difference between the SCF energies for the neutral molecule and positive ion. This difference is 291.0 eV, in close agreement with the ESCA value, 290.7 eV (350). Excellent agreement with experiment is also found for ammonia and water. In fact for

Table V-12. Theoretical and experimental 1s binding energies of the first-row hydrides. Unless indicated, energies are in eV.

Molecule	Neutral SCF Energy (hartrees)	Orbital Energy	Direct SCF Positive Ion Energy (hartrees)	Direct SCF Ionization Potential	Experimental Ionization Potential
BH_3	-26.3779	207.3	-19.1200	197.5	----
CH_4	-40.1812	304.9	-29.4873	291.0	290.7
NH_3	-56.1716	422.8	-41.2626	405.7	405.6
H_2O	-76.0045	559.4	-56.1791	539.4	539.7
HF	-100.0158	715.2	-74.5365	693.3	----
Ne	-128.5159	891.4	-96.5848	868.8	870.2

NH_3 the agreement is so good, 0.1 eV, that it is probably partly fortuitous. That is, there are cancelling effects of the order of 1 eV due to basis set limitations, correlation energy corrections and relativistic effects. Nevertheless the agreement between SCF direct hole state calculations and experiment is truly impressive.

The inner shell binding energy of the BH_3 molecule is likely to be difficult to observe. However, Schwartz points out that the 1s ionization potentials of methane and ethane are nearly identical. Therefore it seems plausible that the predicted BH_3 1s ionization potential will be nearly the same as that in B_2H_6 and other boron hydrides. In one sense, of course, this would be discouraging, since it would not be possible to distinguish between the different boron hydrides on the basis of chemical shifts of core electron binding energies. Of course, substitution of fluorine or some other electronegative atom would be expected to radically change the boron 1s binding energies.

We have discussed the inner shell ionization potentials of a variety of systems including Ne and Ar (section IIA), NO and O_2 (section IIIA), H_2O (section VA), and the polyhydrides considered here. Therefore this is a reasonable point at which to assess the basic theoretical idea used in these calculations. This basic contention, due to Bagus (10), is that single configuration SCF wave functions can provide practical (but not rigorous) upper bounds to the energies of inner-shell hole states. The physical correctness of shell model is thought to be responsible for the reliability of the predicted binding energies. From the practical theoretical experience on the above variety of molecules, it now seems fair to say that Bagus' contention has proved correct. An example may clarify this point. For methane the $^2A_1$ state corresponding to orbital occupancy $1a_1 \; 2a_1^2 \; 1t_1^6$ of course has a very shoft lifetime, since it lies above the energy of the doubly ionized ground state of CH_4^{+2}. In fact the Auger spectrum of methane arises from transitions between the 1s hole state (CH_4^+) and the various states of CH_4^{+2}. Despite the unstable or autoionizing nature of the 1s hole state of CH_4^+, the calculations show that a) this state may be theoretically treated as a stationary state, and b) the orbital picture of molecular structure is sufficient to approximately guarantee orthogonality to all lower states of CH_4^+ of $^2A_1$ symmetry.

The $H_2 + D_2$ exchange reaction. The simplest four-center exchange reaction has long been thought to be

$$H_2 + D_2 \rightarrow 2HD \qquad (V,11)$$

The experimental investigation of this reaction was climaxed by the shock tube work of Bauer and Ossa (26), reported in 1966. Their data may be rationalized by assuming that exchange occurs with high probability only during encounters between H_2-D_2 pairs, one of which is vibrationally excited to approximately 30 kcal/mole. Bauer and Ossa also determined the experimental activation energy to be 42 kcal/mole. Further experimental studies of the same type have been carried out by Burcat and Lifshitz (64) who determined an activation energy of 40 ± 1 kcal, providing support for the "vibrational excitation" mechanism of Bauer and Ossa. The $H_2 + D_2$ reaction is of particular interest to theoreticians a) because of its role as a prototype reaction, and b) due to the fact that several theoretical studies appear to be in conflict with the assumption of a square planar transition state.

Three *a priori* studies of the H_4 potential energy surface have been reported during the past two years. The first, by Wilson and Goddard (398), consisted of full CI calculations with a minimum basis of Slater functions (1s orbital exponent $\zeta = 1.05$). The second series of calculations, also full CI, was done by Rubinstein and Shavitt (327) using a double zeta basis set (two 1s Slater functions on each H atom). The most recent calculations are those of Conroy and Malli (85) and make explicit use of interparticle coordinates. Although this type of calculation appears to be subject to numerical instabilities (see the H_3 discussion in section VA), these errors appear to be of an order of magnitude smaller than the energies of interest for H_4.

The H_4 geometry of greatest interest is the square, since the square corresponds to the transition state (saddle point) along the reaction coordinate (minimum energy path) of the expected bimolecular exchange mechanism. Wilson and Goddard find a minimum energy of -2.06 hartrees, for H-H separation 2.6 bohrs. Since the energy of two separated H_2 molecules is -2.296 hartrees, the predicted barrier height is 148 kcal/mole. Rubinstein and Shavitt calculate an H_4 square energy of -2.075 hartrees for H-H distance 2.47 bohrs, corresponding to a barrier height of 142 kcal/mole. Finally, Conroy and Malli report an energy -2.142 hartrees for a square 2.25 bohrs on a side, which indicates a barrier height of 130 kcal/mole. The surprising result is that all three barrier heights are more than three times the experimental activation energy. The consistency of these results essentially rules out the possibility of the $H_2 + D_2$ reaction proceeding by the square planar exchange mechanism. In fact only 109 kcal are required to break the D_2 bond completely, after which the three-center exchange reaction $H_2 + D \rightarrow H + HD$ could occur. However this mechanism is ruled out by the observed activation energy.

Attempting to find a reaction coordinate corresponding to the experimental activation energy of ~41 kcal, both conventional CI studies (327,398) considered a variety of H_4 geometries. Figure V-3 shows the geometrical arrangements investigated by Wilson and Goddard. For some of these geometries the symmetry of the ground state is not apparent and full CI calculations were carried out for several symmetries. Based on these calculations (327,398), the lowest

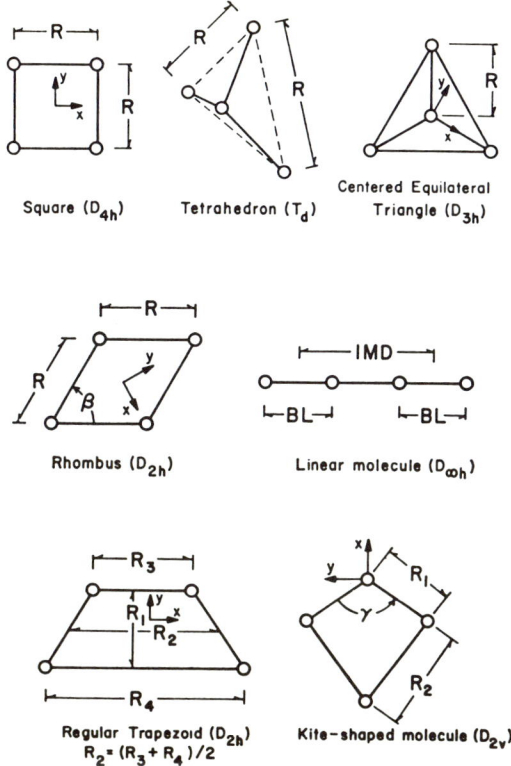

Figure V-3. Possible transition state geometries for the $H_2 + D_2$ reaction.

singlet (triplet states are not relevant since the H_2 and D_2 molecules are initially in their $^1\Sigma_g^+$ electronic ground states) states and corresponding single configuration descriptions for several geometries are

$$\text{Square} \qquad ^1B_{1g} \quad 1a_{1g}^2\ 1e_u^2 \qquad\qquad (V,12)$$

$$\text{Tetrahedron} \qquad ^1E \quad 1a_1^2\ 1t_2^2 \qquad\qquad (V,13)$$

Rhombus	$^1B_{1g}$	$1a_{1g}^2\ 1b_{2u}\ 1b_{3u}$	(V,14)
Rectangle	^1A_g	$1a_{1g}^2\ 1b_{2u}^2$	(V,15)
Linear Equidistant	$^1\Sigma_g^+$	$1\sigma_g^2\ 1\sigma_u^2$	(V,16)

Table V-13 summarizes the energetics predicted by the double-zeta CI. The smaller basis set results of Wilson and Goddard are qualitatively similar. Both calculations predict the tetrahedral arrangement to be highly unfavorable In addition, both calculations show that the lowest electronic state ($^1B_{1g}$) for the square is stable with respect to rhombic distortion. This would appear to rule out the rhombus as a possible transition state. The most favorable approach is seen in Table V-13 to be linear. Clearly, however, this cannot serve as a transition state for the four-center exchange reaction. Rubinstein and Shavitt point out that the linear arrangement is so much lower in energy than the square that some intermediate transition state might be possible. The kite shape and isosceles trapezoid are suggested. Wilson and Goddard have carried out some calculations of this type and are unable to find such a transition state. Therefore, we must conclude that at present, the mechanism of the $H_2 + D_2$ reaction remains a mystery. The H_4 problem is certain to remain a subject of great interest until some way is found to reconcile the theoretical calculations with the low experimental activation energy of 40-42 kcal/mole.

The T and V states of ethylene. The ground state RHF orbital occupancy of ethylene (planar, D_{2h} symmetry) is (using Herzberg's (161) notation)

Table V-13. Results of Rubinstein and Shavitt for possible transition states of the $H_2 + D_2$ reaction.

Geometry	Symmetry	R_e (bohrs)	E(hartrees)	Barrier Height (kcal)
Two H_2 Molecules		1.42	-2.302	0
Linear Equidistant	$^1\Sigma_g^+$	1.70	-2.234	43
Rectangle (ratio of sides constrained to be 4:5)	^1A_g	2.09, 2.61	-2.129	109
Square	$^1B_{1g}$	2.47	-2.075	142
Rhombus (apex angle constrained to be 70°)	$^1B_{1g}$	2.5	-2.061	151
Tetrahedron	^1E	3.8	-2.002	188

$$1a_g^2\ 1b_{1u}^2\ 2a_g^2\ 2b_{1u}^2\ 1b_{2u}^2\ 3a_g^2\ 1b_{3g}^2\ 1b_{3u}^2\ ^1A_g \qquad (V,17)$$

The two lowest excited electronic states are expected to arise from the orbital occupancy in which one of the $1b_{3u}$ orbitals is replaced by a $1b_{2g}$ orbital:

$$1a_g^2\ 1b_{1u}^2\ 2a_g^2\ 2b_{1u}^2\ 1b_{2u}^2\ 3a_g^2\ 1b_{3g}^2\ 1b_{3u}\ 1b_{2g} \quad ^3B_{1u},\ ^1B_{1u} \qquad (V,18)$$

The $^3B_{1u}$ and $^1B_{1u}$ states are generally referred to as the $(\pi\pi^*)$ states of ethylene. Although only the orbitals of linear molecules should be referred to as σ, π, δ, ..., the $1b_{3u}$ orbital is traditionally referred to as π and the

$1b_{2g}$ orbital as π^*. In addition, following Mulliken, the $^3B_{1u}$ state is called the T state and the $^1B_{1u}$ state the V state. These two ($\pi\pi^*$) states are of immense importance since ethylene is the prototype for larger pi-electron systems, the unsaturated hydrocarbons of organic chemistry. Of particular interest here is the controversy which has arisen since the theoretical prediction of Dunning, Hunt, and Goddard (104) that the V state of C_2H_4 is a Rydberg state.

The motivation for the calculations of Dunning, et al. was the fact that earlier SCF calculations on the V ($^1B_{1u}$) state had predicted it to lie at least 9.3 eV above the ground state. Experimentally the vertical excitation energy to the V state is thought to be 7.6 eV (256). Furthermore, the simplest correlation energy arguments would raise the position of the V state predicted in the RHF approximation. The basis set used was of the contracted gaussian type C(9s 5p / 4s 3p) and H(4s / 2s), with diffuse functions added to describe possible Rydberg orbitals (for the V state three $2p\pi$ gaussians with exponents α = 0.0365, 0.0116, and 0.0037 were used). This basis set is sufficiently large so as to give qualitative results which are unlikely to be altered by additional functions.

Table V-14 summarizes the results of Dunning, Hunt, and Goddard. There it is seen that the addition of diffuse or Rydberg functions to the basis set has essentially no effect on the T state results. However, for the V state, the augmentation of the basis set with diffuse functions has radical effects. The most obvious of these is that the excitation energy is reduced from 9.10 to 7.41 eV, in close agreement with the experimental value 7.6 eV. The orbital energies are also changed qualitatively. The V state wave function may be seen to describe a Rydberg state from the value of $<\pi^*|x^2|\pi^*>$, 43 bohrs^2. A value of $<x^2>$ this large can only mean that the π^* orbital is a Rydberg orbital. Note that the value of $<\pi^*|x^2|\pi^*>$ is less than 4 bohrs^2 for the T state. Thus the $1b_{2g}$ or π^* orbital differs greatly between the $^3B_{1u}$ and $^1B_{1u}$ states, both of which arise from the same orbital occupancy, (V,18). These calculations leave little doubt that, within the Hartree-Fock approximation, the lowest state of $^1B_{1u}$ symmetry is a Rydberg state.

Although the calculations of Dunning, Hunt, and Goddard are certainly correct, there has been considerable controversy concerning whether in fact the V ($^1B_{1u}$) state is Rydberg-like. Miron, Ras, and Jortner (261) have reviewed a variety of experimental results relevant to the nature of the V state. In addition they carried out an investigation of the V to ground state electronic transition in solid and liquid rare gases. On the basis of the small solid-liquid spectral shift ($\sim$100 cm^{-1}) Miron, et al. conclude that the V state of C_2H_4 is a normal valence (rather than Rydberg) state.

Additional SCF calculations on the T and V states of ethylene have been carried out by Basch and McKoy (23). The basis set used was similar to but slightly more complete than that of Dunning, Hunt, and Goddard. The interesting feature of the Basch-McKoy calculations involves the consideration of twisted ethylene (90° angle between the two CH$_2$ planes). For this D_{2d} geometry the ($\pi\pi^*$) states arise from electron configuration

$$1a_1^2\ 1b_2^2\ 2a_1^2\ 2b_2^2\ 3a_1^2\ 1e^4\ 2e^2 \qquad\qquad (V,19)$$

Table V-14. Self-consistent-field calculations on the ($\pi\pi^*$) states of ethylene. Unless indicated, properties are in atomic units.

	T ($^3B_{1u}$)		V ($^1B_{1u}$)	
Diffuse Functions in Basis Set?	No	Yes	No	Yes
Energy	-77.8913	-77.8917	-77.6794	-77.7415
Vertical Excitation Energy (eV)	3.33	3.33	9.10	7.41
Orbital Energies				
π	-0.5167	-0.5175	-0.3527	-0.5995
π^*	-0.2422	-0.2419	-0.0113	-0.0593
$\langle \pi | x^2 | \pi \rangle$	2.0592	2.0634	2.1064	2.0294
$\langle \pi^* | x^2 | \pi^* \rangle$	3.6828	3.8468	5.6603	43.0190

The T state is of $^3A_2$ symmetry and the V state of $^1B_2$ symmetry for twisted ethylene. However, orbital occupancy (V,19) is very different from (V,18), since the RHF equivalence restriction forces the π and π^* orbitals (e_x and e_y) to be electronically degenerate. As expected, then, Basch and McKoy find the 2e orbital (and thus the V state) of twisted ethylene to be valence-like. The total SCF energies for the planar (Rydberg) and twisted (valence) V state are -77.74858 and -77.77067. The twisted V state is thus predicted to lie 0.6 eV below the planar. More recent SCF calculations by Hunt, Dunning, and Goddard (176) predict this twisting barrier to be 1.9 eV. These authors also show that the SCF π^* orbital of the V state changes smoothly from Rydberg- to valence-like as the twisting angle goes from 0° to 30° to 60° to 90°. It is experimentally known that the equilibrium (lowest energy) geometry of the C_2H_4 V state occurs for the twisted (90°) form. Merer and Mulliken further state on experimental grounds that the 90°-180° twisting barrier is likely to be 2.0 ± 0.4 eV (256).

The above calculations serve to clarify rather than answer the essential question of theoretical interest: is the V state of ethylene Rydberg-like for planar geometry? Basch and McKoy have suggested that the planar V state is in fact valence-like and that the Rydberg prediction is only due to errors inherent in the Hartree-Fock approximation. This contention may be supported by the fact that the RHF ionization potential of C_2H_4 is about 1.5 eV smaller than experiment. As pointed out in the section VA discussion of the excited states of H_2O, the predicted positions of Rydberg states tend to be in error by the error in the calculated ionization potential. This might have the effect of raising the Rydberg-like, planar, SCF V state of C_2H_4 to ~8.9 eV, and allowing a valence state to become the lowest of $^1B_{1u}$ symmetry. However, such discussions must be considered as speculative in the absence of CI calculations on the planar V state.

Very recently three theoretical groups have independently completed CI calculations on the V state of C_2H_4 (60,33,265). All three calculations give the same qualitative result, that the lowest $^1B_{1u}$ state of planar ethylene remains, as in the SCF approximation, Rydberg-like. The most exhaustive and systematic (33) of the three calculations included 1394 configurations and made use of the iterative natural orbital method. However, Bender et al. (33) find the π* orbital to remain essentially unchanged during the multiconfiguration calculations, which yield a value of $<x^2>$ only 50% smaller than the SCF value. On the basis of their own calculations, Peyerimhoff, et al. (60) have concluded that the V state observed at 7.6 eV is not the (ππ*) singlet ($^1B_{1u}$ state) at all. However, considerably more work, both theoretical and experimental, is needed to provide an adequate description of the low-lying excited states of ethylene.

Approach of two methylenes. The symmetry rules of Woodward and Hoffman (400) have proved immensely useful in the prediction of the course of chemical reactions. The basic idea of Woodward and Hoffman is to construct a correlation diagram showing the orbital energies for reactants and products. If orbital symmetry is conserved between reactants and products, then the reaction is predicted to be concerted, i.e., to proceed with little or no activation energy. One reaction studied in detail by Hoffman and coworkers (169) involves the least motion, coplanar approach of two CH_2 molecules. Our particular interest in this process arises from the interesting ab initio calculations of Basch (24), who has used the multiconfiguration SCF method to study the CH_2 + CH_2 reaction. This study allows a direct comparison between the predictions of a) simple symmetry arguments, b) semi-empirical calculations of the extended Hückel type (167), and c) nonempirical quantum mechanical calculations. Basch's calculations are also of interest because they represent the first application of the MCSCF method to a polyatomic problem of chemical interest.

The ground state SCF orbital occupancy of ethylene, (V,17), may be abbreviated $\sigma^2\pi^2$, omitting the inner twelve electrons. Similarly the orbital occupancy for the $^1A_1$ state of CH_2, (V,7) may be abbreviated σ^2, since organic chemists refer to the $3a_1$ orbital of CH_2 as a σ orbital. At first we consider only the $^1A_1$ state of CH_2, since the extended Hückel studies were limited to closed-shell species. It follows that the orbital occupancy for the approach of two $^1A_1$ methylene is $\sigma^2\sigma*^2$. The reaction may then be visualized as

$$^1A_1 \; CH_2 + {}^1A_1 \; CH_2 \rightarrow {}^1A_{1g} \; C_2H_4$$

$$\sigma^2\sigma*^2 \rightarrow \sigma^2\pi^2 \qquad (V,20)$$

It is clear that orbital symmetry is not conserved in this reaction, and the Woodward-Hoffman rules predict an activation energy or barrier. From the standpoint of the correlation energy diagram the $\sigma*$ and π orbital energies cross in going from 2 CH_2 to C_2H_4. If these symmetry arguments are valid, they imply that there are two low-lying $^1A_{1g}$ electronic states of C_2H_4. There should be an avoided crossing of the potential surfaces for these two states, yielding a hump (in the lower surface) responsible for the activation energy. In their extended Hückel calculations, Hoffmann, et al. find the $\sigma^2\sigma*^2$ configuration to be lowest in energy for C-C separations greater than 3 Å. For C-C separations less than 3 Å, the $\sigma^2\pi^2$ normal C_2H_4 configuration lies lower. Thus the extended Hückel calculations bear out the orbital symmetry predictions and suggest a hump in the potential surface at ~ 3 Å.

To provide an <u>ab initio</u> test to these ideas, Basch has carried out 3-configuration SCF calculations for the least motion coplanar approach of two CH_2 molecules. The three configurations included were $\sigma^2\pi^2$, $\sigma^2\sigma*^2$, and $\sigma^2\pi*^2$.

Table V-15. Three-configuration SCF wave functions for planar ethylene as a function of C-C bond distance (24). All other geometrical parameters are the same as in the $^1A_{1g}$ ground state of C_2H_4. R(C-C) is in Å.

Configuration	$\sigma^2\pi^2$	$\sigma^2\sigma*^2$	$\sigma^2\pi*^2$	3-Configuration SCF Energy
R(C-C)				
9.54	0.0274	0.9992	0.0276	-77.7279
5.31	0.0305	0.9990	0.0324	-77.7252
3.19	0.0485	0.9976	0.0498	-77.7068
2.93	0.7531	0.0179	0.6577	-77.7097
2.66	0.7765	0.0156	0.6299	-77.7471
2.13	0.8566	0.0184	0.5156	-77.8536
1.34 = R_e	0.9760	0.0179	0.2169	-78.0535

The basis set used was adequate for qualitative purposes, C(11s 6p / 5s 3p) and H(5s / 2s). These results are summarized in Table V-15. There it is seen that the $\sigma^2\sigma*^2$ configuration dominates the wave function for C-C distances larger than ~ 3 Å, in essentially perfect agreement with extended Hückel results. At the equilibrium ethylene C-C separation the $\sigma^2\pi^2$ configuration dominates, but between R_e and 2.93 Å, the contribution of the third configuration, $\sigma^2\pi*^2$, becomes increasingly important. The three-configuration SCF treatment predicts a barrier height of about 15 kcal/mole. It can be concluded that symmetry arguments, extended Hückel calculations, and MCSCF calculations all give the same qualitative result.

Despite the consistency of the above results, they do not correspond to physical reality, due to the neglect of the triplet ground state of methylene. For the ethylene equilibrium H-C-H bond angle, 117.6°, the most accurate theoretical work on CH_2 (32,244) predicts the $^3B_1$ state to lie ~0.5 eV below the $^1A_1$ state. Thus the minimum energy path connecting ethylene with two methylenes involves two $^3B_1$ electronic states of CH_2. As pointed out by Hoffman and coworkers (169), the least motion coplanar approach of two $^3B_1$ methylenes does conserve orbital symmetry.

Basch has investigated this problem in a quantitative way using 6-configuration wave functions constructed from the orbitals obtained in the above 3-configuration SCF treatment. The additional three C_2H_4 configurations included in the CI were $\pi^2\pi^{*2}$, $\pi^2\sigma^{*2}$, and $\pi^{*2}\sigma^{*2}$. It is interesting to note that at infinite separation, two $^3B_1$ SCF wave functions are described by the 4-configuration $^1A_{1g}$ C_2H_4 wave function

$$\tfrac{1}{2}\left[\sigma^2\pi^2 + \sigma^2\pi^{*2} + \pi^2\pi^{*2} + \pi^{*2}\sigma^{*2}\right] \qquad (V,21)$$

Basch's calculations show that two $^3B_1$ methylenes approach with no barrier to form the ground state of ethylene. However two $^1A_1$ methylenes approach in a repulsive manner along the second C_2H_4 potential energy surface of $^1A_{1g}$ symmetry. That is, two $^1A_1$ states of CH_2 do not even lie on the same potential surface as the ground state of ethylene. Basch concludes that the dimerization of singlet methylenes by their least motion coplanar approach will not proceed because the reaction path is purely repulsive.

Dimerization energy of BH_3. The first calculations of pair correlation energies for polyatomic molecules were carried out by Ahlrichs and Kutzelnigg (3). These calculations are of particular interest because the pair correlation energies are calculated with respect to localized SCF orbitals. We saw in section IIIB, that for BH, the sum of the localized orbital pair correlation energies gave the most reasonable approximation to the variationally calculated correlation energy. In addition there is great interpretive value in the use of localized pair correlations, since the correlation energy may be discussed in terms of traditional chemical concepts. There will be a certain correlation energy associated with each inner shell pair, bonding pair, lone pair, etc. Although we now know that the interpair correlation energies (e.g., $e(2\sigma,3\sigma)$ in BH) will always be appreciable, the use of localized SCF orbitals is expected to reduce the magnitude of these effects. Thus the localized pair correlation approach is about as close as one can come in an a priori way to the chemist's idea of a molecule being composed of electron pairs.

There is considerable discord between different experimental values of the dimerization energy for the reaction $2BH_3 \rightarrow B_2H_6$ (131). Values of $D(BH_3-BH_3)$ from mass spectrometry vary from 39 to 59 kcal/mole, while those from kinetic data are uniformly lower, 25-38 kcal. From a theoretical point of view, one could hardly expect the Hartree-Fock approximation to provide a reasonable value for this dissociation energy. Therefore, Gelus, Ahlrichs,

Staemmler, and Kutzelnigg (133) have made a theoretical study of this problem including correlation energy via a localized pair correlation treatment. In addition to providing a value for the BH_3 dimerization energy, this work allows some insight into the formation of electron deficient bonds such as those in diborane.

In these calculations a basis set somewhat larger than double-zeta was employed after being carefully contracted from calculations on BH_2^+ and BH_3. The first step in the procedure of Gelus, et al. (133) involved the calculation of SCF wave functions for BH_3 and B_2H_6. The geometry used for B_2H_6 was the experimental one, while that for BH_3 was from earlier theoretical calculations, since the BH_3 geomtery has not been determined experimentally. Then the SCF orbitals were localized, yielding wave functions of the form

$$\psi(BH_3) = c^2 \, t^2 \, t'^2 \, t''^2 \tag{V,22}$$

$$\psi(B_2H_6) = c_A^2 \, c_B^2 \, t_A^2 \, t_A'^2 \, t_B^2 \, t_B'^2 \, b^2 \, b'^2 \tag{V,23}$$

c_A and c_B are the core or essentially 1s orbitals on boron atoms A and B. BH_3 has a planar structure and the localized valence orbitals t, t', and t'' describe the three equivalent B-H terminal bonds. Each boron atom in B_2H_6 is terminally bonded to two hydrogen atoms, and the t_A and t_A' localized orbitals describe the two terminal bonds directed toward B atom A. Finally the two equivalent 3-center BHB bridge bonds are described by localized orbitals b and b'.

The final step in the $2BH_3 \to B_2H_6$ calculations consists of the evaluation of pair correlation energies for the two molecules. In the work of Kutzelnigg and associates, this is done by direct calculation of approximate pseudonatural orbitals for each pair of localized orbitals. These pseudonatural orbitals may then be used to evaluate the pair correlation energies. For example, the t^2 pair correlation energy of BH_3 is obtained by a CI including the SCF plus all orbital occupancies of the type $c^2 \, t'^2 \, t''^2 \, x^2$, where x^2 designates all the pseudonatural orbitals for the t^2 pair. The t^2 pair correlation energy is just the difference between this CI energy and the SCF energy. Put another way, the t^2 pair correlation energy is that part of the correlation energy obtained by all excitations of the type $t^2 \to xy$ with respect to the SCF configuration. For the interpair correlations, e.g., (t,t'), two independent pair correlation energies are obtained. Recall that 4 unpaired electrons give rise to two different singlet spin functions. One of these corresponds to the singlet-coupling of the t and t' orbitals remaining after the tt' → xy double excitation. The other pair correlation energy corresponds to the triplet-coupling of t and t'.

Table V-16 summarizes the results of Gelus, Ahlrichs, Staemmler and Kutzelnigg. On the basis of similar work on smaller systems, these authors have neglected electron correlation involving the core orbitals c_A and c_B in (V,23). In addition, correlation effects between B-H terminal orbitals attached to the two different B atoms are neglected. It is interesting to note that the terminal pair correlation energies e(t,t) and e(t,t') are nearly equal for BH_3 and B_2H_6. As hoped, the interpair correlation energies, though by no means negligible, are smaller (by factors of the order of 5) than the intrapair contributions. However, the intrabond contributions are nevertheless primarily responsible for the correlation contribution to the dimerization

Table V-16. Valence shell pair correlation energies of BH_3 and B_2H_6. See text for a description of the different localized SCF orbitals. Singlet and triplet refer to the two possible spin couplings for the interpair correlation energies.

	Pair	Number of Equivalent Pairs	Correlation Energy
BH_3			
	t - t	3	0.02805
singlet	t - t'	3	0.00329
triplet	t - t'	3	0.00467
	Sum of pairs		0.10804
B_2H_6			
	$t_A - t_A$	4	0.02779
singlet	$t_A - t_A'$	2	0.00301
triplet	$t_A - t_A'$	2	0.00477
	b - b	2	0.02796
singlet	b - b'	1	0.00384
triplet	b - b'	1	0.00770
singlet	b - t_A	8	0.00215
triplet	b - t_A	8	0.00393
	$t_A - t_B$		Neglected
	Sum of Pairs		0.24284

energy. In particular the diborane bond-terminal correlation energies have no counterpart in BH_3. Although each bond-terminal correlation energy is small (0.00608 hartree including both singlet and triplet contributions), the fact that there are eight such equivalent pairs adds up to 0.04864 hartree or 30.5 kcal/mole. It should be pointed out that this contribution is partly compensated for by the fact that there are four fewer (t_A, t_A') pairs in B_2H_6 than in $2BH_3$.

The SCF dimerization energy of BH_3 is calculated to be 8.5 kcal/mole, and the true RHF dimerization estimated energy to be 11.5 kcal/mole. The calculated correlation energy of B_2H_6 is 0.24284 hartree, and that of 2 BH_3 molecules is 0.21608 hartree. The difference, 16.8 kcal/mole, is the calculated correlation energy contribution to the dimerization energy. Inadequacies of the basis set are estimated to add another 7.7 kcal to the correlation contribution. Thus the calculated dimerization energy is 25.3 kcal and the estimated value 36 kcal. The latter prediction is closer to the experimental values obtained from kinetics than those from mass spectrometry. In light of the basis set used, however, it is difficult to place a reliability limit on the theoretical result, but Gelus, et al. estimate 36 ± 5 kcal/mole. Perhaps the most significant result of this work is that the correlation contribution to the dimerization energy is more than twice that obtained at the SCF level.

C. LARGER MOLECULES

<u>Excited states of butadiene</u>. The theoretical results discussed above for the V state of ethylene suggest that π-electron theory may be more difficult than semi-empirical methods (288), for example that of Pariser, Parr,

and Pople, would suggest. Another molecule of great experimental interest for which π-electron theory has been considered very successful is butadiene. Experimental information on the excited singlet states of butadiene has been summarized by Mulliken (268), and the triplet states have been similarly discussed by Evans (115). The first all-electron nonempirical calculations of C_4H_6 were those of Buenker and Whitten (57). They reported calculations on both cis and trans forms, predicting the ^1A_g electronic ground state of the trans form to lie 5 kcal lower. Although the energy difference between the cis and trans forms is not known experimentally, the trans form is known to lie lower in energy. In addition Buenker and Whitten carried out small CI calculations for the excited states of butadiene. Recently, Hosteny, et al. (171) have carried out a truly impressive theoretical study of the excited states of butadiene, and here is given a brief discussion of their work.

The basis set used was of the Dunning contracted gaussian double-zeta type, C(9s 5p / 4s 2p) and H(4s / 2s). However, in light of the controversy concerning the possible Rydberg nature of the V state of ethylene, two diffuse $2p\pi$ (pointing out of the plane of the molecule) functions were added to each carbon atom. All calculations were carried out at the experimental (trans) geometry. With this basis set the ^1A_g ground state SCF energy is -154.8771 hartrees, compared to the Buenker-Whitten result -154.7103 hartrees.

The hallmark of the calculations by Hosteny, et al. is the use of massive configuration interaction. The SCF ground state orbital occupancy for trans butadiene is, excluding the core or carbon 1s orbitals

$$3a_g^2 \; 3b_u^2 \; 4a_g^2 \; 4b_u^2 \; 5a_g^2 \; 5b_u^2 \; 6a_g^2 \; 6b_u^2 \; 7a_g^2 \; 1a_u^2 \; 1b_g^2 \quad ^1A_g \qquad (V,24)$$

The $1a_u$ and $1b_g$ orbitals are usually referred to as the 1π and 2π orbitals. It should be pointed out now that butadiene also has two low-lying orbitals not occupied in (V,24). These are the $2a_u$ and $2b_g$ orbitals, here designated 3π and 4π. Each of the excited-state calculations is based on the SCF wave function for the ground state. A rigorous, full π-electron CI is carried out for all electronic states of interest. This means that full CI is done with the restriction that the ground state SCF orbitals through $7a_g$ are always doubly occupied and only π-type virtual orbitals are used in the construction of the configurations. The symmetries considered (with numbers of configurations in parentheses) were ^1A_g (2752), ^3A_g (3556), ^1B_u (2688), ^3B_u (3584), and ^5A_g (924). These large numbers of configurations arise due to the fact that the basis set includes 14 molecular π orbitals not occupied in the SCF wave function. In addition π-electron CI calculations were carried out including SCF + single excitations, SCF + singles + doubles, and SCF + singles + doubles + triples. The SCF + singles results are of particular interest since this approach has been used frequently in semi-empirical studies.

The results of the butadiene calculations are summarized in Table V-17. The four experimental vertical excitation energies are not the generally accepted values of Mulliken (268) and Evans (115), but rather the very recent results of Mosher, Flicker, and Kuppermann (266). The best theoretical excitation energies for the ^3B_u and ^3A_g states agree well with experiment. For the two singlet states, particularly the ^1B_u state, the agreement is poorer. However these results may be understood by a rather simple model. All the π-electron states of butadiene may be envisaged as arising from the N (ground), T ($\pi\pi^*$ triplet), and V ($\pi\pi^*$ singlet) states of ethylene. The calculations

Table V-17. Vertical excitation energies (in eV) for the π-electron states of butadiene (171). The number preceding each term symbol indicates the first (1), second (2), etc. eigenvalue of the symmetry given by the term symbol.

State	Single Excitation CI	Full CI	Experiment
$1\,^1A_g$	0.0	0.0	0.0
$2\,^1A_g$	7.33	6.77	---
$1\,^1B_u$	6.67	7.05	5.92
$3\,^1A_g$	7.51	7.82	7.25
$1\,^3B_u$	2.77	3.45	3.22
$1\,^3A_g$	4.41	5.04	4.93
$2\,^3B_u$	6.76	7.14	---
$2\,^3A_g$	7.37	7.79	---
^5A_g	---	9.61	---

on ethylene analogous to those discussed here for butadiene predict the T state of ethylene to lie at 4.2 eV, compared to experiment, 4.6 eV. The V state prediction from full π-electron CI is much worse, 8.9 eV, as opposed to 7.6 eV experimentally. We might then expect those states of butadiene which correlate with one V state of C_2H_4 to lie 1.3 eV too high. It turns out that NN (two ethylene molecules in their ground or N states) gives rise to only a single state of butadiene, the ^1A_g state. NT gives rise to two triplet states, ^3B_u and ^3A_g, and these two states are calculated, as expected from the ethylene T state results, only slightly above experiment. The states of butadiene arising from the NV combination of ethylene states are ^1B_u and ^1A_g. These are just the two states which lie too high with respect to experiment, as expected by the fact that the V state of C_2H_4 is calculated to lie too high. Although one does not want to push this type of reasoning too hard, it does seem to provide an explanation for the better agreement with experiment for the triplet than for the singlet states of butadiene.

Table V-18 gives some detailed information concerning the electronic structure of the low-lying states of butadiene. The natural orbital occupation numbers were obtained by diagonalizing the density matrix (I,67) resulting from each wave function. These occupation numbers provide the simplest one-electron explanation for a complicated CI wave function. However, one should be a bit cautious of the numbers in Table V-18 since they do not refer to the full CI, but to the wave functions obtained including all single and double excitations from $1\pi^2\ 2\pi^2$. Each state has been labeled "valence" or "Rydberg" depending on the expectation values $< x^2 >, < y^2 >,$ and $< z^2 >$ evaluated for the most important natural orbitals. The two experimentally observed triplet states are labeled valence states, and the ^3B_u state corresponds, as predicted by simple π-electron theory, to orbital occupancy $1\pi^2\ 2\pi\ 3\pi$. However, the Hosteny description of the ^3A_g state is essentially $1\pi^{1.5}\ 2\pi^{1.5}\ 3\pi^{0.5}\ 4\pi^{0.5}$,

Table V-18. The electronic structure of butadiene as determined by π-electron CI calculations including all configurations differing by one or two orbitals from the SCF ground state $1\pi^2\ 2\pi^2$.

	Natural Orbital Occupation Numbers				
State	1π	2π	3π	4π	Description
1 ^1A_g	1.939	1.893	0.108	0.054	Valence
2 ^1A_g	1.639	1.092	0.849	0.403	Valence
1 ^1B_u	1.954	1.005	0.986	0.008	Rydberg
3 ^1A_g	1.925	1.012	0.113	0.929	Rydberg
1 ^3B_u	1.907	1.070	0.931	0.086	Valence
1 ^3A_g	1.489	1.503	0.510	0.490	Valence
2 ^3B_u	1.957	1.007	0.987	0.009	Rydberg
2 ^3A_g	1.911	1.061	0.088	0.926	Rydberg
^5A_g	1.000	1.000	1.000	1.000	Valence

indicating a nearly equal mixing of the $1\pi^2\ 2\pi\ 4\pi$ and $1\pi\ 2\pi^2\ 3\pi$ configurations. The second triplet states of both B_u and A_g symmetry are predicted to be Rydberg-like. For the unpaired 3π orbital of the 2 ^3B_u wave function, the values of $<x^2>$, $<y^2>$, and $<z^2>$ are 19, 19, and 55 bohrs^2. The excited singlet states thought to be observed experimentally, 1 ^1B_u and 3 ^1A_g, are both predicted to be Rydberg-like.

Of particular interest is the calculated second ^1A_g state of butadiene. Hosteny, et al. (171) do not identify this state with the observed excited state of ^1A_g symmetry lying at 7.25 eV. Rather, they associate the third calculated state of ^1A_g symmetry with the observed state. This is because they identify the calculated 2 ^1A_g state as a doubly-excited state, and the electronic transition between this state and the ground state is expected to be of low, if not vanishing intensity. The calculated 2 ^1A_g state is that which arises from the ππ combination of ethylene states. The 3 ^1A_g state corresponds to the expected orbital occupancy $1\pi^2$ 2π 4π. Figure V-4 provides additional evidence supporting their assignment. The single excitation CI produces only a single valence-like state, the ground state. Only when double excitations (e.g. $2\pi^2 \rightarrow 3\pi^2$) are included does a proper description of the valence-like second state of ^1A_g symmetry become possible. Addition of all triple and quadruple excitations does not qualitatively affect the ^1A_g states.

The results discussed here would appear to indicate the π-electron theory is considerably more complicated than semi-empirical calculations would lead us to believe. It is certainly possible that more complete (including σ electron correlation) theoretical calculations will yield results closer to traditional ways of thinking. However, it is clear that the rigorous application of π-electron theory produces some surprising results, particularly concerning the importance of Rydberg states.

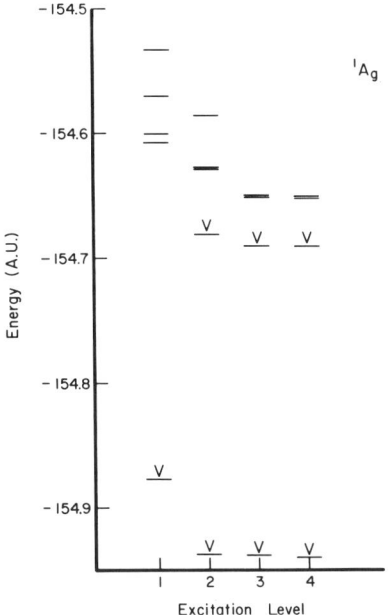

Figure V-4. Effect of different levels of π-electron CI on the ^1A_g states of trans butadiene. V indicates a state identified as valence-like. Other states are Rydberg-like.

The electrocyclic transformation between cyclobutene and cis-butadiene. The prototype electrocyclic reaction (400) involves the isomerization of cis-butadiene to cyclobutene, denoted

$$\begin{array}{c}\text{CH}=\text{CH}_2 \\ | \\ \text{CH}=\text{CH}_2\end{array} \rightleftharpoons \begin{array}{c}\text{CH}-\text{CH}_2 \\ \| \quad | \\ \text{CH}-\text{CH}_2\end{array} \qquad (V,25)$$

The basic difference between these two molecules is that the two terminal methylene (CH_2) groups lie in the plane of the molecule for cis-butadiene, but in cyclobutene each methylene group lies in a plane perpendicular to the plane of the four carbon atoms. Clearly the two CH_2 groups in cis-butadiene must rotate by 90° in order to form cyclobutene. Using the terminology introduced by Woodward and Hoffmann (399), the reaction is called conrotatory if the two CH_2 groups rotate in the same direction, and disrotatory if they rotate in opposite directions. If the CH_2 groups in cis-butadiene are replaced, for example, by CHCl groups, the conrotatory and disrotatory paths will lead to two different substituted cyclobutenes. According to Woodward and Hoffmann, the course of an electrocyclic reaction will be determined by the symmetry of the highest occupied molecular orbital of the open-chain partner. Using a simple prescription, Woodward and Hoffmann were able successfully to predict the stereochemistry of a wide variety of electrocyclic reactions. From a purely theoretical viewpoint, one expects the C_4H_6 isomerization to proceed along the minimum energy path (reaction coordinate) between cis-butadiene and cyclobutene. Here we discuss the work of Hsu, Buenker, and Peyerimhoff (172) who have carried out <u>ab initio</u> calculations aimed at predicting this minimum energy path.

The contracted gaussian basis set used by Hsu, et al. was (10s 5p / 3s 1p) on each carbon atom and (5s / 1s) on each hydrogen. As discussed in chapter I, such a basis set is overcontracted and will in general lead to bond distances too long and dissociation energies too low. However, the use of this basis is certainly understandable in light of the size of the molecule considered and the fact that many points on the potential surface were calculated. The first step in each calculation was a self-consistent-field calculation. CI followed in which the lowest 11 (out of 15) SCF orbitals were held frozen, i.e., were doubly-occupied in all configurations. With this restriction, all single and double excitations, many triple excitations, and a few quadruples were included in the CI, with the total number of configurations being about 250.

Since the C_4H_6 potential surface is 24-dimensional, it is obvious that some chemically reasonable restrictions had to be made concerning the selection of geometries. In all calculations, the following parameters were fixed: methylene CH distances (1.093 Å), other CH distances (1.086 Å), and HCH methylene angles (114°, the average value between the equilibrium values found in butadiene and cyclobutene). Most of the geometry variations involved the rotation θ (both conrotatory and disrotatory) of the CH_2 groups from 0° (cis-butadiene) to 90° (cyclobutene) and the variation of the terminal C-C distance R of cis-butadiene, which becomes a C-C single bond in the cyclic molecule.

Preliminary SCF calculations were carried out to roughly determine the minimum energy path. A primary conclusion of these calculations was that the rotation of the two methyl groups requires much more energy than altering the terminal C-C distance R. In fact, at every R, the lowest energy conformation calculated was that for either $\theta = 0$ or $\theta = 90°$. The lowest energy for $\theta = 45°$ at a given R is usually about 50 kcal/mole higher than the corresponding planar or perpendicular structure. Based on these results, Hsu, Buenker, and Peyerimhoff argue that the methylene rotation takes place in the very narrow region of R where the 0° and 90° conformations are nearly degenerate. The authors refer to this as a stepwise mechanism, as opposed to the linear mechanism (assumed by most earlier workers) in which rotation takes place simultaneously as R varies from 5.51 bohrs (cis-butadiene) to 2.92 bohrs (cyclobutene). According to the calculations, the rotation from $\theta = 0$ to $\theta = 90°$ occurs quite abruptly in the neighborhood of 4.3 bohrs. The stepwise mechanism has an extremely interesting ramification if we consider the reverse of (V,25). Namely, the ring bond connecting the two CH_2 groups of cyclobutene must be completely broken (R increase from 2.92 to 4.3 bohrs) <u>before</u> rotation can proceed with a sufficiently small energy increase to be satisfactorily ascribed to a thermal process.

After pinning down the general features of the C_4H_6 potential surface, CI was used to predict the barrier height more accurately. With CI wave functions the optimum (where the perpendicular and planar energies are the same) R value for the stepwise mechanism is 4.49 bohrs. Figure V-5 summarizes the energy results at this R. First of all, both SCF and CI calculations favor the conrotatory path, in agreement with experiment and the predictions of Woodward and Hoffmann. The origin of the Woodward-Hoffmann rule for this reaction is seen rather dramatically in Figure V-5. Namely, the disrotatory path connects two different electron configurations. Alternatively, the SCF wave function starting along the disrotatory path from cis-butadiene arises from a different orbital occupancy than the SCF wave function of lowest energy starting from cyclobutene.

The energetic and geometry results are summarized in Table V-19. The predicted C-C distances for both cis-butadiene and cyclobutene are qualitatively reasonable. In addition, the CI energy difference between the two isomers is 11.5 kcal/mole compared to experiment, 9.2 kcal. The CI barrier height, 58.8

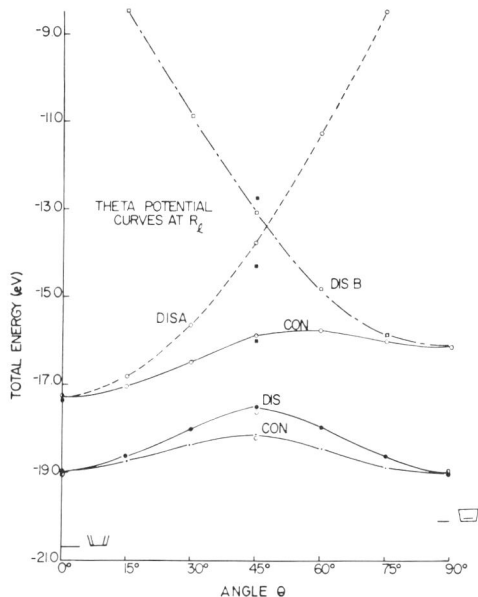

Figure V-5. Total energies for the con- and disrotatory C_4H_6 structures as a function of methylene rotation angle θ at terminal C-C distance 4.49 bohrs. The upper curves are from SCF calculations, the lower from CI. The energies of the equilibrium cis-butadiene and cyclobutene molecules are also indicated.

---

kcal, may be compared with an experimental estimate, 42 kcal (86). As expected, the SCF values for both the exothermicity and barrier height are in poor agreement with experiment. The overall reasonableness of both the geometry and energy differences from CI gives rather strong backing to the interesting stepwise mechanism proposed by the authors.

Electronic ground states of ortho, meta, and para benzyne. Among the most intriguing of organic molecules are the benzynes C_6H_4, seen in Figure

Table V-19. Summary of calculations on the C_4H_6 isomerization. SCF values precede CI values and when available, experimental data is given in parentheses.

	R(bohrs)	θ	E(kcal/mole)
Cis-Butadiene	5.42, 5.39 (5.51)	0°	0
Cyclobutene	3.05, 3.12, (2.92)	90°	1.2, 11.5
Crossing Point	4.320, 4.494	0° and 90°	58.8, 39.0
Conrotatory Mode Maximum	4.320, 4.494	51°, 45°	78.6, 58.8 (~42)
Disrotatory Mode Maximum	4.320, 4.494	47°, 45°	127.8, 72.4

V-6. Of these, the ortho structure is the most interesting since it may be described in terms of a triple bond

The existence of ortho-benzyne as an intermediate was first demonstrated by Roberts and coworkers (315) in 1953. Although none of the benzynes has ever been isolated (in the sense of organic chemistry), some part of the spectra of all three molecules has been observed. The first such work was that of Berry, Spokes, and Stiles (39) on ortho-benzyne, reported in 1962. An interesting and as yet experimentally unresolved problem concerns the determination of the

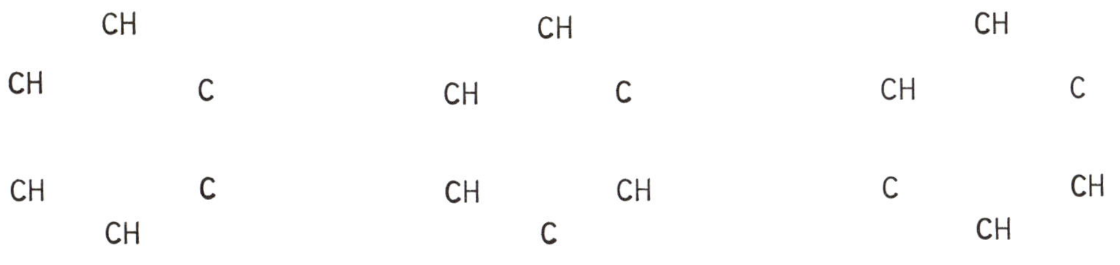

```
         CH                      CH                        CH
  CH          C           CH          C             CH           C
  CH          C           CH          CH             C          CH
         CH                      C                        CH

       ortho                    meta                     para
```

Figure V-6. Ortho-, meta- and para-benzyne.

---

electronic ground states of ortho-, meta-, and para-benzyne. In addition of course there are many qualitative questions to be answered concerning the nature of the chemical bonds in these molecules. These problems have been addressed in a recent theoretical study by Wilhite and Whitten (397).

The basis set used, C(10s 5p / 3s 1p) and H(5s / 1s), is the original contracted gaussian lobe basis of Whitten (395). Although this basis does well for the separated atoms it leaves something to be desired in the way of molecular flexibility. In all calculations the geometry was restricted to be the equilibrium geometry for the ground state of benzene, C_6H_6. Three types of calculations were reported: a) SCF, b) two configuration CI, and c) multi-configuration CI.

Initially only the lowest singlet and triplet states of each molecule were considered. Ignoring the 15 orbitals with lowest orbital energies, the SCF configurations for these states are

$$\text{ortho} \quad 9a_1^2 \; 1b_1^2 \; 1a_2^2 \; 2b_1^2 \; 10a_1^2 \qquad\qquad {}^1A_1 \qquad\qquad (V,26)$$

	$9a_1^2\ 1b_1^2\ 1a_2^2\ 2b_1^2\ 10a_1\ 8b_2$	$^3B_2$	(V,27)
meta	$10a_1^2\ 1b_1^2\ 2b_1^2\ 1a_2^2\ 11a_1^2$	$^1A_1$	(V,28)
	$10a_1^2\ 1b_1^2\ 2b_1^2\ 1a_2^2\ 11a_1\ 7b_2$	$^3B_2$	(V,29)
para	$1b_1^2\ 7b_2^2\ 2b_1^2\ 1a_2^2\ 8b_2^2$	$^1A_1$	(V,30)
	$1b_1^2\ 7b_2^2\ 2b_1^2\ 1a_2^2\ 8b_2\ 10a_1$	$^3B_2$	(V,31)

In general, one would expect the three triplets to have less correlation energy than the closed-shell singlets (V,26), (V,28), and (V,30). The very simplest way to compensate for this is to do two configuration calculations for each of the singlets. For the ortho states, the $10a_1$ and $8b_2$ orbitals can generate two $^1A_1$ configurations, $10a_1^2$ and $8b_2^2$, whereas these same two orbitals yield only one possible $^3B_2$ configuration, that seen in (V,27). The same two configurations are used to describe $^1A_1$ para-benzyne, and for meta-benzyne the $11a_1^2$ and $7b_2^2$ configurations are used to describe the lowest singlet state. The final step in the calculations consisted of moderate CI (of the order of 200 configurations) on all six electronic states.

Figure V-7 summarizes the calculations of Wilhite and Whitten on the lowest singlet and triplet states of benzyne. One interesting result is that in all three cases, the triplet states lie below the singlets in the SCF approximation. For ortho-benzyne, this result is the opposite of that obtained from a variety of semi-empirical calculations predicated on the single configuration approximation. However, the order of the orbital energies is the same for all three molecules as that predicted by Hoffmann, Imamura, and Hehre (168) from an

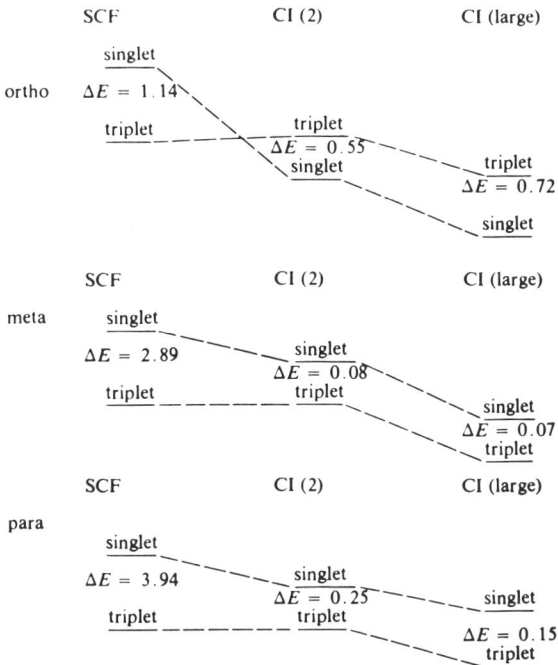

Figure V-7. Diagrammatic presentation of the singlet-triplet separations for each of the three benzynes at the SCF, two configuration, and multiconfiguration levels. Energy differences are in eV.

---

extended Hückel study. For ortho-benzyne, the two-configuration calculation reverses the SCF order and places the singlet state 0.55 eV <u>below</u> the triplet. Additional CI increases this separation to 0.72 eV. The authors thus argue that their calculations predict conclusively that the ground electronic state of ortho-benzyne is a singlet. For the meta and para forms, the SCF calculations predict much larger singlet-triplet splittings, which are drastically reduced by CI. The resulting separations, 0.07 eV for meta-benzyne and 0.15 eV for para-benzyne, are too small to produce definitive triplet ground state predictions. This is particularly true in light of the fact that the benzene ground state geometry was used. The multiconfiguration calculations yield final energies -229.1506 ($^1A_1$ ortho state), -229.1331 ($^3B_1$ meta state), and -229.1342 ($^3B_2$ para state). Thus para-benzyne is predicted to be the most stable of the three C_6H_4 isomers. The SCF calculations yield the same prediction.

Population analyses were carried out for most of the wave functions obtained. Perhaps the most striking result of the population analyses is that the singlet and triplet states, e.g., of ortho-benzyne, are nearly identical in this regard. It is useful to know that the comparable benzene calculation yields 6.31 electrons on each carbon and 0.69 on each hydrogen. For comparison, in para-benzyne there are two types of carbons, the 4 to which H is attached and the other two. The former have population 6.33 electrons and the latter are essentially neutral in a point charge picture, with 6.01 electrons. Similarly carbon atoms in meta-benzyne with no hydrogen attached have population 6.00, and in ortho-benzyne 6.04. For ortho-benzyne, as expected, overlap populations show the migration of electrons into the triple bond region.

For ortho-benzyne only, an attempt was made to predict the electronic spectrum by CI calculations on several low-lying states. The results of these calculations are summarized in Table V-20. The singlet and triplet states are separated, since the probability of observing a singlet-triplet transition for this short-lived species is negligible. Interpretation of the results is simplified by the realization that the 1π, 2π, 3π, and σ orbitals referred to are the $1b_1$, $1a_2$, $2b_1$, and $10a_1$ orbitals of (V,26). The orbital promotions

Table V-20. Predicted positions of the electronic states of ortho-benzyne (397). The most extensive calculations predict the lowest $^3B_2$ state to lie 0.72 eV above the ground $^1A_1$ state.

Electronic State	Orbital Promotion	E(eV)	Wavelength, Å
$^1A_1$	$1\pi^2\ 2\pi^2\ 3\pi^2\ \sigma^2$	0.00	--
$^1B_1$	$\pi \to \sigma^*$	4.03	3071
$^1A_1$	$\pi \to \pi^*$	5.41	2294
$^1A_1$	$\pi\sigma \to \sigma^*\pi^*$	5.93	2094
$^1B_2$	$\sigma \to \sigma^*$	6.04	2056
$^3B_2$	$\sigma \to \sigma^*$	0.00	--
$^3A_2$	$\pi \to \sigma^*$	2.09	5945
$^3B_2$	$\pi \to \pi^*$	3.05	4081
$^3B_2$	$\pi \to \pi^*$	4.41	2816
$^3A_2$	$\sigma \to \pi^*$	5.50	2258
$^3A_2$	$\pi \to \sigma^*$	5.85	2122

were assigned on the basis of the most important configurations in each wave function. The experimental spectrum of photolytically produced ortho-benzyne shows only one very broad transient absorption in the region around 2430 Å (39). Berry, et al. tentatively assign this absorption to a singlet $\sigma \to \sigma^*$ transition. The three singlet excited states at 2294, 2094, and 2056 Å lie in this region, within the expected theoretical reliability, and the last of these corresponds to a $\sigma \to \sigma^*$ transition. The intensity of the 2094 Å transition is likely to be extremely weak since two of the ground state SCF orbitals are replaced. In conclusion, it seems fair to say that theoretical and experimental work are at about an equally preliminary stage of development with respect to these interesting molecules, the benzynes.

Hydrogen bonding in the water dimer and trimer. The importance of the subject of hydrogen bonding need not be labored (297). Theoreticians have paid particular attention to hydrogen bonding in water. Although electron correlation has been included in none of the calculations reported to date, SCF calculations have been carried out with a rather wide variety of basis sets. This discussion weighs heavily on the H_2O dimer results of Diercksen (102) and the dimer and trimer results of Hankins, Moskowitz, and Stillinger (145). A very helpful review of theoretical calculations on hydrogen bonding has been written by Kollman and Allen (208).

The accepted experimental dimerization energy of H_2O is 5.0 kcal/mole, based on second virial coefficient data (326). The geometry of the water dimer has not been determined experimentally. Experimentalists have suggested three possible equilibrium geometries for the H_2O dimer, and these are seen in Figure

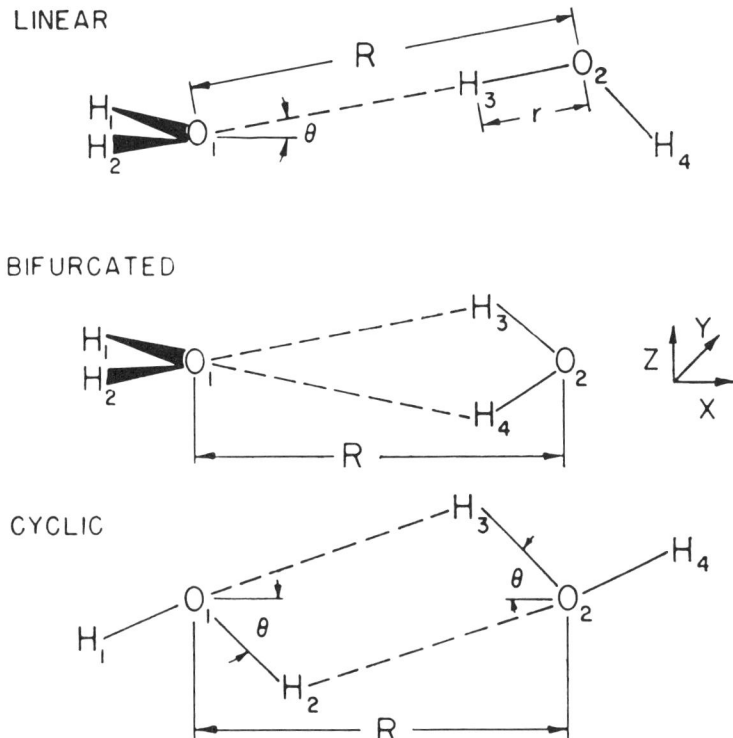

Figure V-8. Three possible geometrical structures for the H_2O dimer (208).

V-8. Morokuma and Pedersen (263) carried out the first nonempirical SCF calculations on the H_2O dimer, using a small uncontracted gaussian basis set. They considered all three geometries in Figure V-8 and found the linear to be most stable, with dimerization energy 12.6 kcal/mole. These calculations also predict the cyclic and bifurcated conformations to be bound, by 7.8 and 9.2 kcal, with respect to two isolated H_2O molecules. Calculations by Kollman and Allen (207) with a larger but overcontracted gaussian basis predicted both the linear and bifurcated structures to be bound by 5.3 kcal/mole and the cyclic to be less stable, 4.0 kcal. The most exhaustive search of H_4O_2 geometries is that of Del Bene and Pople (100), using a minimum basis of Slater functions, each expanded as a linear combination of four gaussians. They found a linear structure to be most stable, with 6.1 kcal binding energy. The conclusion of all three sets of calculations is that the cyclic dimer is less stable than the other two forms.

Diercksen (102) used an extended basis set including polarization functions and investigated both linear and bifurcated geometries. He found the linear dimer to be bound by 4.83 kcal, in good agreement with experiment. Diercksen predicts the bifurcated dimer to lie 1.4 kcal higher in energy than the linear form. The calculated dipole moment for the minimum-energy linear dimer is 4.30 debyes. This is 0.48 debyes lower than the dipole moment calculated for the same system assuming no interaction between H_2O molecules. Diercksen interprets this result as indicating a charge shift in direction from the proton acceptor to the proton donor. Similar extended basis SCF calculations have been reported by Hankins, et al. (145) and yield a dimerization energy of 4.72 kcal/mole.

Table V-21 summarizes the water dimer SCF predictions. The most obvious conclusion to be drawn there is that increasing basis set size ultimately leads to an SCF dimerization energy in close agreement with experiment. A question of great interest which naturally arises concerns whether the RHF approximation can be expected to give an accurate description of hydrogen bonding *in general*. Unfortunately, only a few extended basis comparisons with experiment are possible at present. Diercksen and Kraemers (103) predict a dimerization energy of 4.6 kcal for HF, while experimental estimates of this

Table V-21. Summary of nonempirical calculations on the water dimer for "linear" geometry. R and θ are defined in Figure V-8. Except for the fourth entry (which used a minimum basis set of Slater functions), all calculations employed gaussian functions.

Basis Set	Reference	E(hartrees)	Energy(kcal/mole)	R(O-O) Å	θ
O(5s 3p), H(3s)	263	-151.1188	12.6	2.68	0°
O(10s 5p / 3s 1p) H(5s / 1s)	207	-151.9577	5.3	3.00	25°
O(10s 5p / 4s 2p) H(5s / 2s)	207	-152.0145	7.9	2.85	25°
O(2s 1p), H(1s)	264	-151.4205	6.6	2.76	57°
O(8s 4p / 2s 1p) H(4s / 1s)	100	-151.1100	6.1	2.73	57°
O(11s 7p 1d / 5s 4p 1d) H(6s 1p / 3s 1p)	102	-152.1117	4.84	3.00	40°
O(10s 5p 1d / 5s 3p 1d) H(4s 1p / 2s 1p)	145	-152.0907	4.72	3.00	40°
Experiment	326		5.0		

energy are ~ 7 kcal. A third example is given by the double-zeta-plus-polarization SCF calculations of Clementi, Mehl, and von Niessen (80) on the formic acid dimer. There the predicted dimerization energy is 16.2 kcal, compared to experiment, 14 kcal. Although there is reason for optimism concerning the usefulness of the RHF approximation, at the present time this optimism must be tempered by a reasonable degree of wariness. Another question arising with respect to Table V-21 concerns the reliability of the predicted O-O distances. Although larger basis sets usually shorten predicted bond distances, the opposite is seen to be true for the H_2O dimer. There is really no definitive experimental O-O bond distance, but the infrared spectrum of the matrix-isolated H_2O dimer suggests a bond distance somewhat longer than that in ice, 2.76 Å.

The apparently reasonable SCF description of hydrogen bonding in water suggests the study of the degree of nonadditivity of pair interactions in the H_2O trimer. The study of additivity of pair potentials is particularly important in light of the development of computational methods for the statistical mechanical study of liquids (4). All such methods of course require *a priori* knowledge of the potential surface relating the motions of the various water molecules. The energy of the water trimer may be expressed

$$E(1,2,3) = 3E(1) + V(1,2) + V(1,3) + V(2,3) + V(1,2,3) \qquad (V,32)$$

where $V(1,2,3)$ is the nonadditivity term, which can only be evaluated by calculations on the trimer itself. Hankins, Moskowitz, and Stillinger (145) have investigated the potential surface for the water trimer using an extended basis, O(10s 5p 1d / 5s 3p 1d) and H(4s 1p / 2s 1p). The calculations are analogous to those which predicted the H_2O dimerization energy of 4.72 kcal above.

An obvious problem to be solved, especially in light of the large basis set used, involves the selection of a reasonable number of trimer geometries to investigate. Hankins, et al. chose three geometries observed in ice polymorphs and clathrate hydrates and expected to have significantly different three-body potentials V(1,2,3). Figure V-9 shows these three geometries, referred to as double donor, double acceptor, and sequential. In most of the calculations the hydrogen bond angle was held fixed at 54°44' as seen in Figure V-9. In addition, all trimers calculated involve two O-O bonds of equal length.

Table V-22 shows the most important results of the calculations. Recall that the equilibrium O-O separation in ice is 2.76 Å. An important qualitative feature shown in the table is the variability of V(1,2,3) among the three forms. The sequential trimer is most stabilized with respect to the result expected from the pair potentials (i.e., the dimer energies as a function of O-O distance). On the other hand, the double donor and double acceptor are destabilized by 0.874 and 0.346 kcal/mole at the ice distance. The sequential form of the trimer is lowest in energy and is predicted to have a bond distance of about 3.0 Å, the same as that predicted for the dimer. In tetrahedral clusters (e.g., those thought to occur in ice), the ratio of sequential to double donor to double acceptor forms is 4 to 1 to 1. The authors use the fact that the weighted V(1,2,3) is negative at R = 2.76 Å to rationalize the fact that the dimer bond distance is calculated to be longer than the ice distance. Hankins, et al.

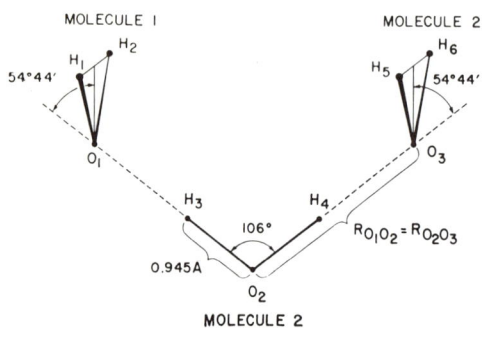

Figure V-9. Geometries considered in SCF calculations on the H_2O trimer (145).

a) Double donor trimer

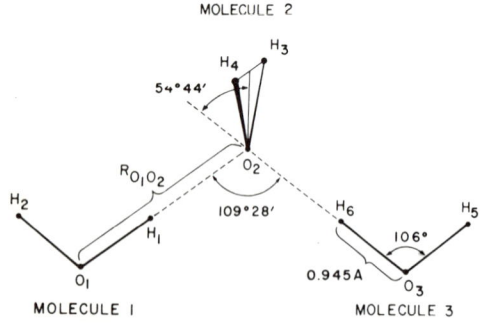

b) Double acceptor trimer

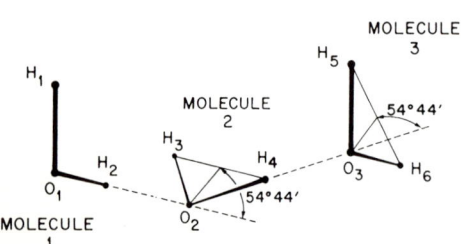

c) Sequential trimer

Table V-22. Hydrogen bond energies and nonadditivities of the water trimer. All energies are in kcal/mole.

Type	R(O-O) Å	Energy Relative to Three Isolated H_2O Molecules	V(1,2,3)
Double Donor	2.76	-2.973	0.874
	3.00	-5.948	0.184
	3.15	-6.141	-0.048
Double Acceptor	2.76	-3.140	0.346
	2.90	-5.294	0.109
	3.00	-5.965	-0.048
	3.15	-6.120	-0.184
	3.39	-5.456	-0.319
Sequential	2.76	-6.766	-1.363
	3.00	-8.293	-0.994
	3.15	-7.949	-0.863

also show that their nonadditivity results cannot, as is sometimes thought, be explained in terms of classical electrostatics, i.e., treating the three molecules as polarizable point particles.

Nickel hexafluoride. Since the classic 1929 paper of Bethe (44), there has been a great deal of activity in the area of crystal field theory. Crystal field theory assumes that the differences in energy levels between a free metal ion (e.g., Ni^{+2}) and an octahedral inorganic complex (e.g., NiF_6^{-4}) are due to changes in the metal ion orbitals due to a point charge environment. In ligand field theory, the orbitals upon which the energy levels depend are allowed to take on some of the character of the ligands (the F^- ions in NiF_6^{-4}). In rigorous molecular orbital theory of course, the orbitals are allowed to take on whatever form yields the lowest energy via the self-consistent-field method. Virtually all calculations reported in the literature on octahedral transition metal complexes are semi-empirical, usually involving rather severe integral approximations (190). The first *ab initio* transition metal complex calculation reported was that of Gladney and Veillard (135) on NiF_6^{-4} and employed a small gaussian basis. More recently, Moskowitz and coworkers (267) have completed work on the same molecule with a larger basis. In both calculations all 86 electrons are explicitly included and all molecular integrals accurately evaluated.

The basis set used by Moskowitz, et al. was designed to be of minimum basis set quality for the inner shell orbitals (nickel 1s, 2s, 3s, 2p, 3p and fluorine 1s and 2s) but double-zeta quality for the valence orbitals (4s, 3d Ni, 2p F). In addition one 4p-like function was included, although this orbital is not occupied in the ground state atomic SCF orbital occupancy. The resulting contracted gaussian basis set is Ni (15s 11p 5d / 5s 3p 2d) and F(8s 5p / 2s 2p). For Ni^{+2} the SCF energy with this basis is -1504.4085 hartrees, as opposed to a near RHF energy -1506.0286 hartrees (75). Despite the rather poor SCF total energy obtained for Ni^{+2}, this basis set is sufficiently flexible in the valence region to give a qualitatively reasonable description of electronic structure. The computation times reported are of interest due to the large size of the ion. 80 minutes of CDC 6600 time were required for evaluation of integrals and 130 minutes for solution of the SCF equations for a given electronic state.

In crystal field theory NiF_6^{-4} is composed of a Ni^{+2} ion and six F^- ions. The ground state SCF orbital occupancy and resulting electronic states of Ni^{+2} are

$$1s^2\ 2s^2\ 2p^6\ 3s^2\ 3p^6\ 3d^8 \qquad {}^3F,\ {}^1D,\ {}^3P,\ {}^1G,\ {}^1S \qquad (V,33)$$

Under octahedral symmetry, however, the 5-fold degenerate 3d orbital splits into triply-degenerate t_{2g} and doubly-degenerate e_g orbitals. The resulting orbital occupancies (all of which correspond to d^8 for the free ion) and electronic states are (17)

$$\begin{array}{ll} t_{2g}^6\ e_g^2 & {}^3A_{2g},\ {}^1E_g,\ {}^1A_{1g} \\ t_{2g}^5\ e_g^3 & {}^3T_{1g},\ {}^3T_{2g},\ {}^1T_{2g},\ {}^1T_{1g} \\ t_{2g}^4\ e_g^4 & {}^3T_{1g},\ {}^1T_{2g},\ {}^1E_g,\ {}^1A_g \end{array} \qquad (V,34)$$

Only for the ${}^3A_{2g}$, ${}^3T_{2g}$, and ${}^1T_{1g}$ states is there an unambiguous single configuration representation. For the other states two configurations are possible and both should be included in electronic structure calculations. The energies of the single-configuration wave functions for the ${}^3A_{2g}$, ${}^3T_{2g}$, and ${}^1T_{1g}$ states may be readily evaluated from Eq.(I,16). If we assume that the ${}^3T_{1g}$ state wave function is the single configuration $t_{2g}^5 e_g^3$, its energy may also be evaluated and expressions for three energy differences obtained in terms of three parameters

$$E(^3T_{2g}) - E(^3A_{2g}) = 10\ Dq$$
$$E(^1T_{1g}) - E(^3T_{2g}) = 12\ B + 2\ C \qquad (V,35)$$
$$E(^3T_{1g}) - E(^3T_{2g}) = 12\ B$$

10 Dq, B, and C are of course just fixed linear combinations of one and two-electron integrals. However, early workers chose to ascribe physical significance to these parameters. 10 Dq is the <u>crystal-field splitting parameter</u>, while B and C are Racah parameters for the Ni^{+2} ion. We now know of course that single configuration wave functions, no matter how accurate, do not usually yield reliable excitation energies. However, since (V,35) represents three equations in three unknowns, the experimental transition energies can be fit to obtain "experimental" values of three empirical parameters 10Dq, B, and C. For NiF_6^{-4} 10 Dq obtained in this way is 7250 cm^{-1}, and B and C are 7% and 13% less than their free ion values. It is, however, important to realize that Hartree-Fock calculations are not expected to reproduce the values of these experimental parameters. Moskowitz and coworkers have calculated SCF wave functions for the $^3A_{2g}$, $^3T_{2g}$, and $^3T_{1g}$ (orbital occupancy $t_{2g}^5 e_g^3$) states of NiF_6^{-4}, yielding energies -2099.1291, -2099.1013, and -2099.0325 hartrees. Subtraction of the $^3A_{2g}$ energy from that of $^3T_{2g}$ gives **0.0278 hartrees or 6089 cm^{-1} for 10 Dq**, in surprisingly good agreement with the experimental value 7250 cm^{-1}. The difference between the $^3T_{1g}$ and $^3T_{2g}$ energies is 1260 cm^{-1}, in qualitative agreement with the experimental value of B, 955 cm^{-1}.

Two approximate calculations of 10 Dq were also carried out. First, the energy of the $^3T_{2g}$ single-configuration was evaluated using the ground state ($^3A_{2g}$) SCF orbitals. This may be referred to as a virtual orbital approximation. The calculated splitting 10 Dq was 33,250 cm^{-1}, much too high since the orbitals used were appropriate for the ground state but much less suited for the $^3T_{2g}$ state. (This type of approximation is analogous to Koopmans' theorem, wherein ground state SCF orbitals for the neutral molecule are used to describe the positive ion.) In the second approximation, the SCF equations were solved for Ni^{+2} surrounded by an array of -1 point charges placed at the F$^-$ ion sites. This is the rigorous equivalent of crystal field theory. The difference between the $^3A_{2g}$ and $^3T_{2g}$ energies is 1512 cm^{-1}, almost a factor of five too small. It would appear that six F$^-$ ions have a much greater effect on the Ni^{+2} energy levels than do six point charges. In conclusion, the direct NiF$_6^{-4}$ calculations yield very reasonable results for 10 Dq but the virtual orbital and crystal field approximations are poor.

Table V-23 indicates the composition of the $^3A_{2g}$ ground state orbitals. This table shows clearly the ionic nature of NiF$_6^{-4}$, as every molecular orbital comes within 2% of having purely atomic character. A population analysis shows the same result. There is essentially no promotion to Ni 4s (population 0.071 electrons) or 4p (population 0.062) orbitals. The nickel 3d orbital population is 8.059, nearly identical to that in the Ni^{+2} free ion.

To investigate the hyperfine interactions at the F$^-$ ligands, Moskowitz, et al. calculated unrestricted SCF wave functions (see section IB) for the ground and first excited states of NiF$_6^{-4}$. The unrestricted energy for the ground state is only 0.0008 hartree lower than the restricted SCF energy.

Table V-23. Orbital energies (hartrees) and percentage composition by population analysis of the SCF orbitals for the ground state of NiF_6^{-4} (267). MO's corresponding to the core orbitals, Ni(1s, 2s, 2p, 3s, 3p) and F(1s) are omitted.

Orbital	Orbital Energy	% Composition Ni^{+2}	F^-
$5a_{1g}$	-0.7283	2(4s)	98(2s)
$4t_{1u}$	-0.7167	1(4p)	98(2s), 1(1s)
$2e_g$	-0.7131	--	100(2s)
$6a_{1g}$	+0.1371	2(4s)	98(2p)
$5t_{1u}$	+0.1661	--	100(2p)
$3e_g$	+0.1970	2(3d)	98(2p)
$6t_{1u}$	+0.2109	--	100(2p)
$4e_g$	+0.6635	98(3d)	2(2p)

However, for the spin density at F^{19} (in units of 10^{-4} cm^{-1}), the restricted SCF value is 26.84, the unrestricted 32.82, and experiment 33.9. On the other hand, the dipole contribution A_σ to the F^{19} hfs is 10.29 for restricted, 13.56 for unrestricted SCF, and 8.8 from experiment. In the restricted SCF approximation, of course, the entire hfs is due to the unfilled e_g orbital. In either the restricted or unrestricted SCF approximations, the predicted hfs parameters are in encouraging agreement with experiment.

XeF_2, XeF_4, and XeF_6. The preparation of the first noble gas compound in 1962 by Bartlett (21) opened a new area of chemistry. Shortly after the announcement of Bartlett's work, Claasen, Selig, and Malm (70) were able to prepare XeF_4, and the discovery of XeF_2 and XeF_6 followed shortly. The existence of xenon fluorides came as a surprise to experimentalists and theoreticians alike, since xenon is a closed shell species, expected to be quite inert. In fact, some experimentalists were quick to say that the existence of the xenon fluorides represented an "embarrassment" to theoretical chemistry. However the discovery of these molecules was in truth only embarassing to those theoretical chemists who believed that all of chemistry could be explained in terms of a few rather naive concepts. Since no rigorous quantum mechanical calculations had been carried out on even XeF_2, in one sense there was no theoretical evidence either for or against the existence of xenon fluorides. Although an accurate treatment of XeF_2 is expected to be completed within the next year or two, it is likely to be several years before reliable a priori calculations of the dissociation energies of XeF_4 and XeF_6 are possible. However, Basch, Moskowitz, Hollister, and Hankins (25) have made an important first step toward the understanding of the electronic structure of the xenon fluorides. They have recently completed ab initio SCF calculations on XeF_2, XeF_4, and XeF_6.

The basis set employed embodies the same philosophy used in the calculations discussed above on nickel hexafluoride. That is, a minimum basis set description of the core orbitals and double-zeta description of the valence orbitals (5s and 5p on xenon, 2p on fluorine) is used. In our notation, the basis set used is Xe(17s 13p 7d / 6s 5p 2d) and F(8s 4p / 2s 2p). The SCF energy for the ground state of xenon lies about 12 atomic units above the RHF energy. However use of this basis set is understandable if one stops to realize that XeF_6 has 108 electrons. The molecular SCF calculations were carried out for the closed-shell ground state electron configurations of linear XeF_2 (bond distance 2.00 Å), square planar XeF_4 (1.95 Å), and octahedral XeF_6 (1.89 Å).

Population analyses obtained from the SCF wave functions are seen in Table V-24. It should be remembered that any scheme (such as a population analysis) for assigning charges to atoms in a molecule is arbitrary. However, comparison of populations analyses for a series of molecules does allow one to make qualitative conclusions concerning changes in electron distribution. For all three molecules in Table V-24, there is a significant shift (with respect to the isolated atoms) in electron density from the Xe to the F atoms. This hardly comes as a surprise, as chemists usually consider fluorine to be the most electronegative element in the periodic table. However, it is interesting to note that 0.38 fewer "electrons" are shifted from Xe to F in XeF_4 than in XeF_2. The same number, 0.38, fewer electrons are shifted to F in going from XeF_4 to XeF_6. Since the total atomic populations on Xe in the three molecules are distinctly noninteger, the calculations seem to warn against the discussion of electronic structure in terms of "oxidation states" of xenon. This is of course quite different from the NiF_6^{-4} results discussed above, where the Ni^{+2} species is clearly identifiable within the molecule.

Table V-24. Population analysis for the valence orbitals of the xenon fluorides (25).

	XeF_2	XeF_4	XeF_6
Xenon			
5s	2.018	2.033	2.058
5p	4.786	3.634	2.622
Net	+1.304	+2.455	+3.458
Fluorine			
2s	1.989	1.991	1.991
2p	5.663	5.625	5.586
Net	-0.652	-0.614	-0.576

An interesting feature of the calculations by Basch and coworkers is the prediction of a variety of molecular properties, seen in Table V-25. Since these quantities depend on the coordinate system used, it should be mentioned that a) for XeF_2 the z axis is the molecular axis, b) for XeF_4, two F atoms lie along the x and y axes and the z axis is perpendicular to the plane of the molecule, and c) for XeF_6, two F atoms lie on each of the x, y, and z axes. Although there are expected to be significant differences due to basis set between the properties in Table V-25 and the RHF values, the reported values should be qualitatively useful. The second moments $<x^2>$,

Table V-25. One-electron properties of the xenon fluorides calculated from SCF wave functions. All properties are in atomic units.

	XeF_2	XeF_4	XeF_6
$<x^2>_{Xe}$	26.32	289.88	279.47
$<y^2>_{Xe}$	26.32	289.88	279.47
$<z^2>_{Xe}$	298.54	31.55	279.47
quadrupole moment			
θ_{zz}	-15.10	13.89	0.00
diamagnetic shielding			
$\sigma^d(Xe)$	323.22	328.05	332.71
$\sigma^d(F)$	42.19	45.97	49.34
electric field gradients			
$q_{xx}(Xe)$	12.83	-14.12	0.0
$q_{yy}(Xe)$	12.83	-14.12	0.0
$q_{zz}(Xe)$	-25.66	28.23	0.0
$q_{xx}(F)$	1.377	-2.601	1.342
$q_{yy}(F)$	1.377	1.059	1.342
$q_{zz}(F)$	-2.755	1.543	-2.685

$< y^2 >$, and $< z^2 >$ indicate that most of the electron density in XeF_2 lies near the molecular axis, and for XeF_4 most of the electron density is contained in a volume slightly above and below the plane of the molecule. It is interesting to note that the quadrupole moment of XeF_2 is close to that estimated by Coulson (90), -14.1 atomic units, from elementary electrostatic considerations. The quadrupole coupling constants $eq_{zz}Q$ have been measured by Mössbauer spectroscopy to be 2490 MHz for $^{129}XeF_2$ and 2620 MHz for $^{129}XeF_4$ (294). If the nuclear-electric quadrupole moment of ^{129}Xe is assumed to be 0.41 barns, one can deduce q_{zz} absolute values at xenon of 25.8 and 27.2 atomic units for XeF_2 and XeF_4. These values are in close agreement with the predicted values in Table V-25, 25.66 and 28.23 atomic units.

Comparisons are also possible between calculated orbital energies and the experimental ionization potentials obtained from photoelectron spectroscopy (195). For example, for the Xe 4d levels, the calculated ionization potential shifts with respect to the atom are 4.20 eV, 8.79 eV, and 13.69 eV for XeF_2, XeF_4, and XeF_6. These are uniformly larger than the experimental shifts 2.95, 5.47, and 7.88 eV. Although the trend is properly predicted, calculations on smaller molecules with larger basis sets (53) have given much better agreement with experiment than this. One suspects that the true RHF binding energy shifts for the xenon fluorides will be in similarly good agreement with experiment. An interesting effect predicted by the calculations is that the shift in ionization potentials decreases slightly on going to higher principle quantum numbers. For those orbital energies identified as corresponding to Xe 3s, the calculated shifts are 4.49, 9.52, and 14.94 eV for XeF_2, XeF_4, and XeF_6. The same shifts for Xe 4s are uniformly less, 4.24 eV, 8.88 eV, and 13.84 eV. For the 2s and 1s orbitals the differences in ionization potentials between Xe and XeF_6 are predicted to be 15.67 and 16.05 eV, both higher than the 3s and 4s shifts.

Basch and coworkers (25) have also attempted to discuss the electronic spectra of the xenon fluorides on the basis of the SCF virtual orbitals. As discussed in section VA with respect to the excited states of H_2O, this is not a very good idea, since the virtual orbitals of a molecule describe the excited states of the negative ion. However, some crude correlations seem possible based on the orbital energies of the lowest unoccupied orbitals, $7\sigma_u$ for XeF_2 ($\epsilon = +0.0498$), $8e_u$ for XeF_4 ($\epsilon = +0.0002$), and $8t_{1u}$ for XeF_6 ($\epsilon = -0.0398$).

Naphthalene and Azulene. In January of 1969, a paper was published in Chemical Physics Letters which foretold the important role to be played by ab initio calculations in organic chemistry. In that paper (59), Buenker and Peyerimhoff reported nonempirical SCF calculations on naphthalene and azulene, two $C_{10}H_8$ isomers. Although a rather inflexible basis set, C(10s 5p / 3s 1p) and H(5s / 1s), was used, most chemists who read the paper were surprised that any type of rigorous calculation could be carried out for a molecule with eighteen atoms. The present author was reminded at the time of a seminar at M.I.T. during 1965 when a leading organic chemist stated that ab initio calculations on the much smaller benzene molecule would not be possible for many years to come. In light of the fact that 0.89 billion electron repulsion or two-electron integrals were calculated in each of the naphthalene and azulene calculations, it is not difficult to understand earlier predictions of failure. It is just this type of experience which makes the present author hesitant to state that certain types of calculations will not be possible for the next 5, 10, or 20 years.

The SCF calculations of Buenker and Peyerimhoff were carried out for the experimental geometries of naphthalene and azulene, depicted in Figure V-10. The calculated total energy for naphthalene was -382.7883 hartrees and that for azulene -382.7082 hartrees. Thus naphthalene is predicted to lie lower in energy by 50.3 kcal/mole. The experimental heats of formation confirm this prediction that naphthalene is the more stable, but the true energy difference is only 29.5 kcal/mole. This SCF error is rationalized by the authors using the fact that the lowest virtual orbital of naphthalene, $2b_{2g}$ has $\epsilon = +0.0771$ hartree while for azulene the lowest unoccupied orbital energy $(4b_1)$ is 0.0204 hartree. Since the first virtual orbital of azulene lies lower than that of naphthalene, the authors argue that azulene should have more correlation energy and that the introduction of CI would thus lower azulene with respect to naphthalene. The Koopmans' theorem ionization potentials for naphthalene and azulene, 8.26 eV and 9.20 eV, are in reasonable agreement with the experimental values 7.41-7.72 and 8.5 eV.

Particular emphasis is placed on the interpretation of the calculated orbital energies, seen in Table V-26. Naphthalene has D_{2h} symmetry and azulene C_{2v}, and each orbital seen in Table V-26 is doubly-occupied in one of the two SCF wave functions. Both molecules have 68 electrons. As we

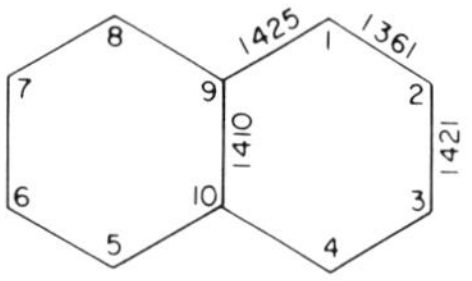

Naphthalene

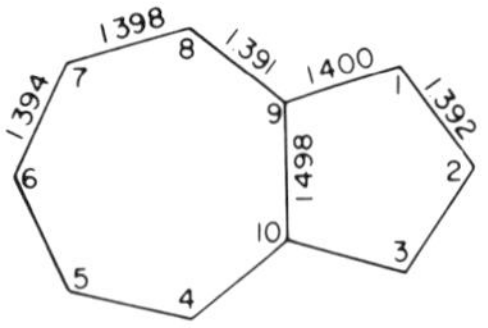

Azulene

Figure V-10. Experimental ground state geometries of napthalene and azulene. Bond lengths are given in Å.

---

have seen, electron distributions in large molecules are usually illustrated by Mulliken population analyses. However, Buenker and Peyerimhoff suggest that inner shell orbital energies for the various atoms give the most direct information on the charge distributions. This interpretation is consistent with the ideas of experimental workers in the area of x-ray photoelectron spectroscopy. There, experimentally measured core electron binding energies are used to assign charges on the atoms in a molecule (350). If one uses this argument, the carbon atoms with lowest (largest in absolute magnitude) orbital energy are assigned the smallest charge. For naphthalene the $1a_g$ and $1b_{2u}$

orbitals correspond to two fusion carbon atoms, 9 and 10 in Figure V-10. Since $1a_g$ and $1b_{2u}$ have the lowest orbital energies, carbon atoms 9 and 10 are predicted to be the most positive, i.e., have the least charge or number of electrons in a point-charge picture. On the basis of orbital energy, the α positions (1,4,5, and 8) are expected to be the most negative, although the difference between α and β (2,3,6, and 7 in Figure V-10) orbital energies is less than 0.1 eV. Buenker and Peyerimhoff point out that this point charge picture is consistent with the experimental finding that the α position is most susceptible to electrophilic substitution. For electrophilic substitution to occur, a positive intermediate species must first be formed, and the easiest position from which to remove an electron is the more negative α position.

In azulene, the predicted separation of 1s binding energies is much greater than in naphthalene. A general observation is that the orbital energies identified with the five-membered ring lie significantly higher than those identified with the seven-membered ring. In a point charge picture, there are more electrons per carbon atom on the five-membered ring. This picture is consistent with the early semi-empirical calculations of Brown (52). On the basis of his semi-empirical electron density, Brown was able to correctly predict that for azulene, electrophilic substitution is most likely to proceed at the more negative 1 or 3 positions.

A final point investigated concerns the predicted π <u>bond orders</u> (88) for the two molecules. At the outset it should be made clear that bond order is not an observable but rather a purely theoretical construct. Naphthalene has 5 so-called π orbitals, $1b_{1u}$, $1b_{2g}$, $1b_{3g}$, $2b_{1u}$, and $1a_u$. Azulene also

Table V-26. Orbital energies, in hartrees, for the electronic ground states of naphthalene and azulene. Numbering of the inner shell ε's indicates the atomic centers at which the orbitals are localized. See Figure V-10 for the numbering of atomic centers. Center 1 is indicated four times in the naphthalene calculation since there are three other equivalent carbon atoms of this type.

Naphthalene		Azulene	
$1a_g$	− 11.3926 (9)	$1a_1$	− 11.3985 (4)
$1b_{2u}$	− 11.3916 (9)	$1b_2$	− 11.3985 (4)
$2a_g$	− 11.3628 (2)	$2a_1$	− 11.3942 (6)
$1b_{3u}$	− 11.3628 (2)	$3a_1$	− 11.3821 (9)
$1b_{1g}$	− 11.3620 (2)	$2b_2$	− 11.3815 (9)
$2b_{2u}$	− 11.3620 (2)	$4a_1$	− 11.3703 (5)
$2b_{3u}$	− 11.3599 (1)	$3b_2$	− 11.3703 (5)
$3a_g$	− 11.3599 (1)	$5a_1$	− 11.3423 (2)
$2b_{1g}$	− 11.3599 (1)	$4b_2$	− 11.3193 (1)
$3b_{2u}$	− 11.3598 (1)	$6a_1$	− 11.3192 (1)
$4a_g$	− 1.2133	$7a_1$	− 1.2074
$3b_{3u}$	− 1.1455	$8a_1$	− 1.1447
$4b_{2u}$	− 1.0898	$5b_2$	− 1.1037
$5a_g$	− 1.0466	$9a_1$	− 1.0372
$3b_{1g}$	− 1.0249	$6b_2$	− 0.9945
$4b_{3u}$	− 0.8866	$10a_1$	− 0.9187
$5b_{2u}$	− 0.8719	$7b_2$	− 0.8847
$6a_g$	− 0.8512	$11a_1$	− 0.7879
$4b_{1g}$	− 0.7484	$12a_1$	− 0.7795
$5b_{3u}$	− 0.7447	$8b_2$	− 0.7466
$7a_g$	− 0.7178	$13a_1$	− 0.7193
$6b_{2u}$	− 0.6689	$14a_1$	− 0.6752
$8a_g$	− 0.6616	$9b_2$	− 0.6707
$5b_{1g}$	− 0.6347	$10b_2$	− 0.6201
$6b_{3u}$	− 0.6305	$15a_1$	− 0.5986
$7b_{2u}$	− 0.6094	$16a_1$	− 0.5778
$1b_{1u}$	− 0.5758	$11b_2$	− 0.5678
$7b_{3u}$	− 0.5673	$1b_1$	− 0.5678
$6b_{1g}$	− 0.5300	$12b_2$	− 0.5546
$9a_g$	− 0.5221	$17a_1$	− 0.5176
$1b_{2g}$	− 0.4966	$2b_1$	− 0.4961
$1b_{3g}$	− 0.4355	$1a_2$	− 0.4549
$2b_{1u}$	− 0.3750	$3b_1$	− 0.3470
$1a_u$	− 0.3418	$2a_2$	− 0.3035

has 5 π orbitals, $1a_1$, $2a_1$, $1a_2$, $3b_1$, and $2a_2$. Bond order is most easily defined if there is a single atomic orbital (e.g., a π orbital, that is a p orbital pointing out of the plane of the molecule) centered on each atom. Then the partial bond order (between atoms A and B) for the ith molecular orbital is just $c_{iA} c_{iB}$, where c_{iA} is the coefficient of the function centered on A in the ith molecular orbital. The total bond order, or just bond order, is then the sum of the $c_{iA} c_{iB}$ over all the occupied molecular orbitals i. The π bond orders are obtained for naphthalene by allowing the summation in i to go over only the five π orbitals. Traditionally a large bond order has been taken to imply a short bond distance. For example, ethane, ethylene, and acetylene are frequently assumed to have bond orders of 1, 2, and 3, to go with the observed bond distances 1.54, 1.33, and 1.20 Å (370). For naphthalene, the calculations of Buenker and Peyerimhoff predict the following bond orders: 0.420 (9-10), 0.307 (1-9), 0.577 (1-2), and 0.315 (2-3). Comparison with Figure V-10 shows that the 1-2 bond, with highest bond order, is in fact the shortest. Perhaps surprisingly, the order of bond orders perfectly predicts the observed order of bond distances. For azulene, the ordering of bond distances predicted in this way is (with calculated bond orders in parentheses): 1-2 (0.448), 1-9 (0.446), 4-5 (0.437), 5-6 (0.435), 4-10 (0.411), and 9-10 (0.075). Unfortunately, the observed order of bond distances is: 4-10, 1-2, 5-6, 4-5, 1-9, and 9-10. Far from being the second longest, the 4-10 bond turns out to be the shortest. These results would seem to indicate that the use of bond orders to predict bond distances is not advisable.

The guanine-cytosine base pair. Most high school students are familiar with the term DNA, or deoxyribonucleic acid, the hereditary substance believed to carry the genetic message in the cell. As determined by Watson and Crick in the early 1950's, DNA consists of a double helix held together by hydrogen bonds between pairs of bases on the two helices. In particular adenine is hydrogen bonded to thymine and guanine is hydrogen bonded to cytosine. For concreteness, Figure V-11 shows the geometry of the guanine-cytasine (G-C) pair according to Spencer (365). There it is seen that there are three hydrogen bonds connecting the two bases. There is reason to believe that the potential energy surface describing the motion of the three bridging hydrogen atoms (H23, H24, and H25 in Figure V-11) is of fundamental biological importance. It is usually assumed that a potential curve with a double minimum governs

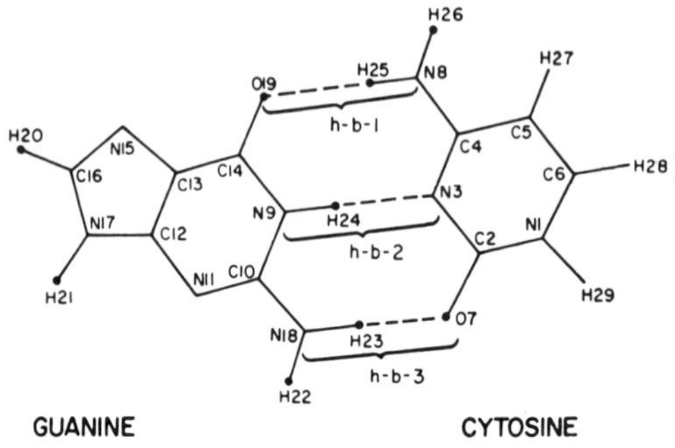

Figure V-11. Geometry of the guanine-cytosine base pair, with illustration taken from the paper of Clementi, Mehl, and von Niessen (80).

the movement of, for example, H24 from N9 to N3 in the G-C pair. If, for example, H25 moved to a second equilibrium position closer to O19 than N8 and simultaneously H24 moved closer to N3, a tautomeric form of guanine-cytosine results and is referred to as G*-C*. An interesting discussion has been given by Löwdin (236) of the possible importance of quantum mechanical tunneling for the interconversion between different tautomeric forms.

It has been pointed out earlier that SCF calculations using extended basis sets appear to describe hydrogen bonding in a qualitatively correct way for the handful of cases where comparison with experiment has been possible. Thus it comes as no surprise that theoreticians would like to carry out SCF calculations on the base pairs. However, G-C, for example, has 136 electrons, a formidable barrier to any *ab initio* procedure. Nevertheless Clementi, Mehl, and von Niessen (80) have been able to perform SCF calculations on G-C for 27 different geometries. The basis set used was (7s 3p / 2s 1p) on C, N, and O and (3s / 1s) on each H atom. The feasibility of these calculations was in part the result of a feature (79) of Clementi's program IBMOL which allowed each calculation on G-C after the first to be completed in about one-tenth the computation time required for the first calculation. Only those integrals involving basis functions centered on atoms whose positions change from one geometry to the next are recomputed. This feature is clearly desirable if an entire potential energy surface is to be investigated. Each complete calculation after the first required somewhat less than 5 hours of IBM 360/195 computer time.

Three different types of geometries were considered. In the first, hydrogen atom 25 (see Figure V-11) was moved between oxygen atom 19 and nitrogen atom 8. In this process, the positions of all 28 atoms except H25 remained fixed in their experimental positions. A second series of calculations varied the position of H24 between N9 and N3, and the final 12 calculations varied the position of the third hydrogen bridge, H23. The surprising result of these calculations was that in none of the three cases was a double minimum potential predicted. The potential curve for the motion of H25 is seen in Figure V-12. As known experimentally, the calculation predicts the first hydrogen to be associated with cytosine. Although there is a noticeable shoulder in Figure V-12 where the second minimum might be expected, there is no second minimum. Clementi and coworkers cite three possible sources of error in their prediction of the lack of a double minimum. The first of these is the possibility that the G-C experimental geometry of Spencer is not correct. Changing the distance between guanine and cytosine is expected to alter the potential curves for the three hydrogen bridges. Secondly, simultaneous motion of two hydrogen bridges was not considered. In going from G-C to G*-C*, hydrogen bridges 1 and 2 are expected to simultaneously move in opposite directions. Third, the small gaussian basis set used may yield SCF results very different from the true Hartree-Fock potential curves.

To test the second and third possible sources of error, Clementi and coworkers carried out calculations on the formic acid dimer, which has two hydrogen bonds. First only the motion of a single hydrogen bridge was considered. Using the G-C basis set only a single minimum was found. If a much larger double-zeta-plus-polarization basis is used, a very pronounced shoulder

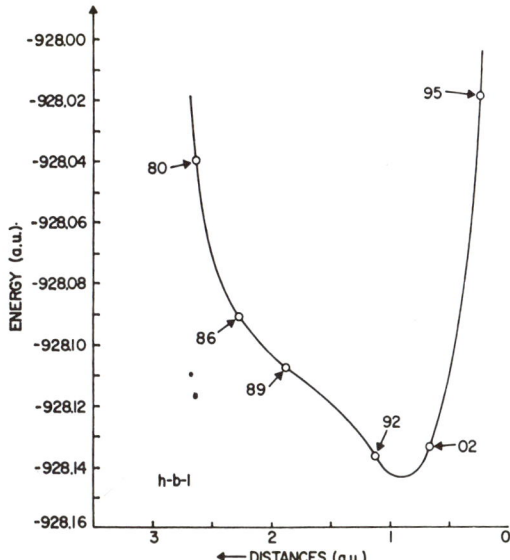

Figure V-12. Potential energy curve for hydrogen bridge 1 in guanine-cytosine (80). Point 80 is about the same distance from O19 as point 95 is from N8.

is predicted, but still no double minimum. Then the coupled motion of the two hydrogen bridges was considered. Using the G-C basis, a pronounced shoulder was again predicted. However, the use of both double-zeta and double-zeta-plus-polarization basis sets yielded a double minimum. In the latter SCF calculation the dimerization energy is predicted to be 16.2 kcal/mole, in good agreement with the experimental value 14 kcal. These formic acid calculations seem to show rather clearly that a) simultaneous motion of the hydrogen bridges in G-C should be considered, and b) any basis set smaller than double-zeta may yield spurious results in studies of hydrogen bonding.

For the experimental geometries, Figure V-13 shows the predicted orbital energies for G-C and the two isolated base pairs. A point stressed in earlier papers by Clementi is that the σ and π orbital energies are interwoven. In π-electron theory, the σ orbitals are neglected and only π orbitals taken into account. Most semi-empirical workers had assumed that the π orbital energies always lie above the highest σ orbital energy. For G-C, the 1π orbital energy is seen in Figure V-13 to lie nearly 10 eV <u>below</u> the 56σ orbital. This would appear to raise serious questions concerning the suitability of the π-electron approximation.

D. OTHER TOPICS

Here are discussed briefly two additional topics which did not appear to fit neatly into sections A, B, or C of this chapter.

<u>Barriers to rotation and inversion</u>. Interest in barriers to internal rotation began in 1936 when Kemp and Pitzer (202) postulated on the basis of thermodynamic data the existence of a barrier of about 3 kcal/mole in ethane. However, the first direct quantum mechanical calculation of a rotational barrier was not reported until 1963. During the past five years, a great deal of theoretical effort has been spent in the calibration of methods for the prediction of rotation and inversion barriers (6). The overall conclusion drawn from these calculations is that the magnitudes of barriers to rotation

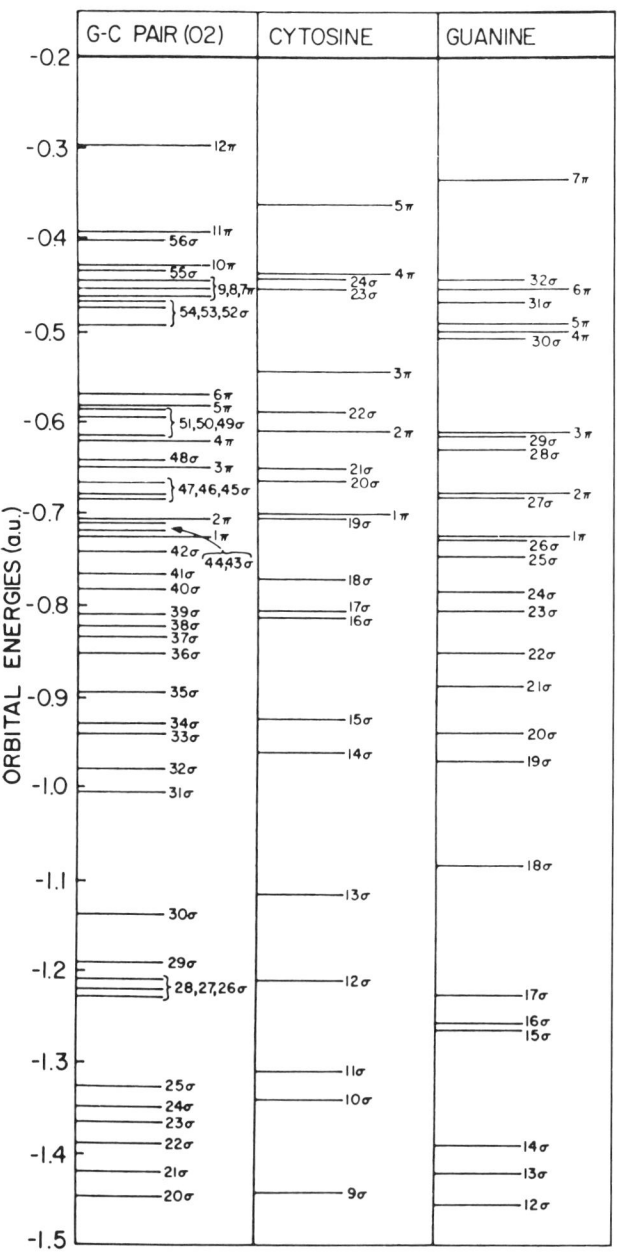

Figure V-13. Orbital energies for guanine, cytosine, and the G-C base pair from <u>ab initio</u> SCF calculations.

and inversion are quite accurately predicted within the single configuration Hartree-Fock approximation. Here we discuss the three barriers which have attracted the most interest.

The first nonempirical barrier calculation was that of Pitzer and Lipscomb (298) for ethane, C_2H_6. They carried out SCF calculations using a minimum basis set of Slater functions. The geometry used (except for the relative orientation of the two methyl groups) was taken from experiment. For staggered ethane, the SCF energy was -78.99115 hartrees, and for eclipsed ethane -78.98593 hartrees was the result. Thus staggered ethane was predicted to be the more stable by 0.00522 hartree or 3.3 kcal/mole. Since the best experimental value at the time was 3.03 ± 0.30 kcal, the theoretical prediction was within experimental error. Many persons who read the Pitzer-Lipscomb article in 1963 probably thought that the agreement between theory and experiment for this small energy difference was a fluke, due to some fortuitous combination of computational errors, deficiency of basis set, and neglect of electron correlation. All three of these objections have since been invalidated. Using a completely different set of computer programs, Stevens (367) has recently verified the essential correctness of the Pitzer-Lipscomb results. Again using a minimum basis set, Stevens optimized (for both staggered and eclipsed forms) both orbital exponents and geometrical parameters. These fully optimized minimum basis set C_2H_6 results yield staggered and eclipsed energies of -79.0999 and -79.0946 hartrees. Although the total energies are more than 0.1 hartree lower than those of Pitzer and Lipscomb, the barrier is the same, 3.3 kcal.

To test whether the above agreement with experiment was only due to a poor basis, numerous SCF calculations on ethane have been carried out with larger basis sets. Of these, the calculation by Veillard (379) seems to be the most comprehensive. Veillard used a large contracted gaussian basis set including polarization functions, C(11s 7p 1d / 5s 3p 1d) and H(6s 1p / 3s 1p). First Veillard performed calculations on the staggered and eclipsed forms using the experimental geometry of ethane. SCF energies obtained were -79.2377 and -79.2319. The predicted barrier is 3.65 kcal, only slightly higher than the minimum basis result. Next the C-C bond distances were optimized individually for the staggered and eclipsed forms, and a barrier of 3.47 kcal obtained. Finally the HCH angles were optimized for both forms, yielding 107.31° for the staggered and 106.97° for the eclipsed. For comparison the experimental HCH angle (which refers to the staggered form) is 109.3°. Subtraction of these final two energies yields a barrier of 3.07 kcal/mole. This theoretical prediction may be compared to the most recent experimental value, 2.93 kcal (392). A survey of several other nonempirical calculations of the ethane barrier shows that all of these predict values between 2.5 and 3.6 kcal/mole. We must conclude that the ethane barrier is both insensitive to basis set and properly described in the RHF approximation.

A rather different story is found for the inversion barrier in ammonia. The experimental difference between the energies of pyramidal and planar NH_3 is 5.8 kcal/mole or 0.0092 hartree (371). The first conventional SCF calculations of the NH_3 barrier were those of Kaldor and Shavitt (191). Using a minimum basis set of Slater functions, they predicted a barrier twice as large as experiment. By going to a double-zeta basis (4s 2p) on nitrogen, a worse result was obtained, with planar NH_3 actually predicted to lie below the pyramidal form in energy. The first calculation using gaussian functions was that of Clementi (78), using a poorly contracted N(11s 7p / 5s 3p) and H(6s 1p / 3s 1p) basis. Clementi predicted NH_3 to be pyramidal, but with an inversion barrier of 0.0166 hartree, almost a factor of 2 too large. Similarly discouraging results were obtained for the inversion barrier by Fink and Allen (120), who used a N(10s 5p / 3s 1p) and H(5s / 1s) basis. Their predicted barrier was much too small, 0.0013 hartree. The above four calculations made it clear that the ammonia inversion barrier is much more difficult to predict than the ethane internal rotation barrier.

The first calculation of the NH_3 barrier which could be considered "successful" was that of Body, McClure, and Clementi (47). They used a contracted gaussian basis set of N(11s 7p 1d / 7s 3p 1d) and H(6s 2p / 3s 2p). This calculation yielded a total energy of -56.2109 hartrees for the pyramidal form and a barrier of 0.0116 hartree, only about 20% greater than experiment. Furthermore, Body, et al. explained why a basis set including only s and p functions will usually yield an inversion barrier too small. The forms of the SCF wave functions for pyramidal and planar NH_3 are

$$C_{3v} \quad 1a_1^2 \; 2a_1^2 \; 1e_1^4 \; 3a_1^2$$
$$D_{3h} \quad 1a_1'^2 \; 2a_1'^2 \; 1e_1'^4 \; 1a_2''^2 \qquad (V,36)$$

d functions centered on nitrogen would in general be expected to make significant contributions to the two highest occupied orbitals of NH_3. And for pyramidal NH_3, the $1e_1$ and $3a_1$ orbitals are found to have significant d components. However, for point group D_{3h}, none of the five (corresponding to m_ℓ = -2, -1, 0, 1, and 2) d functions are of a_2'' symmetry. And only those d functions in the plane of the molecule have components of e_1' symmetry. Thus one can argue by symmetry that d orbitals will be more important for pyramidal than planar geometry. This in turn explains the computationally observed result that otherwise adequate basis sets which exclude d functions consistently give an inversion barrier which is too small.

Recently two SCF calculations using very extended basis sets have been reported for NH_3. The first of these, by Rauk, Allen, and Clementi (311) used a basis of N(13s 8p 2d / 8s 5p 2d) and H(8s 2p / 4s 2p). The equilibrium geometry was predicted by varying both bond angle and bond distance. The results, R = 1.00 Å and θ = 107.2°, agree well with experiment, 1.01 Å, θ = 106.7°. The calculated dipole moment is 1.66 debyes, compared to experiment 1.48. The total energy at the predicted equilibrium geometry is -56.2219 hartrees. Similarly, for the planar geometry the bond distance was varied and thus predicted to be slightly shorter, 0.984 Å. The inversion barrier is predicted to be 0.0081 hartree or 5.1 kcal, in good agreement with experiment. It is interesting to note that by excluding d functions from their basis set, Rauk, et al. (311) calculate a barrier of only 0.0019 hartree. A population analysis of this near RHF wave function shows

7.639 electrons on nitrogen and 0.787 on each hydrogen. The predicted bending and stretching force constants, 1.828 md/Å and 22.07 millidynes/Å, are in good and excellent agreement with experiment, 1.48 and 21.78.

A nearly equally complete SCF calculation has been carried out by Stevens (369) using a large, N(5s 4p 1d) and H(2s 1p), set of Slater functions. The geometries of both pyramidal and planar NH_3 were optimized, and the predicted geometries agreed very closely with those of Rauk, et al. At the predicted ammonia equilibrium, Stevens' SCF energy is -56.2212 hartrees, or 0.0007 hartree above the extended gaussian result. The present author estimates the true Hartree-Fock energy of NH_3 to be no lower than -56.225 hartrees. The barrier predicted by Stevens is 0.0094 hartree, very close to experiment. These last two calculations are sufficiently close to the RHF limit that it is apparent that the barrier is well described in the single configuration approximation. However, the barrier is not properly predicted until one approaches the RHF limit quite closely. Stevens has also predicted the vibrational spectrum of the bending or inversion mode of NH_3, and this is seen in Table V-27. There the agreement with experiment (371) is seen to be quite satisfactory.

The most recalcitrant of barrier problems has been the hydrogen peroxide problem. H_2O_2 has an experimental geometry between the cis (0°) and trans (180°) forms, namely at dihedral angle 111.5°. The most recent analysis (116) of experimental data (173) indicates that the cis form lies 2649 cm^{-1} or 7.6 kcal above the energy at 111.5° and the trans form lies a mere 386 cm^{-1} or 1.1 kcal/mole above the energy at equilibrium. Among the earliest calculations of

Table V-27. Near Hartree-Fock and experimental vibrational energy levels for the γ_2 mode of ammonia. Energies are in wave numbers (cm^{-1}).

	Vibrational Energy Levels	
	Computed	Experimental
	NH_3	
0_S	0	0
0_A	1.19	0.7935
1_S	991.24	932.51
1_A	1043.09	968.32
2_S	1690.55	1597.6
2_A	2041.82	1910
3_S	2595.15	2383.46
3_A	3174.78	2895.48
	ND_3	
0_S	0	0
0_A	0.068	0.053
1_S	804.682	745.7
1_A	809.498	749.4
2_S	1451.217	1359
2_A	1543.892	1429
3_S	1960.615	1830
3_A	2285.958	2106.6

the H_2O_2 barriers were those of Palke and Pitzer (286). Using a minimum basis selected by Slater's rules, their SCF calculations predicted a reasonable cis barrier of 9.4 kcal, but no trans barrier at all. Palke and Pitzer obtained a much lower total energy by using a minimum basis optimized for H_2O, but a poorer cis barrier (13.1 kcal) and again no trans barrier. By optimizing orbital exponents and geometry for both cis and trans forms, Stevens (367) obtained the limit of the minimum basis approximation, a cis barrier of 9.4 kcal and still no trans barrier. Several calculations with small gaussian basis sets gave large cis barriers (13 - 16 kcal) and no trans barrier. The first calculation to predict a trans barrier appears to be that of Fink and Allen (120). They used a double-zeta contracted gaussian basis and found a cis barrier of 13.2 kcal and a trans barrier of 0.3 kcal. In some ways, the most discouraging calculation of all was that of Veillard (380). He used an O(11s 7p 1d / 5s 3p 1d) and H(6s 1p / 3s 1p) basis, which should be adequate although it is somewhat overcontracted. Some geometry optimization was carried out for cis and trans forms and the two barriers were predicted to be 10.9 and 0.6 kcal/mole. Veillard concluded that the remaining discrepancies with experiment were probably due to electron correlation.

Happily, the H_2O_2 problem has now been resolved, due to a very careful series of calculations by Dunning and Winter (109) using an optimally contracted gaussian basis, O(9s 5p 1d / 4s 3p 1d) and H(4s 1p / 2s 1p). The SCF energy at the trans configuration was -150.82016, compared to that obtained by Veillard, -150.7983 hartrees. The other key to the success of the Dunning-Winter calculation was the complete variation of geometry. For each of four dihedral angles, all geometrical parameters were optimized. Table V-28

Table V-28. Comparison of calculated and experimental barriers to internal rotation and equilibrium geometries for hydrogen peroxide.

Barriers	Near Hartree-Fock	Experiment
$V(cis)$, cm^{-1}	2921	2649
$V(trans)$	384	386
Geometry		
ϕ_0	113.7°	111.5°
θ_{HOO}	102.5°	94.8°
R_{OH}, a.u.	1.788	1.795
R_{OO}	2.632	2.787

summarizes their results. The agreement with experiment is excellent and a considerable surprise in light of earlier failures. The fact that the O-O predicted bond distance is 0.155 bohrs (0.08 Å) shorter than experiment is a bit surprising, but near RHF calculations on O_2 predict a bond distance 0.06 Å too short, and this error can only be accounted for by inclusion of electron correlation (342). In conclusion, the present status of barrier problems is that every barrier which has been carefully studied in the SCF approximation using a large basis set has been correctly predicted.

Localized Orbitals. In several sections of this book localized orbitals have been mentioned. In all cases these localized orbitals refer to single configuration wave functions and are related to the ordinary or canonical SCF orbitals by a unitary transformation. From the properties of determinants, it is easy to show(317) that a single determinant wave function for a closed-shell system is invariant to a unitary transformation among the orbitals. That is, the total wave function does not change under such a transformation, e.g., Eq.(III,24). More generally it is true that any single configuration wave function (which may be a linear combination of determinants) is invariant to a unitary transformation among the doubly-occupied orbitals. For molecules, the canonical SCF orbitals (which are optimum for describing ionization processes) tend to be delocalized or spread over the entire molecule. It is not surprising therefore that chemists have looked for unitary transformations which would <u>localize</u> the SCF orbitals, i.e., produce a set of orbitals which relates to chemical ideas about electron-pair bonds and the like.

Perhaps the most straightforward and nonarbitrary method for obtaining localized orbitals is that of Edmiston and Ruedenberg (111). The essence of this approach is to minimize the sum of exchange integrals $(ij|ji)$, where i and j go over the occupied SCF orbitals. If a unitary transformation could be found which made <u>all</u> the exchange integrals zero, then the antisymmetrizer would be unnecessary and the wave function could be written as a simple product of spin orbitals. This would correspond to a classical electrostatic description of the electronic structure. Edmiston and Ruedenberg provide a very simple prescription for the minimization of the sum of the $(ij|ji)$. Two by two rotations, such as that in Eq.(III,24), of each pair of SCF orbitals i and j are carried out until the desired sum is a minimum. For a molecule the size of diborane, it is typically found that 5 cycles (through all ij pairs) are required to obtain minimization to within 10^{-5} hartrees. For all SCF wave functions, constructed from adequate basis sets, to which the Edmiston-Ruedenberg method has been applied to date, the localized orbitals are in agreement with qualitative chemical concepts. The only drawback of the method is that it is quite time-consuming. The first step in the process requires the transformation of the two-electron integrals over basis functions to two-electron integrals over SCF orbitals. For large basis sets, the latter process may require significantly more computer time than the entire SCF calculation.

An alternate scheme, proposed by Boys (50), obtains localized orbitals by maximizing the product of the distances between the centroids of charge of the various orbitals. This method has the distinct computational advantage of requiring only one-electron integrals involving the dipole moment operator $\vec{r}$. Thus the Boys method may be readily applied to any molecule for which an SCF calculation can be carried out. In most cases the localized orbitals obtained by the Boys procedure are nearly the same as the Edmiston-Ruedenberg orbitals. However there are some exceptions, the most obvious perhaps being that the Boys method cannot distinguish between two s orbitals on the same atom, since the center of charge for both is the nucleus in question.

An interesting molecule for which localized orbitals have been obtained is diborane, discussed earlier with respect to the dimerization energy of BH_3. Switkes, Stevens, Lipscomb, and Newton (372) have calculated a minimum basis set SCF wave function (with optimized Slater function exponents) for B_2H_6 and then obtained localized orbitals by the Edmiston-Ruedenberg method. For diborane, the effect of localization is pronounced, leading to nearly a 20-fold reduction in the sum of the exchange integrals. In terms of the canonical SCF orbitals the sum of the exchange integrals is 4.4339 hartrees, while after localization this sum is only 0.2688 hartree. The eight canonical SCF orbitals become three sets of equivalent orbitals upon localization. As expected there are two boron core orbitals, four B-H terminal bond orbitals, and two B-H-B three-center orbitals. The localized valence orbitals are seen in Figure V-14. A population analysis shows that the B-H terminal orbital is made up of 0.936 part boron and 1.082 part terminal hydrogen. Similarly each bridge bond has a population of 1.016 on the bridging hydrogen and 0.507 on each of the two boron atoms. Thus the H atoms in B_2H_6 are nearly neutral in a point charge model, whereas the H atoms in hydrocarbons are substantially positively charged.

A second interesting study of localized orbitals is that of Rothenberg (321) on simple saturated hydrocarbons. For this study SCF calculations were carried out on methane, ethane, and methanol. For methane the sum of exchange integrals decreases from 0.9933 to 0.2957 hartree upon localization, while for ethane the change is more dramatic, 5.7197 to 0.6048. For CH_3OH the decrease is from 2.8859 to 0.9798 hartrees. The interesting question addressed by Rothenberg concerns the transferability of the CH bond orbital. CH_4 has four tetrahedrally oriented localized orbitals, C_2H_6 has six such bond orbitals and CH_3OH has three. Table V-29 summarizes the properties of these Edmiston-Ruedenberg localized CH bond orbitals. This table suggests that the properties of the CH bond orbital are very insensitive both to basis set and to the saturated system in which the orbital resides. This conclusion is also supported by the fact that the overlap integrals between the five different C-H bond

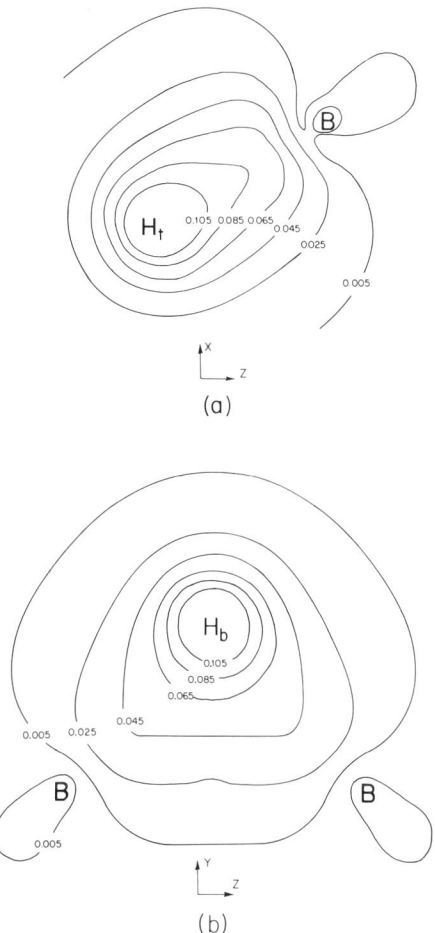

Figure V-14. Diborane localized SCF orbitals (372). Figure a) is the localized orbital representing the B-H terminal bond, and b) represents the three-center or bridge bond.

orbitals are all greater than 0.996. These results seem to provide quantitative verification of one of the simplest notions in organic chemistry. A point of particular interest concerns the dipole moment of the CH bond. Chemists have been very successful at transferring "bond dipole moments" from molecule to

Table V-29. Theoretical C-H bond orbital properties in atomic units.

Molecule	Basis Set	Kinetic Energy	$<4/r_C>$	$<1/r_H>$	Dipole Moment
CH_4	C(10s 5p / 3s 1p) H(5s / 1s)	0.8690	2.6012	0.9296	1.341
CH_4	C(10s 5p / 4s 2p) H(5s / 2s)	0.8680	2.6024	0.9344	1.375
C_2H_6	C(10s 5p / 3s 1p) H(5s / 1s)	0.8860	2.6097	0.9314	1.344
CH_3OH	C,O(10s 5p / 3s 1p) H(5s / 1s)	0.8890	2.6149	0.9319	1.347
CH_3OH	C,O(10s 5p / 4s 2p) H(5s / 2s)	0.9120	2.6573	0.9457	--

molecule. For saturated systems the C-H bond dipole moment is thought to be about 0.4 debye. Although the electronic contributions to the dipole moment seen in Table V-29 are nearly constant, they are much too large, being about 3.4 debyes. In addition these values must be multiplied by two since two electrons occupy the bond orbital. The problem of course is that the nuclear contribution to the bond moment has been neglected. Unfortunately it is not possible to rigorously evaluate the nuclear contribution, since that would necessitate [see Eq.(I,29)] assigning charges to the C and H atoms within the molecule. However, Rothenberg concludes that a plausible choice of the nuclear contribution gives reasonable agreement with the "experimental" value, 0.4 debye.

An alternate interpretation of the experimental CH bond dipole moment has been given by Pritchard and Kern (309). They studied the bond moments of CH_4, C_2H_4, C_2H_6, and C_2H_2 using localized orbitals from minimum basis set Slater function calculations. Like Rothenberg, Pritchard and Kern obtained bond dipole moments significantly larger than "experiment." However, they rationalized these results by noting that most of the calculated bond dipole moment is "atomic" in nature, being due to the carbon sp^3 hybrid orbital. Since the four methane sp^3 hybrid orbitals have a net dipole moment of zero, Pritchard and Kern argue that this "atomic" effect must be excluded if one wishes to compare a theoretical bond dipole moment with experiment. When a plausible value for this atomic sp^3 bond moment is subtracted from the ab initio bond moment of the localized CH orbital, reasonable agreement with the experimental bond moment is found.

REFERENCES

1. R. Ahlrichs, W. Kutzelnigg, and W.A. Bingel, Theoret. Chim. Acta. $\underline{5}$, 291, 305 (1966).
2. R. Ahlrichs and W. Kutzelnigg, J. Chem. Phys. $\underline{48}$, 1819 (1968).
3. R. Ahlrichs and W. Kutzelnigg, Theoret. Chim. Acta $\underline{10}$, 377 (1968); Chem. Phys. Letters $\underline{1}$, 651 (1968).
4. B. J. Alder and T. E. Wainwright, J. Chem. Phys. $\underline{33}$, 1439 (1960).
5. L.C. Allen, E. Clementi, and H. M. Gladney, Rev. Mod. Phys. $\underline{35}$, 465 (1963).
6. L. C. Allen, Ann. Rev. Phys. Chem. $\underline{20}$, 315 (1969).
7. L. C. Allen, unpublished.
8. I. Amdur and J. E. Jordan, Advan. Chem. Phys. $\underline{10}$, 29 (1966).
9. S. Aung, R. M. Pitzer, and S. I. Chan, J. Chem. Phys. $\underline{49}$, 882 (1968).
10. P. S. Bagus, Phys. Rev. $\underline{139}$, A619 (1965).
11. P. S. Bagus and B. Liu, Phys. Rev. $\underline{148}$, 79 (1966).
12. P. S. Bagus and T. L. Gilbert, reported in Argonne National Laboratory Report 7271 (January, 1968).
13. P. S. Bagus, B. Liu, and H. F. Schaefer, Phys. Rev. $\underline{A2}$, 555 (1970).
14. P. S. Bagus and H. F. Schaefer, J. Chem. Phys. $\underline{55}$, 1474 (1971).
15. P. S. Bagus and H. F. Schaefer, J. Chem. Phys. $\underline{56}$, 224 (1972).
16. B. Bak, E. Clementi, and R. N. Kortzeborn, J. Chem. Phys. $\underline{52}$, 764 (1970).
17. C. J. Ballhausen, <u>Introduction to Ligand Field Theory</u> (McGraw-Hill, New York, 1962).
18. E. A. Ballik and D. A. Ramsay, Astrophys. J. $\underline{137}$, 84 (1963).
19. M. P. Barnett, Methods in Comput. Phys. $\underline{2}$, 95 (1963).
20. T. L. Barr and E. R. Davidson, Phys. Rev. $\underline{A1}$, 644 (1970).
21. N. Bartlett, Proc. Chem. Soc. $\underline{1962}$, 218.
22. N. Bartlett and F. O. Sladky, <u>The Chemistry of Krypton, Xenon, and Radon</u>, University of California Radiation Laboratory Report 19658, June, 1970.
23. H. Basch and V. McKoy, J. Chem. Phys. $\underline{53}$, 1628 (1970).
24. H. Basch, J. Chem. Phys. $\underline{55}$, 1700 (1971).
25. H. Basch, J. W. Moskowitz, C. Hollister, and D. Hankins, J. Chem. Phys. $\underline{55}$, 1922 (1971).
26. S. H. Bauer and E. Ossa, J. Chem. Phys. $\underline{45}$, 434 (1966).

27. C. F. Bender and E. R. Davidson, J. Phys. Chem. $\underline{70}$, 2675 (1966).
28. C. F. Bender and E. R. Davidson, J. Chem. Phys. $\underline{47}$, 360 (1967).
29. C. F. Bender and E. R. Davidson, Phys. Rev. $\underline{183}$, 21 (1969).
30. C. F. Bender and H. F. Schaefer, J. Am. Chem. Soc. $\underline{92}$, 4984 (1970).
31. C. F. Bender and H. F. Schaefer, unpublished.
32. C. F. Bender and H. F. Schaefer, unpublished.
33. C. F. Bender, H. F. Schaefer, T. H. Dunning and W. A. Goddard, unpublished.
34. W. S. Benedict and E. K. Plyer, Can. J. Phys. $\underline{35}$, 1235 (1957).
35. H. G. Bennewitz, H. Busse, and H. D. Dohmann, Chem. Phys. Letters $\underline{8}$, 235 (1971).
36. J. Berkowitz and W. A. Chupka, Chem. Phys. Letters $\underline{7}$, 447 (1970).
37. R. A. Bernheim, H. W. Bernard, P. S. Wang, L. S. Wood, and P. S. Skell, J. Chem. Phys. $\underline{53}$, 1280 (1970).
38. R. B. Bernstein and J. T. Muckerman, Advan. Chem. Phys. $\underline{12}$, 389 (1967).
39. R. S. Berry, G. N. Spokes, and M. Stiles, J. Am. Chem. Soc. $\underline{84}$, 3570 (1962).
40. R. Bersohn, Y. H. Pao, and H. L. Frisch, J. Chem. Phys. $\underline{45}$, 3184 (1966).
41. P. J. Bertoncini, G. Das, and A. C. Wahl, J. Chem. Phys. $\underline{52}$, 5112 (1970).
42. P. Bertoncini and A. C. Wahl, Phys. Rev. Letters $\underline{25}$, 991 (1970).
43. N. Bessis, H. Lefebvre-Brion, C. M. Moser, A. J. Freeman, R. K. Nesbet, and R. E. Watson, Phys. Rev. $\underline{135}$, A588 (1964).
44. H. Bethe, Ann. Physik $\underline{3}$, 133 (1929).
45. R. J. Blint, W. A. Goddard, R. C. Ladner, and W. E. Palke, Chem. Phys. Letters $\underline{5}$, 302 (1970).
46. R. J. Blint and W. A. Goddard, unpublished.
47. R. G. Body, D. S. McClure, and E. Clementi, J. Chem. Phys. $\underline{49}$, 4916 (1968).
48. R. Bonaccorsi, E. Scrocco, and J. Tomasi, J. Chem. Phys. $\underline{52}$, 5270 (1970).
49. S. F. Boys, Proc. Roy. Soc. (London) $\underline{A200}$, 542 (1950).
50. S. F. Boys, in <u>Quantum Theory of Atoms, Molecules and the Solid State</u>, Edited by P. O. Lowdin (Academic Press, New York, 1966).
51. L. Brillouin, Actualites Sci. Ind. No. 71 (1933); No. 159 (1934).
52. R. D. Brown, Trans. Faraday Soc. $\underline{44}$, 984 (1948).
53. C. R. Brundle, M. B. Robin, and H. Basch, J. Chem. Phys. $\underline{53}$, 2196 (1970).
54. A. Büchler, J. Stauffer, W. Klemperer, and L. Wharton, J. Chem. Phys. $\underline{39}$, 2299 (1963).

55. A. D. Buckingham, Quart. Rev. 13, 183 (1959).
56. R. J. Buenker and S. D. Peyerimhoff, J. Chem. Phys. 45, 3682 (1966).
57. R. J. Buenker and J. L. Whitten, J. Chem. Phys. 49, 5381 (1968).
58. R. J. Buenker and S. D. Peyerimhoff, Theoret. Chim. Acta 12, 183 (1968).
59. R. J. Buenker and S. D. Peyerimhoff, Chem. Phys. Letters 3, 37 (1969).
60. R. J. Buenker, S. D. Peyerimhoff, and H. L. Hsu, Chem. Phys. Letters 11, 65 (1971).
61. A. Bunge and C. F. Bunge, Phys. Rev. A1, 1599 (1970); A. Bunge, J. Chem. Phys. 53, 20 (1970).
62. C. F. Bunge, Phys. Rev. 168, 92 (1968).
63. C. F. Bunge and E. M. A. Peixoto, Phys. Rev. A1, 1277 (1970).
64. A. Burcat and A. Lifshitz, J. Chem. Phys. 47, 3079 (1967).
65. P. E. Cade, K. D. Sales, and A. C. Wahl, J. Chem. Phys. 44, 1973 (1966).
66. P. E. Cade and W. M. Huo, J. Chem. Phys. 47, 614, 649 (1967).
67. K. D. Carlson, K. Kaiser, C. Moser, and A. C. Wahl, J. Chem. Phys. 52, 4678 (1970).
68. R. W. Carr, T. W. Eder, and M. G. Topor, J. Chem. Phys. 53, 4716 (1970).
69. A. Carrington and A. D. McLachlan, Introduction to Magnetic Resonance (Harper and Row, New York, 1967).
70. H. H. Claasen, H. Selig, and J. G. Malm, J. Am. Chem. Soc. 84, 3593 (1962).
71. E. Clementi, C. C. J. Roothaan, and M. Yoshimine, Phys. Rev. 127, 1618 (1962).
72. E. Clementi, J. Chem. Phys. 38, 2248 (1963).
73. E. Clementi and D. L. Raimondi, J. Chem. Phys. 38, 2686 (1963); E. Clementi, D. L. Raimondi, and W. P. Reinhardt, J. Chem. Phys. 47, 1300 (1967).
74. E. Clementi, J. Chem. Phys. 40, 1944 (1964); E. Clementi, R. Matcha, and A. Veillard, J. Chem. Phys. 47, 1865 (1967).
75. E. Clementi, Tables of Atomic Wave Functions, a supplement to IBM J. Res. Develop. 9, 2 (1965).
76. E. Clementi and A. Veillard, J. Chem. Phys. 44, 3050 (1966).
77. E. Clementi and D. R. Davis, J. Comput. Phys. 2, 223 (1967).
78. E. Clementi, J. Chem. Phys. 46, 3851 (1967).
79. E. Clementi, Intern. J. Quantum Chem. 1S, 307 (1967).
80. E. Clementi, J. Mehl, and W. von Niessen, J. Chem. Phys. 54, 508 (1971).
81. A. J. Coleman, Rev. Mod. Phys. 35, 668 (1963).

82. G. A. D. Collins, D. W. J. Cruickshank, and A. Breeze, Chem. Comm. 1970, 884.
83. E. U. Condon and G. H. Shortley, The Theory of Atomic Spectra (McGraw-Hill, New York 1960).
84. H. Conroy and B. Bruner, J. Chem. Phys. 47, 921 (1967).
85. H. Conroy and G. Malli, J. Chem. Phys. 50, 5049 (1969).
86. W. Cooper and W. D. Walter, J. Am. Chem. Soc. 80, 4220 (1958).
87. F. A. Cotton, Chemical Applications of Group Theory (Wiley-Interscience, New York, 1971).
88. C. A. Coulson, Proc. Roy. Soc. (London) A169, 413 (1939).
89. C. A. Coulson, Valence (Oxford University Press, London, 1961).
90. C. A. Coulson, J. Chem. Soc. 1964, 1442.
91. C. A. Coulson, J. Chem. Phys. 44, 468 (1966).
92. I. G. Csizmadia, M. C. Harrison, J. W. Moskowitz, and B. T. Sutcliffe, Theoret. Chim. Acta 6, 191 (1966).
93. I. G. Csizmadia, J. C. Polanyi, A. C. Roach, and W. H. Wong, Can. J. Chem. 47, 4097 (1969).
94. G. Das and A. C. Wahl, J. Chem. Phys. 44, 876 (1966).
95. G. Das, J. Chem. Phys. 46, 1568 (1967).
96. G. Das and A. C. Wahl, Phys. Rev. Letters, 24, 440 (1970).
97. E. R. Davidson and C. F. Bender, J. Chem. Phys. 49, 465 (1968); Chem. Phys. Letters 3, 33 (1969).
98. D. W. Davies, The Theory of the Electric and Magnetic Properties of Molecules (John Wiley & Sons, New York, 1967).
99. D. W. Davis and D. A. Shirley, J. Chem. Phys. 56, 669 (1972).
100. J. Del Bene and J. A. Pople, J. Chem. Phys. 52, 4858 (1970).
101. M. J. S. Dewar, The Molecular Orbital Theory of Organic Chemistry (McGraw-Hill, New York, 1969).
102. G. H. F. Diercksen, Chem. Phys. Letters 4, 373 (1969).
103. G. H. F. Diercksen and W. P. Kraemers, Chem. Phys. Letters 6, 419 (1970).
104. T. H. Dunning, W. J. Hunt, and W. A. Goddard, Chem. Phys. Letters 4, 147 (1969).
105. T. H. Dunning, J. Chem. Phys. 53, 2823 (1970).
106. T. H. Dunning, unpublished.
107. T. H. Dunning, J. Chem. Phys. 55, 3958 (1971).

108. T. H. Dunning and N. W. Winter, J. Chem. Phys. 55, 3360 (1971).
109. T. H. Dunning and N. W. Winter, Chem. Phys. Letters 11, 194 (1971).
110. N. C. Dutta, C. Matsubara, R. T. Pu, and T. P. Das, Phys. Rev. 177, 33 (1969).
111. C. Edmiston and K. Ruedenberg, Rev. Mod. Phys. 34, 457 (1963).
112. C. Edmiston and M. Krauss, J. Chem. Phys. 45, 1833 (1966).
113. C. Edmiston, J. Doolittle, K. Murphy, K. C. Tang, and W. Willson, J. Chem. Phys. 52, 3419 (1970).
114. W. C. Ermler and C. W. Kern, J. Chem. Phys. 55, 4851 (1971).
115. D. F. Evans, J. Chem. Soc. 1960, 1735.
116. C. S. Ewig and D. O. Harris, J. Chem. Phys. 52, 6268 (1970).
117. H. Eyring and M. Polanyi, Z. Physik. Chem. B12, 279 (1931).
118. H. Eyring, J. Walter, and G. E. Kimball, Quantum Chemistry (John Wiley & Sons, Inc., New York, 1944).
119. W. E. Falconer, J. R. Morton, and A. G. Streng, J. Chem. Phys. 41, 902 (1964).
120. W. H. Fink and L. C. Allen, J. Chem. Phys. 46, 2276 (1967).
121. W. H. Flygare and R. C. Benson, Mol. Phys. 20, 225 (1971).
122. V. Fock, Z. Physik 61, 126 (1930).
123. J. M. Foster and S. F. Boys, Rev. Mod. Phys. 32, 303 (1960).
124. J. M. Foster and S. F. Boys, Rev. Mod. Phys. 32, 305 (1960).
125. S. Fraga and G. Malli, Many-Electron Systems: Properties and Interactions (W. B. Saunders, Philadelphia, 1968).
126. K. F. Freed, Phys. Rev. 173, 1 (1968); see also Ann. Rev. Phys. Chem. 21, 313 (1971).
127. A. J. Freeman and R. E. Watson, in Magnetism, edited by G. T. Rado and H. Suhl (Academic Press, New York, 1965).
128. C. Froese, Can. J. Phys. 41, 1895 (1963).
129. A. A. Frost, J. Chem. Phys. 47, 3707 (1967).
130. A. A. Frost and R. A. Rouse, J. Am. Chem. Soc. 90, 1965 (1968).
131. P. S. Ganguli and H. A. McGee, J. Chem. Phys. 50, 4658 (1969).
132. A. G. Gaydon, Dissociation Energies and Spectra of Diatomic Molecules (Chapman and Hall, London, 1968).
133. M. Gelus, R. Ahlrichs, V. Staemmler, and W. Kutzelnigg, Chem. Phys. Letters 7, 503 (1970).

134. Z. Gershgorn and I. Shavitt, Intern. J. Quantum Chem. $\underline{1S}$, 403 (1967).
135. H. M. Gladney and A. Veillard, Phys. Rev. $\underline{180}$, 385 (1969).
136. W. A. Goddard, Phys. Rev. $\underline{182}$, 48 (1969).
137. J. Goldstone, Proc. Roy. Soc. (London) $\underline{A239}$, 267 (1957).
138. D. A. Goodings, Phys. Rev. $\underline{123}$, 1706 (1961).
139. S. Green, J. Chem. Phys. $\underline{54}$, 827 (1971); $\underline{56}$, 739 (1972).
140. F. Grimaldi, A. Lecourt, and C. Moser, Intern. J. Quantum Chem. $\underline{1S}$, 153 (1967).
141. S. R. Gunn, J. Phys. Chem. $\underline{71}$, 2934 (1967).
142. J. Guy, J. Tillieu, and J. Baudet, Compt. Rend. $\underline{246}$, 574 (1958).
143. S. Hagstrom and H. Shull, Rev. Mod. Phys. $\underline{35}$, 624 (1963).
144. M. L. Halberstadt and J. R. McNesby, J. Am. Chem. Soc. $\underline{89}$, 3417 (1967).
145. D. Hankins, J. W. Moskowitz, and F. H. Stillinger, J. Chem. Phys. $\underline{53}$, 4544 (1970).
146. F. E. Harris and H. H. Michels, Intern. J. Quantum Chem. $\underline{1S}$, 329 (1967).
147. F. E. Harris and H. H. Michels, Intern. J. Quantum Chem. $\underline{2S}$, 21 (1968).
148. J. F. Harrison and L. C. Allen, J. Am. Chem. Soc. $\underline{91}$, 807 (1969).
149. H. Hartmann and E. Clementi, Phys. Rev. $\underline{133}$, A1295 (1964); E. Clementi, J. Mol. Spectry. $\underline{12}$, 18 (1964).
150. D. R. Hartree, Proc. Cambridge Phil. Soc. $\underline{24}$, 89 (1928).
151. D. R. Hartree, W. Hartree, and B. Swirles, Phil. Trans. Roy. Soc. (London) $\underline{A238}$, 229 (1939).
152. D. R. Hartree, The Calculation of Atomic Structures (John Wiley & Sons, New York, 1957).
153. R. A. Hegstrom and W. N. Lipscomb, Rev. Mod. Phys. $\underline{40}$, 354 (1968).
154. W. J. Hehre, R. F. Stewart, and J. A. Pople, J. Chem. Phys. $\underline{51}$, 2657 (1969).
155. T. G. Heil and H. F. Schaefer, Astrophys. J. $\underline{163}$, 425 (1971); J. Chem. Phys. $\underline{54}$, 2573 (1971).
156. W. H. Henneker and H. E. Popkie, J. Chem. Phys. $\underline{54}$, 1763, 4597 (1971).
157. G. Herzberg, Spectra of Diatomic Molecules (Van Nostrand Reinhold, New York, 1950).
158. G. Herzberg, Can. J. Phys. $\underline{31}$, 657 (1953).
159. G. Herzberg, Proc. Roy. Soc. (London) $\underline{A262}$, 291 (1961).
160. G. Herzberg and J. W. C. Johns, Proc. Roy. Soc. (London) $\underline{A295}$, 107 (1966).
161. G. Herzberg, Electronic Spectra of Polyatomic Molecules (D. Van Nostrand, Princeton, N. J., 1967).

162. G. Herzberg, *The Spectra and Structures of Simple Free Radicals* (Cornell University Press, Ithaca, N.Y., 1971).
163. G. Herzberg and J. W. C. Johns, J. Chem. Phys. $\underline{54}$, 2276 (1971).
164. M. M. Hessel, Phys. Rev. Letters $\underline{26}$, 215 (1971).
165. F. B. Hildebrand, *Methods of Applied Mathematics* (Prentice-Hall, Englewood Cliffs, N.J., 1965).
166. J. O. Hirschfelder, C. F. Curtiss, and R. B. Bird, *Molecular Theory of Gases and Liquids* (John Wiley & Sons, New York, 1954).
167. R. Hoffman, J. Chem. Phys. $\underline{39}$, 1397 (1963).
168. R. Hoffmann, I. Imamura, and W. J. Hehre, J. Am. Chem. Soc. $\underline{90}$, 1499 (1968).
169. R. Hoffmann, R. Gleiter, and F. B. Mallory, J. Am. Chem. Soc. $\underline{92}$, 1460 (1970).
170. R. P. Hosteny, R. R. Gilman, T. H. Dunning, A. Pipano, and I. Shavitt, Chem. Phys. Letters $\underline{7}$, 325 (1970)
171. R. P. Hosteny, T. H. Dunning, R. R. Gilman, A. Pipano, and I. Shavitt, unpublished.
172. K. Hsu, R. J. Buenker, and S. D. Peyerimhoff, J. Am. Chem. Soc. $\underline{93}$, 2117 (1971).
173. R. H. Hunt, R. A. Leacock, C. W. Peters, and K. T. Hecht, J. Chem. Phys. $\underline{42}$, 1931 (1965).
174. W. J. Hunt and W. A. Goddard, Chem. Phys. Letters $\underline{3}$, 414 (1969).
175. W. J. Hunt, T. H. Dunning, and W. A. Goddard, Chem. Phys. Letters $\underline{3}$, 609 (1969).
176. W. J. Hunt, T. H. Dunning, and W. A. Goddard, unpublished.
177. W. J. Hunt, P. J. Hay, and W. A. Goddard, J. Am. Chem. Soc. $\underline{94}$, 638 (1972).
178. W. Huo, J. Chem. Phys. 43, 624 (1965).
179. W. M. Huo, K. F. Freed, and W. Klemperer, J. Chem. Phys. $\underline{46}$, 3556 (1967).
180. W. M. Huo, J. Chem. Phys. $\underline{49}$, 1482 (1968).
181. A. C. Hurley, J. Lennard-Jones, and J. A. Pople, Proc. Roy. Soc. (London) $\underline{A220}$, 446 (1953).
182. S. Huzinaga, J. Chem. Phys. $\underline{42}$, 1293 (1965).
183. S. Huzinaga and Y. Sakai, J. Chem. Phys. $\underline{50}$, 1371 (1969).
184. S. Huzinaga and C. Arnau, J. Chem. Phys. $\underline{53}$, 451 (1970).
185. E. A. Hylleraas, Z. Physik $\underline{48}$, 469 (1928).
186. M. H. Johnson, Phys. Rev. $\underline{39}$, 197 (1932).

187. H. S. Johnston, Gas Phase Reaction Rate Theory (Ronald Press, New York, 1966).
188. J. E. Jordan and I. Amdur, J. Chem. Phys. 46, 165 (1967).
189. P. C. H. Jordan and H. C. Longuet-Higgens, Mol. Phys. 5, 121 (1962).
190. C. K. Jorgensen, Modern Aspects of Ligand Field Theory (North-Holland, Amsterdam, 1971).
191. U. Kaldor and I. Shavitt, J. Chem. Phys. 45, 888 (1966).
192. U. Kaldor and F. E. Harris, Phys. Rev. 183, 1 (1969).
193. U. Kaldor, Phys. Rev. A1, 1586 (1970).
194. U. Kaldor, Phys. Rev. A2, 1267 (1970).
195. S. F. Karlsson, K. Siegbahn, and N. Bartlett, reported in reference 350; C. R. Brundle, G. R. Jones, and H. Basch, J. Chem. Phys. 55, 1098 (1971).
196. J. M. Keller and W. L. Taylor, J. Chem. Phys. 51, 4829 (1969).
197. H. P. Kelly, Phys. Rev. 131, 690 (1963).
198. H. P. Kelly, Phys. Rev. 144, 39 (1966).
199. H. P. Kelly, Phys. Rev. 173, 142 (1968); 180, 55 (1969).
200. H. P. Kelly, Advan. Chem. Phys. 14, 129 (1969).
201. P. S. Kelly, Astrophys. J. 140, 1247 (1964).
202. J. D. Kemp and K. S. Pitzer, J. Chem. Phys. 4, 749 (1936).
203. C. W. Kern and M. Karplus, J. Chem. Phys. 40, 1374 (1964).
204. C. W. Kern and R. L. Matcha, J. Chem. Phys. 49, 2081 (1968).
205. N. R. Kestner and O. Sinanoglu, J. Chem. Phys. 45, 194 (1966).
206. N. R. Kestner, J. Chem. Phys. 48, 252 (1968).
207. P. A. Kollman and L. C. Allen, J. Chem. Phys. 51, 3286 (1969).
208. P. A. Kollman and L. C. Allen, Chem. Rev. (in press).
209. W. Kolos and C. C. J. Roothaan, Rev. Mod. Phys. 32, 219 (1960).
210. W. Kolos and L. Wolniewicz, J. Chem. Phys. 46, 1426 (1967).
211. W. Kolos and L. Wolniewicz, J. Chem. Phys. 50, 3228 (1969).
212. T. Koopmans, Physica 1, 104 (1933).
213. M. Kotani, Table of Molecular Integrals (Maruzen Company, Tokyo, 1955).
214. J. Kouba and Y. Öhrn, Intern. J. Quantum Chem. 3, 513 (1969).
215. M. Krauss, J. Res. Nat. Bur. Stand., Sect. A, 68, 635 (1964).
216. M. Krauss, "Compendium of ab initio Calculations of Molecular Energies and Properties," Nat. Bur. Stand. (USA) Tech. Note 438 (1967).
217. M. Krauss, Ann. Rev. Phys. Chem. 20, 39 (1970).

218. Krishnaji and V. Prakash, Rev. Mod. Phys. **38**, 690 (1966).
219. R. C. Ladner and W. A. Goddard, J. Chem. Phys. **51**, 1073 (1969).
220. A. Lagerqvist, "Investigations of the Band-Spectrum of Beryllium Oxide," thesis, University of Stockholm, 1948.
221. P. W. Langhoff, M. Karplus, and R. P. Hurst, J. Chem. Phys. **44**, 505 (1966).
222. E. A. Laws, R. M. Stevens, and W. N. Lipscomb, Chem. Phys. Letters **4**, 159 (1969); J. Chem. Phys. **54**, 4269 (1971).
223. R. M. Lees, R. F. Curl, and J. G. Baker, J. Chem. Phys. **45**, 2037 (1966); P. D. Foster, J. A. Hodgeson, and R. F. Curl, J. Chem. Phys. **45**, 3760 (1966).
224. J. A. Lely and J. M. Bijvot, Rec. Trav. Chim. **61**, 244 (1952).
225. W. A. Lester, J. Chem. Phys. **53**, 1511 (1970).
226. M. Levy, W. J. Stevens, H. Shull, and S. Hagstrom, J. Chem. Phys. **52**, 5483 (1970).
227. S. P. Liebmann and J. W. Moskowitz, J. Chem. Phys. **54**, 3622 (1971).
228. C. C. Lin, Phys. Rev. **116**, 903 (1959).
229. B. Liu and H. F. Schaefer, J. Chem. Phys. **55**, 2369 (1971).
230. B. Liu, Intern. J. Quantum Chem. **5S**, 123 (1971).
231. B. Liu, P. S. Bagus, and H. F. Schaefer, unpublished.
232. F. London, Z. Elektrochem. **35**, 552 (1929).
233. P. O. Löwdin, Phys. Rev. **97**, 1474 (1955).
234. P. O. Löwdin, Advan. Chem. Phys. **2**, 207 (1959).
235. P. O. Löwdin, Rev. Mod. Phys. **32**, 328 (1960); **36**, 966 (1964).
236. P. O. Löwdin, Rev. Mod. Phys. **35**, 724 (1963).
237. P. O. Löwdin and B. Pullman, editors, <u>Molecular Orbitals in Chemistry, Physics, and Biology</u> (Academic Press, New York, 1964).
238. J. K. L. MacDonald, Phys. Rev. **43**, 830 (1933).
239. B. H. Mahan, Accounts Chem. Res. **3**, 393 (1970).
240. H. Margenau and N. R. Kestner, <u>Theory of Intermolecular Forces</u> (Pergamon Press, Oxford, 1969).
241. R. L. Matcha, J. Chem. Phys. **47**, 4595, 5295 (1967); **48**, 335 (1968); **49**, 1264 (1968).
242. G. H. Matsumoto, C. F. Bender, and E. R. Davidson, J. Chem. Phys. **46**, 402 (1967).
243. D. McKee and N. Bartlett, unpublished.

244. D. R. McLaughlin, C. F. Bender, and H. F. Schaefer, unpublished.
245. A. D. McLean, J. Chem. Phys. $\underline{32}$, 1595 (1960).
246. A. D. McLean and M. Yoshimine, J. Chem. Phys. $\underline{45}$, 3467 (1966).
247. A. D. McLean and M. Yoshimine, J. Chem. Phys. $\underline{45}$, 3676 (1966).
248. A. D. McLean and M. Yoshimine, J. Chem. Phys. $\underline{46}$, 3682 (1967).
249. A. D. McLean and M. Yoshimine, J. Chem. Phys. $\underline{47}$, 3256 (1967).
250. A. D. McLean and M. Yoshimine, Intern. J. Quantum Chem. $\underline{1S}$, 313 (1967).
251. A. D. McLean and M. Yoshimine, Tables of Linear Molecule Wave Functions, a supplement to IBM J. Res. Develop. $\underline{12}$, 206 (1968).
252. A. D. McLean and S. Seung, unpublished, 1968.
253. A. D. McLean, in Proceedings of the Conference on Potential Energy Surfaces in Chemistry, Publication RA 18, IBM Research Laboratory, San Jose, California.
254. R. McWeeny and B. T. Sutcliffe, Methods of Molecular Quantum Mechanics (Academic Press, New York, 1969).
255. E. L. Mehler, K. Ruedenberg, and D. M. Silver, J. Chem. Phys. $\underline{52}$, 1174, 1181 (1970).
256. A. J. Merer and R. S. Mulliken, Chem. Rev. $\underline{69}$, 639 (1969).
257. W. Meyer, J. Chem. Phys. $\underline{51}$, 5149 (1969).
258. W. Meyer, Intern. J. Quantum Chem. $\underline{5S}$, 341 (1971).
259. J. Miller, R. H. Friedman, R. P. Hurst, and F. A. Matsen, J. Chem. Phys. $\underline{27}$, 1385 (1957).
260. W. H. Miller, Accounts Chem. Res. $\underline{4}$, 161 (1971).
261. E. Miron, B. Raz, and J. Jortner, Chem. Phys. Letters $\underline{6}$, 563 (1970).
262. C. Møller and M. S. Plesset, Phys. Rev. $\underline{46}$, 618 (1934).
263. K. Morokuma and L. Pedersen, J. Chem. Phys. $\underline{48}$, 3275 (1968).
264. K. Morokuma and J. R. Winick, J. Chem. Phys. $\underline{52}$, 1301 (1970).
265. K. Morokuma, unpublished.
266. O. Mosher, W. M. Flicker, and A. Kuppermann, unpublished.
267. J. W. Moskowtiz, C. Hollister, C. J. Hornback, and H. Basch, J. Chem. Phys. $\underline{53}$, 2570 (1971).
268. R. S. Mulliken, Rev. Mod. Phys. $\underline{14}$, 765 (1942).
269. R. S. Mulliken, J. Chem. Phys. $\underline{23}$, 1833, 1841 (1955).
270. R. K. Nesbet, Proc. Roy. Soc. (London) $\underline{A230}$, 312 (1955).

271. R. K. Nesbet, Rev. Mod. Phys. $\underline{32}$, 272 (1960).
272. R. K. Nesbet, unpublished work at Boston University, 1962.
273. R. K. Nesbet, J. Chem. Phys. $\underline{40}$, 3619 (1964).
274. R. K. Nesbet, J. Chem. Phys. $\underline{43}$, 311 (1965).
275. R. K. Nesbet, Phys. Rev. $\underline{155}$, 51, 56 (1967).
276. R. K. Nesbet, Phys. Rev. $\underline{175}$, 2 (1968).
277. R. K. Nesbet, T. L. Barr, and E. R. Davidson, Chem. Phys. Letters $\underline{4}$, 203 (1969).
278. R. K. Nesbet, Advan. Chem. Phys. $\underline{14}$, 1 (1969).
279. R. K. Nesbet, Phys. Rev. $\underline{A2}$, 1208 (1970).
280. R. K. Nesbet, Phys. Rev. $\underline{A3}$, 87 (1971).
281. D. B. Neumann and J. W. Moskowitz, J. Chem. Phys. $\underline{49}$, 2056 (1968).
282. D. B. Neumann and J. W. Moskowitz, J. Chem. Phys. $\underline{50}$, 2216 (1969).
283. M. D. Newton, W. A. Lathan, W. J. Hehre, and J. A. Pople, J. Chem. Phys. $\underline{52}$, 4064 (1970).
284. M. D. Newton, unpublished.
285. S. V. O'Neil, H. F. Schaefer, and C. F. Bender, J. Chem. Phys. $\underline{55}$, 162 (1971).
286. W. E. Palke and R. M. Pitzer, J. Chem. Phys. $\underline{46}$, 3948 (1967).
287. J. H. Parker and G. C. Pimentel, J. Chem. Phys. $\underline{51}$, 91 (1969).
288. R. G. Parr, <u>Quantum Theory of Molecular Electronic Structure</u> (W. B. Benjamin, New York, 1964).
289. L. Pauling, J. Am. Chem. Soc. $\underline{53}$, 1367 (1931).
290. L. Pauling and E. B. Wilson, <u>Introduction to Quantum Mechanics</u> (McGraw-Hill, New York, 1935).
291. R. Pauncz, <u>Alternate Molecular Orbital Theory</u> (W. B. Saunders, Philadelphia, Philadelphia, 1967).
292. C. L. Pekeris, Phys. Rev. $\underline{126}$, 1470 (1962).
293. R. A. Penneman, J. B. Mann, and C. K. Jørgensen, Chem. Phys. Letters $\underline{8}$, 321 (1971).
294. G. J. Perlow, chapter 7 in <u>Chemical Applications of Mössbauer Spectroscopy</u> (Academic Press, New York, 1968).
295. P. E. Phillipson, Phys. Rev. $\underline{125}$, 1981 (1962).
296. F. L. Pilar, <u>Elementary Quantum Chemistry</u> (McGraw-Hill, New York, 1968).
297. G. C. Pimentel and A. L. McClellan, <u>The Hydrogen Bond</u> (W. H. Freeman, San Francisco, 1960).

298. R. M. Pitzer and W. N. Lipscomb, J. Chem. Phys. $\underline{39}$, 1995 (1963).
299. R. M. Pitzer and S. Aung, unpublished.
300. O. R. Platas and H. F. Schaefer, Phys. Rev. $\underline{A4}$, 33 (1971).
301. J. M. Pochan, R. G. Stone, and W. H. Flygare, J. Chem. Phys. $\underline{51}$, 4278 (1969).
302. J. C. Polanyi and D. C. Tardy, J. Chem. Phys. $\underline{51}$, 5717 (1969).
303. J. C. Polanyi, in *Proceedings of the Conference on Potential Energy Surfaces in Chemistry*, Publication RA 18, IBM Research Laboratory, San Jose, California.
304. H. E. Popkie and W. H. Henneker, J. Chem. Phys. $\underline{55}$, 617 (1971).
305. J. A. Pople and R. K. Nesbet, J. Chem. Phys. $\underline{22}$, 571 (1954).
306. J. A. Pople and D. L. Beveridge, *Approximate Molecular Orbital Theory* (McGraw-Hill, New York, 1970).
307. R. K. Preston and J. C. Tully, J. Chem. Phys. $\underline{54}$, 4297 (1971).
308. H. Preuss, Z. Naturforsch. $\underline{11}$, 823 (1956).
309. R. H. Pritchard and C. W. Kern, J. Am. Chem. Soc. $\underline{91}$, 1631 (1969).
310. B. J. Ransil, Rev. Mod. Phys. $\underline{32}$, 245 (1960).
311. A. Rauk, L. C. Allen, and E. Clementi, J. Chem. Phys. $\underline{52}$, 4133 (1970).
312. W. P. Reinhardt and J. D. Doll, J. Chem. Phys. $\underline{50}$, 2767 (1969).
313. W. G. Richards, G. Verhaegen, and C. M. Moser, J. Chem. Phys. $\underline{45}$, 3226 (1966).
314. W. G. Richards, T. E. H. Walker, and R. K. Hinkley, *Bibliography of ab initio Calculations* (Oxford University Press, London, 1971).
315. J. D. Roberts, H. E. Simmons, L. A. Carlsmith, and C. W. Vaughan, J. Am. Chem. Soc. $\underline{75}$, 3290 (1953).
316. B. Roos and P. Siegbahn, Theoret. Chim. Acta. $\underline{17}$, 199 (1970).
317. C. C. J. Roothaan, Rev. Mod. Phys. $\underline{23}$, 69 (1951).
318. C. C. J. Roothaan, Rev. Mod. Phys. $\underline{32}$, 179 (1960).
319. C. C. J. Roothaan and P. S. Bagus, Methods in Comput. Phys. $\underline{2}$, 47 (1963).
320. A. Rotenberg, J. Chem. Phys. $\underline{39}$, 512 (1963).
321. S. Rothenberg, J. Chem. Phys. $\underline{51}$, 3389 (1969).
322. S. Rothenberg and H. F. Schaefer, J. Chem. Phys. $\underline{53}$, 3014 (1970).
323. S. Rothenberg, P. Kollman, M. E. Schwartz, E. F. Hayes, and L. C. Allen, Intern. J. Quantum Chem. $\underline{3S}$, 715 (1970).
324. S. Rothenberg and H. F. Schaefer, J. Chem. Phys. $\underline{54}$, 2765 (1971).
325. S. Rothenberg and H. F. Schaefer, Chem. Phys. Letters $\underline{10}$, 565 (1971).

326. J. A. Rowlinson, Trans. Faraday Soc. $\underline{47}$, 120 (1951).
327. M. Rubinstein and I. Shavitt, J. Chem. Phys. $\underline{51}$, 2014 (1969).
328. K. Ruedenberg, J. Chem. Phys. $\underline{19}$, 1459 (1951); F. E. Harris, J. Chem. Phys. $\underline{32}$, 3 (1960).
329. N. Sabelli and J. Hinze, J. Chem. Phys. $\underline{50}$, 684 (1969).
330. L. M. Sachs and M. Geller, Intern. J. Quantum Chem. $\underline{1S}$, 445 (1967).
331. L. M. Sachs, M. Geller, and J. J. Kaufman, J. Chem. Phys. $\underline{51}$, 2771 (1969).
332. H. F. Schaefer and F. E. Harris, Phys. Rev. $\underline{167}$, 67 (1968).
333. H. F. Schaefer and F. E. Harris, J. Chem. Phys. $\underline{48}$, 4946 (1968).
334. H. F. Schaefer and F. E. Harris, J. Comput. Phys. $\underline{3}$, 217 (1968).
335. H. F. Schaefer, R. A. Klemm, and F. E. Harris, Phys. Rev. $\underline{176}$, 49 (1968).
336. H. F. Schaefer, R. A. Klemm, and F. E. Harris, Phys. Rev. $\underline{181}$, 137 (1969).
337. H. F. Schaefer, R. A. Klemm, and F. E. Harris, J. Chem. Phys. $\underline{51}$, 4643 (1969).
338. H. F. Schaefer, J. Chem. Phys. $\underline{52}$, 6241 (1970).
339. H. F. Schaefer, J. Comput. Phys. $\underline{6}$, 142 (1970).
340. H. F. Schaefer, D. R. McLaughlin, F. E. Harris, and B. J. Alder, Phys. Rev. Letters $\underline{25}$, 988 (1970); D. R. McLaughlin and H. F. Schaefer, Chem. Phys. Letters $\underline{12}$, 244 (1971).
341. H. F. Schaefer and S. Rothenberg, J. Chem. Phys. $\underline{54}$, 1423 (1971).
342. H. F. Schaefer, J. Chem. Phys. $\underline{54}$, 2207 (1971).
343. H. F. Schaefer, J. Chem. Phys. $\underline{55}$, 176 (1971); S. V. O'Neil, P. K. Pearson, and H. F. Schaefer, Chem. Phys. Letters $\underline{10}$, 404 (1971).
344. T. P. Schafer, P. E. Siska, J. M. Parson, F. P. Tully, Y. C. Wong, and Y. T. Lee, J. Chem. Phys. $\underline{53}$, 3385 (1970).
345. B. Schiff and C. L. Pekeris, Phys. Rev. $\underline{134}$, A638 (1964).
346. M. E. Schwartz, Chem. Phys. Letters, $\underline{5}$, 50 (1970).
347. I. Shavitt, R. M. Stevens, F. L. Minn, and M. Karplus, J. Chem. Phys. $\underline{48}$, 2700 (1968).
348. I. Shavitt, J. Chem. Phys. $\underline{49}$, 4048 (1968).
349. I. Shavitt, J. Comput. Phys. $\underline{6}$, 124 (1970).
350. K. Siegbahn, C. Nordling, G. Johansson, J. Hedman, P. F. Heder, K. Hamrin, U. Gelius, T. Bergmark, L. O. Werme, R. Manne, and Y. Baer, <u>ESCA Applied to Free Molecules</u> (North-Holland, Amsterdam, 1969).
351. D. M. Silver, K. Ruedenberg, and E. L. Mehler, J. Chem. Phys. $\underline{52}$, 1206 (1970).
352. H. J. Silverstone and O. Sinanoglu, J. Chem. Phys. $\underline{44}$, 1899, 3608 (1966).

353. H. J. Silverstone and M. L. Yin, J. Chem. Phys. $\underline{49}$, 2026 (1968).
354. Since the appearance of Bunge's work on Be (62), an even lower variational energy, -14.66654 hartrees, have been reported by J. S. Sims and S. Hagstrom, Phys. Rev. $\underline{A4}$, 908 (1971). The Sims-Hagstrom calculation involves the explicit use of interparticle coordinates r_{ij}.
355. O. Sinanoglu, J. Chem. Phys. $\underline{36}$, 706, 3198 (1962).
356. O. Sinanoglu, Advan. Chem. Phys. $\underline{6}$, 315 (1964).
357. O. Sinanoglu and I. Oksuz, Phys. Rev. Letters $\underline{21}$, 507 (1968).
358. O. Sinanoglu, Advan. Chem. Phys. $\underline{14}$, 237 (1969).
359. A. K. Q. Siu and E. R. Davidson, Intern. J. Quantum Chem. $\underline{4}$, 223 (1970).
360. J. C. Slater, Phys. Rev. $\underline{35}$, 210 (1930).
361. J. C. Slater, Phys. Rev. $\underline{36}$, 57 (1930).
362. J. C. Slater, Quantum Theory of Atomic Structure, Volumes I and II (McGraw-Hill, New York, 1960).
363. J. C. Slater, Quantum Theory of Molecules and Solids, Volume 1 (McGraw-Hill, New York, 1963).
364. L. C. Snyder and H. Basch, J. Am. Chem. Soc. $\underline{91}$, 2189 (1969).
365. M. Spencer, Acta Cryst. $\underline{12}$, 59 (1959).
366. a) R. M. Stevens and W. N. Lipscomb, J. Chem. Phys. $\underline{42}$, 3666 (1965); b) R. A. Hegstrom and W. N. Lipscomb, J. Chem. Phys. $\underline{45}$, 2378 (1966).
367. R. M. Stevens, J. Chem. Phys. $\underline{52}$, 1397 (1970).
368. R. M. Stevens, E. Switkes, E. A. Laws, and W. N. Lipscomb, J. Am. Chem. Soc. $\underline{93}$, 2603 (1971).
369. R. M. Stevens, J. Chem. Phys. $\underline{55}$, 1725 (1971).
370. A. Streitwieser, Molecular Orbital Theory for Organic Chemists (John Wiley & Sons, New York, 1961).
371. J. D. Swalen and J. A. Ibers, J. Chem. Phys. $\underline{36}$, 1914 (1962).
372. E. Switkes, R. M. Stevens, W. N. Lipscomb, and M. D. Newton, J. Chem. Phys. $\underline{51}$, 2085 (1969).
373. H. S. Taylor and J. K. Williams, J. Chem. Phys. $\underline{42}$, 4063 (1965).
374. R. E. Trees, Phys. Rev. $\underline{92}$, 308 (1953).
375. D. W. Turner, C. Baker, A. D. Baker, and C. R. Brundle, Molecular Photoelectron Spectroscopy (Wiley-Interscience, New York, 1970).
376. J. H. Van Vleck, The Theory of Electric and Magnetic Susceptibilities (Oxford University Press, London, 1932).

377. A. Veillard and E. Clementi, J. Chem. Phys. $\underline{49}$, 2415 (1968).
378. A. Veillard, Theoret. Chim. Acta $\underline{12}$, 405 (1968).
379. A. Veillard, Chem. Phys. Letters $\underline{3}$, 128 (1969).
380. A. Veillard, Theoret. Chim. Acta $\underline{18}$, 21 (1970).
381. J. W. Viers, F. E. Harris, and H. F. Schaefer, Phys. Rev. $\underline{A1}$, 24 (1970).
382. A. J. H. Wachters, J. Chem. Phys. $\underline{52}$, 1033 (1970).
383. A. C. Wahl, P. E. Cade, and C. C. J. Roothaan, J. Chem. Phys. $\underline{41}$, 2578 (1964).
384. A. C. Wahl, J. Chem. Phys. $\underline{41}$, 2600 (1964).
385. A. C. Wahl and G. Das, Advan. Quantum Chem. $\underline{5}$, 261 (1970).
386. J. M. Walter and S. Barratt, Proc. Roy. Soc. (London) $\underline{A119}$, 257 (1928).
387. E. Wasserman, V. J. Kuck, R. S. Hutton, and W. A. Yager, J. Am. Chem. Soc. $\underline{92}$, 7491 (1970).
388. R. E. Watson, Phys. Rev. $\underline{119}$, 170 (1960).
389. F. Weinhold, J. Chem. Phys. $\underline{54}$, 1874 (1971).
390. A. W. Weiss, Phys. Rev. $\underline{162}$, 71 (1967).
391. A. W. Weiss, Phys. Rev. $\underline{119}$, 170 (1960).
392. S. Weiss and G. E. Leroi, J. Chem. Phys. $\underline{48}$, 962 (1968).
393. D. White, K. S. Seshadri, D. F. Deves, P. E. Mann, and M. J. Leneusky, J. Chem. Phys. $\underline{39}$, 2463 (1963).
394. D. R. Whitman and C. J. Hornback, J. Chem. Phys. $\underline{51}$, 398 (1968).
395. J. L. Whitten, J. Chem. Phys. $\underline{44}$, 359 (1966).
396. J. H. Wilkinson, The Algebraic Eigenvalue Problem (Oxford University Press, London, 1965).
397. D. L. Wilhite and J. L. Whitten, J. Am. Chem. Soc. $\underline{93}$, 2858 (1971).
398. C. W. Wilson and W. A. Goddard, J. Chem. Phys. $\underline{51}$, 716 (1969).
399. R. B. Woodward and R. Hoffmann, J. Am. Chem. Soc. $\underline{87}$, 395 (1965).
400. R. B. Woodward and R. Hoffmann, The Conservation of Orbital Symmetry (Academic Press, New York, 1970).
401. M. Yoshimine, J. Phys. Soc. Japan $\underline{25}$, 1100 (1968).

MOLECULE INDEX

Al	89,100	C_2NH	242,244
AlF	248	C_2N_2	238
AlH	165	C_3H_6, cyclopropane	60,328
Ar	88,89,129	C_4H_6, butadiene	352,359
Ar^+	89	C_6H_4, benzyne	363
ArF_2	89	C_6H_6, benzene	388
As	100	$C_9N_8O_2H_{10}$, guanine-cytosine	394
B	100,102,123,140,195	$C_{10}H_8$, naphthalene and azulene	388
B_2H_6	348,409,411	CF_2	326
BF	80,248	CF_4	330,331
BF^+	168	CH	19,184,201
BH	9,26,158,188,191,200,245	CHF_3	330,331
BH_2	37	CH_2	19,290,309,344
BH_3	331,348	CH_2F_2	331
BH_3^+	331	CH_3	19,33,35
BH_4^+	290	CH_3F	331
BH_4^-	290	CH_3OH	410
BO	168	CH_4	249,290,328,330,331,410
Be	87,111,123,129	CH_4^+	331
BeF	168	CN	168,200
BeH	188,200	CO	6,30,78,163,189,197,203,246
BeO	159,222	CO^+	168
Br	100	CO_2	238,240,244
C	19,60,65,72,74,100,102,111,122,140	CS	200
		CaO	159
C^+	105	Cl	100
C_2	223	ClCN	242,244,248
C_2^-	168	Cl_2	73
C_2ClH	242,244,248	ClH_2	30
C_2FH	242,244,248	Co	100
C_2H_2	239,328	Cr	100
C_2H_4	32,328,338,345,352	Cu	100
C_2H_6	289,328,398,410	F	100,123

Molecule Index

FCN	242,244,248	KF	153
F_2	73,146,159,205,283	KrF	146,232
FH_2	272	KrF^+	146,232
F_2N_2	330	KrF_2	281
F_2O_2	329	$KrSbF_{12}$	235
Fe	100	Li	86,100,195
Ga	100	LiCN	259
Ge	100	LiCl	153
H	65,76	LiF	80,153
HCN	64,241,244,247	LiH	17,53,145,158,183,190,201
$(HCOOH)_2$	374,396	Li_2	227
HCl	158,248	Li_2O	250
HF	80,146,183,201,248,275,319,331	LiNC	259
HF^+	331	LiOH	250
$(HF)_2$	372	Mg	130
H_2	43,247	MgO	159
H_3	272,290,316	Mn	100
H_3^+	266	N	23,30,96,100,105,123,140,195
H_4	272,335	N^+	105
H_2CO	14,65,249,289,330	NF_3	330
H_2O	7,70,75,78,195,249,290,298,304,319,328,331,370	NH	184,190,201,203
H_2O^+	289,307,331	NH_2	22,290
$(H_2O)_2$	370	NH_3	12,64,330,331,401
$(H_2O)_3$	370	NH_3^+	331
H_2O_2	195,330,404	N_2	75,146,158,163,177,203
H_2S	298	N_2^+	146,158,168
He	40	N_2O	242,244,246
He_2	216	NO	176
He_2^+	48	NO^+	176
HeH_2^+	266	NO_2	290
HeO_2^+	271	Na	100,103
K	100	Na_2	227
		NaCl	153

NaF	153	PN	249
NaLi	202,227	S	58,100
NaLi$^+$	229	SCN$^-$	242,248
Ne	6,29,33,34,51,89,111,122, 134,319,332	SCO	242,244
		SO$_2$	294
Ne$^+$	89,332	Sc	100
Ni	100,378	Se	100
NiF$_6^{-4}$	378	Si	100
O	60,65,68,100,105,123,140	SeO	203,248
OCN$^-$	242,248	SrO	159
OH	183,201	Ti	99
O$_2$	30,34,36,159,181,200,205, 283	V	99
O$_2^+$	181	XeF$_2$	238,383
O$_3$	298	XeF$_4$	383
P	99	XeF$_6$	383

SUBJECT INDEX

acceleration form of oscillator strength	102	bond distances	74,152,159, 196,205,215, 221,226,230, 236,250,261, 269,275,285, 301,317,325, 339,364,373 403,407
activation energy	30,237,267, 272,318,335, 362		
active electron approximation	170		
additivity of pair correlation energies	130,134,183, 323	bond orbirals	197,327,349, 411
		bond order	391
alkali halides	153	Born-Oppenheimer approximation	1
angular momentum	33		
anharmonicity corrections	152,158,196, 215,236	Brillouin's theorem	10,140
		charges on atoms	263,298,368, 381,384,390
antishielding	164	cartesian gaussians	57
antisymmetrizer	4	chemical reactions	30,237,266, 272,316,344, 359
atomic correlation energies	29,86,101,111, 129,134,140		
atomic populations	263,298,368, 381,384,390	chemical shifts of inner shell electron binding energies	331,387
atomic SCF calculations	86	computer programs	290
		concerted reactions	344,359
atomic units	3	configuration	32,49,135, 239,354
atoms	85		
Auger spectrum	334	configuration interaction	30
barrier height for a chemical reaction	30,237,267, 272,318,335, 362	conrotatory mode of reaction	359
		contour maps for potential energy surfaces	277,320
basis sets	56,147,153, 177,251,406		
bond angles	250,300,309, 325,339,364, 373,403,407	contracted functions	70
		core polarization	122
bond dipole moments	411	core electron ionization potentials	89,176,325, 331,387

Subject Index

correlation diagram	345
correlation energy	27
coulomb integrals	9
coupled Hartree-Fock method	161
crystal field splitting parameter	380
crystal field theory	378
cubic force constants	300
current density induced by a magnetic field	169
degeneracy effects in electronic structure	114
degenerate irreducible representations	35
density matrix	41
determinant	4
diagmanetic shielding	297,304,386
diamagnetic susceptibility	161,296,304
diatomic molecules	145
dimerization energy of BH_3	348
dimers of water	370
dipole-dipole interaction	292
dipole moments	13,30,81,154, 197,236,240, 255,265,295
dipole polarizability	243
direct diagonalization of symmetry operators	35
direct hole state calculations	89,176,325, 331
disrotatory mode of reaction	359
dissociation behavior of Hartree-Fock wave functions	148,158,208, 214,230,271, 284,347
dissociation energies	30,74,80,148, 157,205,207, 225,229,232, 240,255,276, 285,348
DNA, deoxyribonucleic acid	394
d orbitals, importance of	282,298,402
double excitation	37
double minimum	394
double-quantum light scattering	243
double zeta basis set	63,252
double zeta plus polarization basis sets	77,252
doubly excited configuration	37
eigenvalue problem	30
electric field gradients	81,120,155, 236,256,297, 386
electrocyclic reaction	336
electron configuration	33
electron correlation	27
electron density diagrams	15,16,63,411
electron pair function	192
electron paramagnetic resonance (EPR)	312
electrostatic Hamiltonian	2
energy of a single-determinant wave function	8
exchange integrals	9,408
exchange reaction	335
exchange splitting	176
excited state Hartree-Fock wave functions	19

excitation energies	110,150,200, 222,309,327, 352,367,378, 388	Hamiltonian matrix elements	37
		Hamiltonian, quantum mechanical	2,239
expectation values	13	harmonic force constants	300
exponential functions	57		
extended basis set	63,238,252	Hartree-Fock approximation	4
Fermi contact interaction	30,95,119,292	Hartree-Fock Hamiltonian	305
first natural configuration	45	heats of reaction	29,269,273
first-order electronic wave functions	122,140,210, 224,234,273, 285,312,325	hole states	89,176,325, 331,387
		hybrid orbitals	27,413
		hydrogen bonding	370,394
floating spherical gaussian orbital method	326	hyperfine structure	30,95,119, 290,381
force constants	152,301,414	hyperpolarizability	243
forces on nuclei in a molecule	258,296	improved virtual orbitals	306
frozen orbital approximation	179	inner shell electron binding energies	89,176,325, 331,387
f values	102,168	interatomic correlation energy	220,348
gauge invariance	162		
gaussian functions	57,61,67,70	internal barriers	398
gaussian lobe functions	58	interpair correlation energies	192,324,348
geometry predictions	74,152,159, 196,205,215, 221,226,230, 236,250,261, 267,275,285, 300,309,317, 325,339,361, 371,395,403, 407	ionic basis set	253
		ionization potential	14,89,176, 325,331
		ion-molecule reactions	266
		isomers of LiCN	259
		iterative natural orbital method	52,183,212, 273,312
g-factor, rotational	257		
group theory	7	Koopmans' theorem	14,89,148, 179,305
guanine-cytosine base pair	394	Lagrange multipliers	14

least motion coplanar approach of two methylenes	344	noble gas molecules	232,383
least-squares expansion of Slater functions	61	nonadditivity of pair interactions in the water trimer	374
length form of oscillator strength	102	nonorthogonality problem	8,104,168, 321
ligand field theory	378	normal coordinates	302
linear polyatomic molecules	237	occupation numbers	42,357
localized hole states of molecules	182	octupole moment, molecular	296,303
localized orbitals	190,327,348, 407	one electron bond	229
		one-electron integrals	9
L-S eigenfunctions	135	one-electron properties	13
MacDonald's theorem	32	open-shell systems	19,140
magnetic hyperfine structure	29,95,119, 290,381	optimized valence configurations, method of	208,220,228
magnetic susceptibility	160	orbital	5
many-body perturbation theory	127,129	orbital contribution to atomic hyperfine structure	119
matrix diagonalization	32	orbital energies	17,255,389, 399
matrix elements	37		
microwave spectroscopy	290	orbital exponents	57
minimum basis set	58,202,238, 252,328	orbital occupancy	32
		orthogonality	8
minimum energy path	267,273,317, 336,361	oscillator strengths	102,168
molecular quadrupole moments	81,236,256, 281,294,386	overcontracted basis sets	72
		pair correlation approximations	129,134,140, 182,219,323, 348
Møller and Plesset, theorem of	13		
multiconfiguration Hartree-Fock	40	pairwise additivity of intermolecular forces	374
multiconfiguration SCF method	40	paramagnetism	160
natural orbitals	41	partitioning of degenerate spaces	118

perturbed Hartree-Fock method	161	quadrupole polarizability	244
photoelectron spectroscopy	95,176,331, 387,390	Racah parameters	380
		reaction coordinate	237,267,273, 317,336,361
pi-electron theory	338,352,363, 388,394	reaction mechanism	335,361
polarizabilities	243,256	relativistic corrections	86,94
polarization basis functions	75,252,281	restricted Hartree-Fock wave function	7
polarization wave function	122	Roothaan procedure	21,70,238
		rotational g factor	257
polyatomic molecules, nonlinear	289	Rydberg-Klein-Rees potential energy curves	148,214
population analysis	263,298,368, 381,384,390	Rydberg states	304,340,353
potential curves	146,153,173, 200,205,216, 222,231,234	saddle point	275,318,335, 364
		Schrödinger equation	1
potential energy surfaces	266,272,299, 315,316,335, 344,359,394	second moments of the electronic charge distribution	296
		secular equation	32
potential maximum	286,345	self-consistent-field method	4
primitive gaussian functions	71	semi-empirical calculations	58,311,344, 352,366,378, 391
projected unrestricted Hartree-Fock wave function	24		
		separated pair approximation	190
projection operators	24,35	single excitation	37,199
pseudonatural orbital method	48,106,267, 322	singlet-triplet splittings	309,367
quadrupole coupling constants	127,156,298, 303,387	singly-excited configuration	37,186
		Slater determinant	4
quadrupole moments, molecular	81,236,256, 281,294,386	Slater functions	57,58,65
		Slater's rules	59
quadrupole moments, nuclear	127,156,298, 387		

Subject Index

spectroscopic constants	146, 157, 196, 205, 215, 226, 230, 236, 263, 405
spin-adapted pair correlation energies	184, 323, 350
spin density at a nucleus	30, 95, 119, 292
spin-dipolar contribution to atomic hyperfine structure	120
spin-extended Hartree-Fock approximation	24
spin-optimized SCF method	25
spinorbital pair correlations	129, 136
spinorbitals	6
spin projection operator	24
spin rotation constants	165
strong orthogonality	192
symmetry	7
symmetry-adapted linear combination of determinants	33
symmetry-adapted pair correlations	134, 140, 184, 323, 350
symmetry and equivalence restrictions	6
symmetry operators	35
term symbols	7, 19, 147
three- and four-center integrals	238, 289
total orbital energy	262
transition moments	103, 171
transition probabilities	102, 168
transition state geometries	275, 318, 335, 364
trimers of water	370
triple and quadruple excitations, importance of	117
tunneling	395
two-center integrals	145
two-electron integrals	9, 58, 85, 237, 389
unitary transformation	189, 408
unrestricted Hartree-Fock approximation	23, 95
van der Waals interaction	216, 234, 319
variational principle	3
velocity form of oscillator strength	102
vibrational frequencies	152, 158, 197, 205, 215, 226, 235, 263, 404
vibrationally averaged properties	298
virial ratio, V/T	146, 255
virtual orbital approximation	171
virtual orbitals	305
Wigner-Witmer rules	204
Woodward-Hoffman rules	344, 359
Zeeman effect, rotational	294
zero-point vibrational effects	298